UPDATES IN VOLCANOLOGY – A COMPREHENSIVE APPROACH TO VOLCANOLOGICAL PROBLEMS

Edited by **Francesco Stoppa**

Updates in Volcanology – A Comprehensive Approach to Volcanological Problems
Edited by Francesco Stoppa

Published by InTech

Janeza Trdine 9, 51000 Rijeka, Croatia

Publishing Process Manager Niksa Mandic

Technical Editor Teodora Smiljanic

Cover Designer InTech Design Team

Additional hard copies can be obtained from orders@intechopen.com

Updates in Volcanology – A Comprehensive Approach to Volcanological Problems,
Edited by Francesco Stoppa
p. cm.
ISBN 978-953-307-434-4

For EU product safety concerns:
IN TECH d.o.o., Prolaz Marije Krucifikse Kozulić 3, 51000 Rijeka, Croatia,
info@intechopen.com or visit our website at intechopen.com.

Meet the editor

Professor Francesco Stoppa, PhD, is a full professor of Volcanology and Geochemistry at the G. d'Annunzio University of Chieti, Italy. He specializes in the alkaline, mafic-ultramafic volcanism with particular emphasis on its geodynamic properties. He maintains long-term collaboration with famous English petrologists, such as Ken Bailey, Keith Bell, and Alan Woolley, and leads high-profile scientific exchange and research with Russian and Chinese specialists, experimenting in new approaches to the major challenges that the volcanic phenomenon poses to humans. He discovered a number of carbonatites in Italy, publishing his work in widely circulated international journals. He is an activist in organizing international conferences and excursions on the alkaline volcanoes of Italy and Spain. He is also engaged in conservation programs of volcanic areas and in the mitigation of seismic and volcanic risk. He devotes much of his research in Italy to the monogenic intra-Apennine ultra-alkaline volcanoes, as well as to Mount Vulture Volcano and to the Roman province. His main research areas outside of Italy are Uganda, Canary Islands, and Calatrava in Castilla-La Mancha.

Contents

Preface

Volcanism witnesses every major change of our planet and other planets.

In Advances in Volcanology, scientists from highly active volcanic countries, such as Japan, Italy, and New Zealand, as well as others from Germany, Portugal, and Russia, debate less commonplace themes. Topics from classic field volcanology, including practical problems with volcanic stratigraphy in oil exploitation, to the most modern techniques related to tomographic studies are discussed. The question about the role of hydro-volcanism as a modifying factor versus juvenile gases as the primary engine of volcanism is discussed in full. The complex geodynamic meaning of the large basaltic province versus large alkaline provinces is analyzed by means of large scale examples, using geochemical, tectonic, and stratigraphic demonstrations. Tectonic modification related to collisional-extensional volcanic environments, which puzzle structural geologists, is also considered. This is germane to a modern conception of volcanology as a typical multi-scale, multi-method discipline.

Field methods in volcanology

Chapter 1 by Stoppa, Rosatelli, Schiazza, and Tranquilli, and chapter 2 by Németh provide excellent examples to understand the volcanic facies and the distribution of monogenetic volcanoes that cluster in intra-continental settings. Large monogenetic volcanic fields in western Hungary and central Spain are presented in detail with the aim of characterizing their pyroclastic successions and chemistry, and inferring their eruptive mechanisms. In Chapter 3, Umeda and Ban provide a compilation of the distribution of 139 volcanic centres depicting eruption style, magma compositions, and eruptive volume related to change from the condition of a neutral stress regime with low crustal strain rate to compression along major thrust faults associated with uplift in a volcanic front. It is widely assumed that magma cannot rise so easily in compressional settings, and the distribution of volcanic centres is controlled mostly by local extensional dislocations and gravitational instability. However, in this chapter, the reason why an increase in erupted magma volume may be related to the subduction rate and to the lowering of differential stress by thermal effects is discussed.

Large igneous provinces

Volcanism is spread and distributed well at the surface of the Earth in the form of large plumes that last for long geological periods, affecting big areas. For several years, this concept has been debated, and the existence of the plumes themselves is questioned. Thus, the following chapters are devoted to this problem.

In Chapter 4, Sushchevskaya, Belyatsky, and Laiba show that remote volcanic provinces, which are interpreted as the manifestation of the Karoo–Maud plume in Antarctica and Africa, have a considerable duration and multistage character. Derbeko, in chapter 5, depicts the bimodal petrochemical series of the Mongol-Okhotsk orogenic belt in the interval 119 – 97 Ma. The mantle source composition is characterized by trace element enrichment/depletion in terms of LILE/HFSE ratios and related to their tectonic position. Adam, in Chapter 6, and Lavecchia and Bell, in Chapter 7, consider a large-scale analysis of regional geochemistry, volcanology, and tectonics of famous igneous provinces, such as those of the Mediterranean and French Polynesia regions, which is discussed in a broad comparative analysis that brings us back to the mystery of the planet dynamics. French Polynesia is characterized by a great concentration of volcanism on the South Pacific Superswell. The description of this area provides a fairly accurate image of the mantle underneath this region, demonstrating that a direct link exists between the mantle convection and the surface observation, and can bring new insight to the plume debate. In Chapter 7, Lavecchia and Bell take inspiration from the Mediterranean potassic series paradoxes, due to a peculiar coexistence, sometimes within the same location and at the same age, of SiO2-oversaturated rock-types (calcalkaline to high-K calcalkaline products and, more rarely, leucite-free lamproites) and of SiO2-undersaturated potassic to ultra-potassic rock-types (leucite-phonolites to leucitites, melilitites, and kamafugites) and Na-rich series. Strangely enough, volcanic products, although clearly belonging to the same magmatotectonic domain, the Mediterranean wide-rift basin, are attributed in the literature to contrasting geodynamic environments. The first of these being anorogenic and intra-plate, and the second being orogenic and subduction-related. The discussion mainly concerns the nature of the metasomatic component, which might result from pressure-related dehydration of the subducting slabs, or from upwelled deep mantle components. When not a priori forced to fit all the available multidisciplinary source elements within a subduction view, other interesting scenarios can be opened, which also allows a unifying interpretation of the overall Mediterranean and peri-Mediterranean magmatism.

Applied volcanology

Chapter 8 by Klarner and Klarner makes us aware of the role of pyroclastics and epiclastics when exploring hydrocarbon reservoirs. These rocks may produce practical problems, due to complex diagenetic overprints and lateral seals or migration barriers, which produce both positive and negative impacts on the petroleum system. It is therefore essential to understand the distribution of volcanics in the vicinity of the reservoir. In Chapter 9, Koulakov demonstrates the capacity of tomographic methods for studying magma sources in different areas of volcanic provinces at different scales. Tomographic data are considered in a multidisciplinary context together with geological, geophysical, and geochemical data.

All the authors stress that modern volcanology is a young science, but the interest in volcanoes is perhaps as old as human beings. It is thus necessary to place the arguments presented in this book in a historical light, which will help readers to understand the basis of many volcanological arguments.

A fundamental step in volcanological history was the eruption of Vesuvius in 79 AD. Mount Vesuvius became active after a secular resting, and it destroyed Pompeii and other neighbouring towns. Pliny the Elder (c. 23 AD – 79 AD), led primarily by his curiosity for natural history, tried to get close to the volcano, but lost his life during the final phase of the eruption. However, his nephew Pliny the Younger (c. 61–114 AD) provided the first direct accurate description of a volcanic eruption. Paradoxically, this eruption, which marks the maximum advance of knowledge of volcanoes, was the last opportunity for the discussion of these topics before the medieval stagnation. For two millennia, scientists continued to place emphasis on the lava effusions, neglecting the explosive processes. The resumption of studies in volcanology ideally coincided with the formation of Monte Nuovo (1538) in the Phlegraean Fields. Although a relatively small-scale eruption, it brought into question almost all of the medieval dogmas on the creation of the Earth. The eruption of Vesuvius in 1631 came after centuries of stagnation, and prompted European scientists to come to the first formulation of the modern geological theories. From that point on until the beginning of 1900, Vesuvius, with its continuous activity and its proximity to the city of Naples, a capital of arts and culture, was considered the prototype of all volcanoes. Aristotle said that man, because of his limited perception of the flow of events, erroneously attributes to disasters and the power to change the course of nature, when in fact, they are part of a constantly changing Earth (Meteorologica, Book II). Catastrophism, the theory adopted by Christianity, dominated Western countries' thinking for many centuries. Extreme isolation and harsh living conditions prevented visits by scholars or the birth of scientific schools in areas outside of an active volcanic continental Europe. For these reasons, the current dispute between the various scientific communities continued, until recently, to be based on an unrepresentative number of more strategically placed volcanoes, such as Vesuvius.

With Newton's influence, the focus slowly shifted to the idea of a planet resulting from the balance of constantly active forces. During the eighteenth century, the dispute between Neptunists and Plutonists offered the opportunity to break the deadlock that was already present in Aristotle's thoughts concerning the dichotomy between fire (central heat, actualistic, and evolutionary vision of the earth) and water (diluvian vision, catastrophic, and "chemical" volcanism). The sudden formation of Ferdinandea Island, in the Strait of Sicily in 1831, dispelled the last doubts about the nature of the volcanic phenomena. For these reasons, Italy is considered the cradle of volcanology.

The geologists of the nineteenth century had a clearer view of the fact that volcanism was not randomly distributed or conditioned by local phenomena. However, this view of the global distribution of volcanism was not aware of the existence of volcanic mid-ocean ridges and African rift volcanoes, which were still very poorly understood. Substantial progress has occurred in our century. The vision of volcanism is now framed in the global tectonic theory, although not always completely circumscribed by a lithosphere formed of plates. Some eruptions have been of great importance for the impetus given to the progress of volcanological studies: the eruptions of Krakatau in the Sunda Strait in 1883, Mount Pelée in Martinique in 1902, Bezimianny Kamchakta in 1956, and St Helens in the Cascades Range in 1980.

Nowadays, volcanology is trying to escape from the extreme fragmentation and specialization that has occurred in recent years. It is rapidly gaining renewed interest among geologists and geophysicists. Its global significance becomes clearer when one tries to tackle the open questions that geology still poses. It is understood that only comprehensive and comparative study of large volcanic provinces and their peculiarities can form a consistent picture of the dynamics of the planet. Advances in Volcanology is a good opportunity to open our minds about volcanoes and the problems with their interpretation in a multicultural world-wide approach.

Prof. Francesco Stoppa
Earth Sciences Department, Gabriele d'Annunzio University, Chieti, Italy

Part 1

Field Methods in Volcanology

1

Hydrovolcanic vs Magmatic Processes in Forming Maars and Associated Pyroclasts: The Calatrava -Spain- Case History

F. Stoppa, G. Rosatelli, M. Schiazza and A. Tranquilli
Università Gabriele d'Annunzio, Dipartimento di Scienze, Chieti
Italy

1. Introduction

The Calatrava Volcanic Field (CVF) of Castilla-La Mancha is characterised by numerous monogenetic volcanic centres, that erupted mainly foidites, melilitites and carbonatites (ultra-alkaline rock-association sensu, Le Bas, 1981) carrying abundant mantle xenoliths. At CVF, carbonatites have been described by Bailey et al. (2005) and Stoppa et al. (2011). Along with the volcanic field of Eifel of Germany, Limagne basin of France and Intra-mountain Ultra-alkaline Province (IUP) of Italy, the CVF encompasses the most numerous Pliocene-Quaternary extrusive carbonatites in Western Europe in terms of dimension, number and size of volcanoes (Bailey et al., 2005; Bailey et al., 2006). Similar volcanic fields are Toro-Akole and Bufumbira in Uganda (Bailey & Collier, 2000), the Avon district in Missouri (Callicoat et al., 2008), Mata da Corda in Brazil (Junqueira-Brod et al., 1999) and West Qinling in Gansu Province, China (Yu et al., 2003). In spite of abundant local studies (González Cárdenas et al., 2010; Peinado et al., 2009), the CVF has been mostly neglected by the international audience, although Bailey (2005) outlined the need for a long-term research program on CVF. This work focuses on the role of deep CO_2 at CVF, which is considered an intrinsic component of carbonatitic mantle magmatism (Hamilton et al., 1979). Previous, studies of CVF volcanoes considered that the hydrovolcanism is a necessary and sufficient condition to explain the CVF volcanological features, and, as a corollary that the carbonate present in the pyroclastic rocks is remobilised limestones (e.g., López-Ruiz et al., 2002). We propose an alternative hypotheses based on CO_2 violent exolution and expansion germane to diatremic propagation of ultra-alkaline melts towards the surface and to dry-magmatic origin of the maars (Mattsson & Tripoli, 2011; Stoppa, 1996; Stoppa & Principe, 1998).

2. Volcano-tectonic setting

The CVF volcanoes occur in a circular area of about 3000 km^2, at the western termination of the SSW-NNE elongated Guadiana valley (Fig. 1), which is one of the largest tectonic basins in central-southern Spain. Most of the CVF centres are nested in the Palaeozoic rocks of the Calatrava and Almagro massifs, composed of quartzite, slate and lesser granite, deformed in E-W and N-S vertical, flexural folds (De Vicente et al., 2007). The massifs are cut by faults striking NW-SE and E-W, which determine a low profile, "horst and graben" -type morphology. The CVF has been subject to a generalised uplift that produced erosion of the

Neogene alluvial and lacustrine sediments filling the "grabens". This erosional phase was followed by paleosol-caliche formation during Lower Pliocene (Peinado et al., 2009). The uplift shortly predates the main volcanic phase. Post-volcanic lacustrine sedimentation, composed of travertine plus epiclastites and diatomite with bioturbation and slumps, has been observed in some maars such as Casa de los Cantagallos, Vega de Castellanos, Hoya de los Muertos (Peña, 1934; Portero García et al., 1988). It is likely that post-volcanic travertines are related to magmatic CO_2 dissolved in the ground-water and/or carbonatite weathering and remobilisation. Lacustrine travertines from Granátula de Calatrava gave C isotopes ratios averaging -5.73‰ $\delta^{13}C_{PBD}$ (average of 4 analyses data unpublished courtesy of M. Brilli CNR, Roma) in agreement with values measured from CO_2 emission at Calatrava.

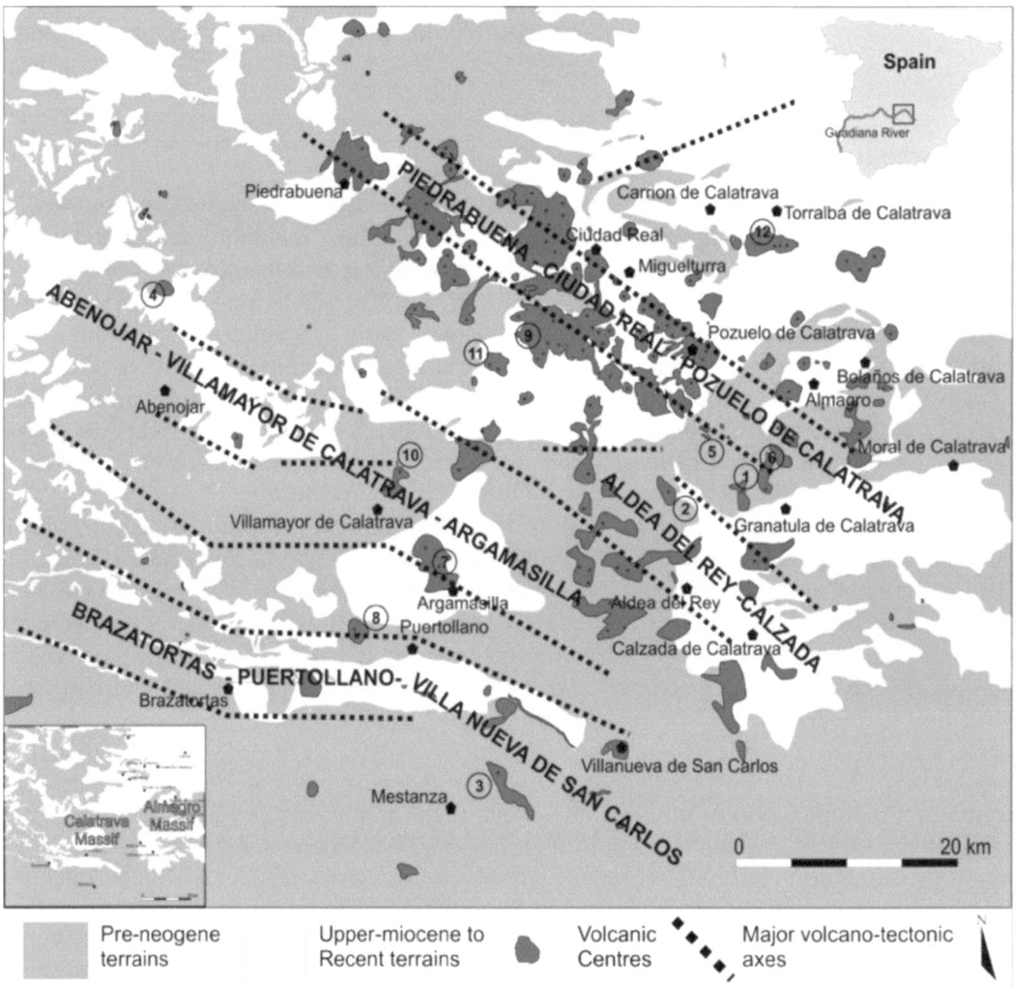

Fig. 1. Geological sketch map of CVF. Left top N 39°10'4.70" W 4°31'10.70" and the right bottom N 38°30'4.80" W 3°31'10.80". 1) La Sima, 2) Hoya de la Cervera, 3) Laguna de la Alberquilla, 4) Laguna Los Michos, 5) La Nava, 6) Cerro Gordo - Barondillo, 7) Laguna Blanca; 8) Laguna Almodovar del Campo, 9) Poblete, 10) Morron de Villamayor, 11) Cabezo Segura II, 12) Cerro San Marcos.

Volcanoes and CO_2 emissions are aligned NW-SE (Fig. 1). This direction corresponds to the elongation of the four major "grabens": a) Piedrabuena-Ciudad Real-Pozuelo de Calatrava, b) Aldea del Rey-Calzada de Calatrava, c) Abenojar-Villamayor de Calatrava-Argamasilla, d) Brazatortas-Puertollano-Villanueva de San Carlos (González Cárdenas & Gosálvez Rey, 2004; Poblete Piedrabuena, 1997). Some seismic activity has been identified east of the CVF. It is very weak, with 2-3 events per year and an average Mw of 2.7. A maximum event of Mw 5.1 occurred in Pedro Muñoz at the NE termination of the upper Guadiana basin, on August 12, 2007. The focal mechanism is compatible with a right, lateral strike-slip fault oriented ENE (data of Instituto Geográfico Nacional de España). The seismological evidence is in agreement with recent stress field estimates in western Spain, indicating pure strike-slip faulting conditions (De Vicente et al., 2007). The volcanic activity has been intense and relatively continuous over a few million years in the CVF (Ancochea, 1982; Cebriá et al., 2011). The subcontinental lithosphere, metasomatised by a rising asthenospheric diapir, has been considered the CVF melt source (Cebriá & López-Ruiz, 1995). However, deep seismic sounding studies on regional scale do not show any notable crustal thinning or upper-mantle upwelling confirming works based on Bouguer anomalies (Bergamín & Carbo, 1986; Díaz & Gallart, 2009; Fernàndez et al., 2004). If CVF activity is not driven by lithosphere tectonic it could be consequence of a hot finger detached by the megaplume active between the Canary Islands, Azores Islands and the western Mediterranean Sea (Hoernle et al., 1995).

3. CO_2 emissions and hydrothermalism

CO_2-bubbling springs, locally known as "hervideros" (Poblete Piedrabuena, 1992; Yélamos & Villarroya Gil, 1991), and CO_2 vents (mephites), lethal for animals, are frequent in the CVF. $^{13}C/^{12}C$ determination at Granátula de Calatrava and Puertollano CO_2-rich springs gave $\delta^{13}C_{PBD}$ between -4.9‰ and -5.6‰ similar to primitive mantle values (Redondo & Yélamos, 2005). Mephites at La Sima and Granátula de Calatrava are associated with sporadic H_2S emissions and historical thermal anomalies (Calvo et al., 2010; Gosálvez et al., 2010). Past hydrothermal activity seems to have deposited relatively conspicuous Mn(Co-Fe) concretionary cryptomelane $K(Mn^{4+}, Mn^{2+})_8O_{16}$ and litioforite $(Li_6Al_{14}Mn_{21}O_{42}(OH)_{42})$. These ores are found in La Zarza and El Chorrillo (Fig. 1), about 2 km SSW of Pozuelo de Calatrava (Crespo & Lunar, 1997).

The seismic crisis of August 2007 produced a dramatic increase in gas emissions at La Sima (Peinado et al., 2009). Before the shock of August 12, the CO_2 values were about 0.03 kg/m^2 per day. After the earthquake new CO_2 vents opened with apparent damage to the surrounding vegetation. A constant increase in the CO_2 emission, up to 324 kg/m^2 per day and a grand total of 4,86 kg per day only in the La Sima emission area was recorded (González Cárdenas et al., 2007; Peinado et al., 2009).

In CVF shallow well drillings have caused exceptional escapes of CO_2 in Los Cabezos, El Rosario and Añavete. Abrupt large emissions of gas-water are frequent in the area even if not lasting more than a few days. The "chorro" of Granátula de Calatrava in the Granátula-Moral de Calatrava graben has recently released gas, water and debris. After this event, a geophysical study identified a positive gravimetric and thermal anomalies (EPTISA, 2001). On March 2011, the "geyser" of Bolaños de Calatrava swamped an area of about 90,000 m^2 and issued up to 40 tonnes of CO_2 per day for several days. It spontaneously arose in a vineyard emitting 50,000 cubic meters of water propelled by gases composed 90% vol. of carbon dioxide plus sulphur compounds (H_2S and HgS). An estimate of the temperature and pressure of the deep seated hydrothermal system is about 118 °C and 63 bar pressure (data Grupo de Investigación GEOVOL de la Universidad de Castilla-La Mancha). These localised activities have been interpreted as ephemeral gas releases along deep fractures. Well-eruption due to drilling confirm that CO_2 is locally accumulated at shallow level (<1km) and any perturbation, either natural or artificial, might lead to the violent release of gas producing water-debris currents. Evidence for a Holocene discrete phreatic eruption, which produced no juvenile ejecta, is recorded in the stratigraphy of the La Columba volcano (González Cárdenas et al., 2007). Future volcanic scenarios can be considered including diatreme formation, volcanian-like explosion, phreatic events, primary lahars, local volcano-seismic crises due to fluids/melt intrusion, potentially fatal CO_2-H_2S rapid emissions. All these phenomena are triggered by the abundant presence of juvenile gases in the magmatic system of Calatrava.

4. CVF magma composition

The entire CVF activity produced no less than 15 km^3 of alkaline mafic/ultra-mafic rocks. Rock type occurrences at 33 investigated volcanoes (Fig. 1) are 36% nephelinite, 30% olivine melilitite, 21% leucite nephelinite (leucitite *s.l.*), 6% tephritic nephelinite, 3% melilite nephelinite and 3% carbonatites. It is not possible to calculate rock type in term of individual volume due to their complicate distribution and stratigraphy. However, carbonatite largely dispersed as ash-tuff are probably dominant in volume. Some of the CVF rocks are somewhat similar to ugandite or kamafugite having larnite in the CIPW norms, strong SiO_2 undersaturation and a potassic character with agpaitic index (Na+K/Al) of about 0.9. High K content of nepheline suggest that kalsilite, a key mineral for kamafugites, or kaliophyllite occurrence is possible. Worldwide association of melilitite and carbonatite is noteworthy (e.g., Hamilton et al., 1979; Stoppa et al., 2005). This association can be found in many place worldwide and it covers 50% of the occurrences of extrusive carbonatite outcrops (Woolley & Church, 2005). Approximately 50% of the CVF outcrops contain mantle nodules. Plagioclase-bearing rocks are subordinate in all these districts, and in the CVF modal tephrite and basanite are notably absent. CVF nephelinites are depleted in ^{87}Sr and enriched in ^{143}Nd, whereas leucitite-melilitite and carbonatites are enriched in ^{87}Sr and depleted in ^{143}Nd (Cebriá & López-Ruiz, 1996). In the CVF peridotitic nodules are spinel-lherzolite to amphibole-lherzolite equilibrated up to 20 kbar and a temperature of 956-1382 °C (Villaseca et al., 2010). Possibly different magma sources in the CVF may explain rock associations with different geochemical characteristics: I - melilitite and carbonatite; II - nephelinite-tephritic nephelinite. A level intensely metasomatized with amphibole-carbonate and with phlogopite veins would form the thermal boundary layer. These two components would produce, due to a slightly different partial melting point, the CVF magmatic spectrum. A similar feature has also been found in Italian carbonatites and kamafugites (Stoppa & Woolley, 1997) and is possibly related to reaction of alkali carbonatite with spinel or garnet lherzolite (Rosatelli et al., 2007).

5. Volcanology

Volcanic activity started on the western side of the CVF with the emplacement of melilite leucite foidites. This early phase is mostly represented at Volcano Morrón de Villamayor (located N 38°49'20'' W 4°07'30''). K/Ar ages are inconsistent, giving a range of 8.7-6.4 Ma for the same lava cooling-unit of this volcano (Bonadonna & Villa, 1986). Some other deeply eroded emission centres located in the Tirteafuera area may tentatively be related to this first magmatic phase. In the CVF activity lasted till the Quaternary (Ancochea & Ibarrola, 1982; Cebriá & López-Ruiz 1995). However, the dating by K/Ar methods available so far might not be perfectly suitable for the most challenging problems of Recent volcanism in CVF (Balogh et al., 2010). Due to the hundreds of vents, maars, cones and their multiple-clustered pattern in the CVF, it is important to give a general view of the volcanological features of this area and examples of the dominant volcanic forms. As for the definition of volcanic forms as vent, maar, diatrema and scoria cone, we conform to the definition of tab. 2 in White & Ross (2011). We assume, however, that the diatreme "feeder-dike" is very deep and located in the mantle (Stoppa et al., 2011, Fig. 3). In addition, we prefer the term tuffisite (Cloos, 1941) instead of peperite as the latter implies a magma/wet-sediment interaction. For specific discussion about tuffisite definition see Stoppa et al. (2003). CVF volcanoes density ranges from 10 to 15 per 100 km^2, leading to an estimation of about 250-300 volcanoes in the whole area (Fig. 1). However, volcanic landforms, eruption style and chemical composition are repetitive and are well represented by describing a limited number of volcanoes. Two main areas, located NW and SE of the city of Ciudad Real, can be identified, in terms of density of volcanoes and volume of the deposits. All the other volcanoes are scattered and decrease in size and density with distance from these areas. At least 150 of them have names and are now recognised by local people as volcanoes. The local idiom is very precise and distinguishes between different volcanic forms. *Peña* indicates a cone having a summit covered by blocks, *cabezo* is a small and isolated cone, *laguna* is a maar containing water and *nava* is a maar with a flat dry surface or a marsh, without trees, surrounded by hills, *hoya* and *pozo* are names for a large, deep diatreme. The presence of polygenic volcanoes (Becerra-Ramírez et al., 2010) is questionable from a stratigraphic point of vieiw. In fact, the paleosols delineate the overlap of products of the adjacent volcanoes, rather than polygenic activity. However, co-eruptive vents are frequent and are represented by coalescent, multiple and/or nested vents associated with volcanic complexes of cones and maars. So we prefer the term polyphasic to polygenic. In some cases vents are aligned along NW-SE fissures some kilometres long such as in Miguelturra-Pozuelo de Caltrava (Fig. 1). Exposures of feeding dykes are lacking along these alignments and in general in the CVF.

5.1 Maar/diatreme systems

At CVF there are no geophysical data or exposure allowing direct observation of diatremes. Circular depressions sharply excavated into the Palaeozoic crystalline hard-rocks with steep internal escarpment and without significant accumulation of volcanics outside the rims, are considered here as the surface expression of eroded diatremic conduits. The few remnants of volcanics may indicate the presence of a former maar. When a pyroclastic ring is found around these depressions, it is classified as a maar (Martín-Serrano et al., 2009). La Hoya de la Cervera (Figs. 1, 2a,b) is located north of the Aldea del Rey, along a NW-SE alignment, which links with the CO_2-rich springs. Less than 3 km NE is the large maar of Finca la Nava (Fig. 3). La Hoya de la Cervera is a depression with diameter of about 300 m, totally excavated into the hard Palaeozoic rocks. The depression bottom is at about 675 m a.s.l., while the rim is between 750 and 825 m above sea level (a.s.l.). Sparse remains of lapilli tuffs and breccias are exposed along the depression rim in the NE section. These deposits are characterised by occurrence of tuffs, containing concentric-shelled cored lapilli, along with heterolithic breccias which are hardened by carbonate matrix (Fig. 8c). The diatreme of La Hoya de La

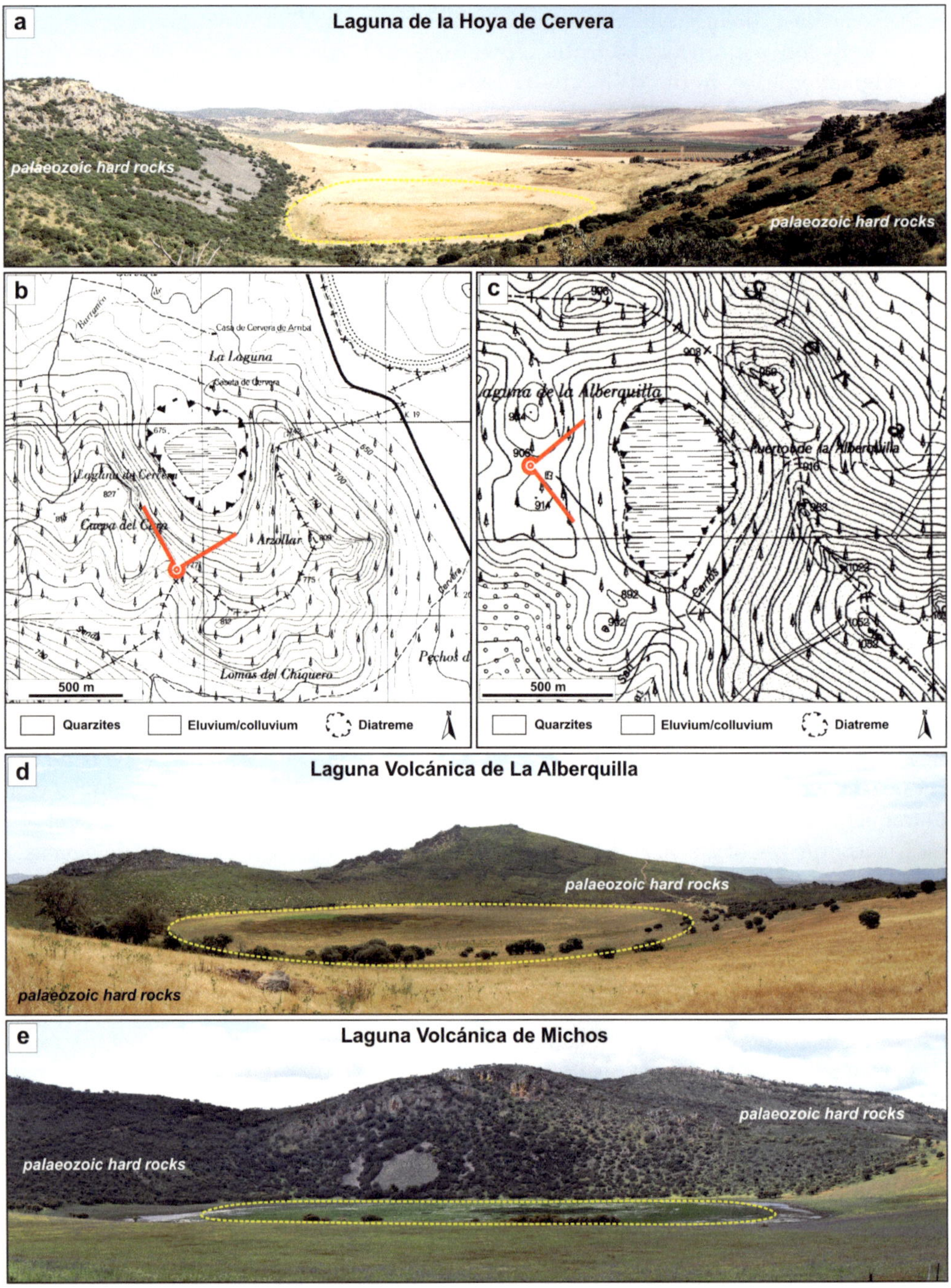

Fig. 2. Surface expression of CVF diatremes: a) general view of Hoya de la Cervera and b) geological sketch map, c) geological sketch and d) general view of Laguna de la Alberquilla, e) general view of Los Michos diatreme. Red symbol in this and other figures are outlook points from where the pictures were token. In the yellow dashed line is indicated the rim of the eroded diatreme.

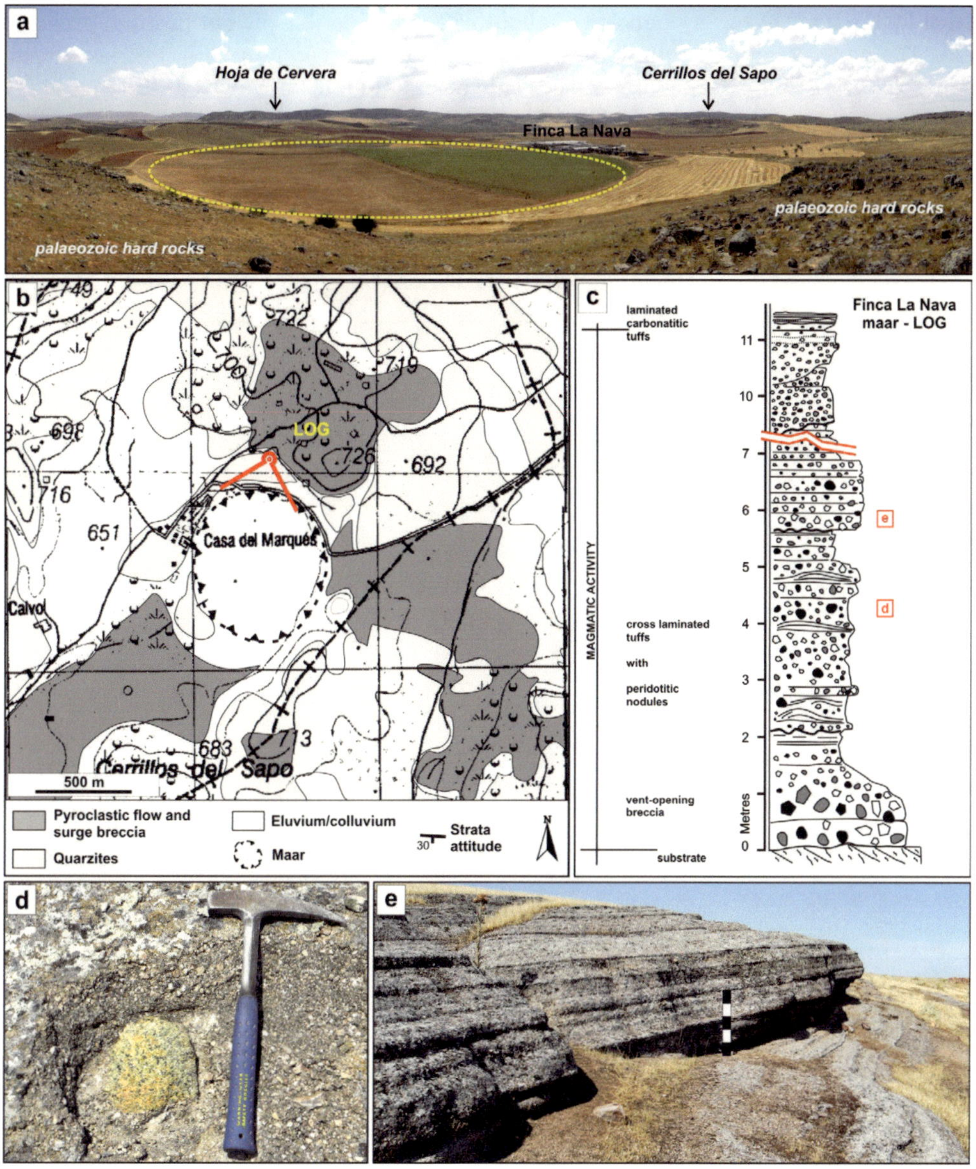

Fig. 3. a) General view of Finca la Nava maar and, on the skyline, Hoya de la Cervera and Cerrillos del Sapo indicated by an arrow, b) geological sketch of Finca la Nava maar, c) stratigraphy of La Nava volcanic products (log position located on the map and written in yellow), d) discrete mantle nodule in the La Nava tuffs (hammer is 30 cm long) and e) picture showing the intermediate part of the La Nava volcanic sequence (the bar scale is 1 m long).

Alberquilla (Figs. 1, 2c,d) is located on the escarpment of the Puertollano graben, and is part of a NW-SE elongated cluster of volcanoes, east of the village of Mestanza. The graben shoulders are made of Palaeozoic quartzite. The depression hosts a laguna at 865 m a.s.l., while the rim is at 950-1000 m a.s.l.; the shape of the depression is elliptical (500 x 300 m). The Los Michos diatreme (Figs. 1, 2e) is one of the best preserved and hosts a temporary lake at 700 m a.s.l.. It has a diameter of about 450 m and is sharply cut through crystalline rocks. There are no above ground, pyroclastics rocks preserved around it. La Nava maar is about 1 km wide and is located on the NE side of the Río Jabalón valley, at 620 m a.s.l., and is excavated in the Palaeozoic hard-rock substrate (Figs. 1, 3). Pyroclastic rocks outcrop discontinuously around the maar, especially at the north side, where a maximum thickness of about 11 metres was observed. They are mostly roughly layered, cross-laminated strata about 7 m thick overlying 2 m thick, vent opening breccia. Vent opening breccias are pyroclastic rocks composed largely of country rock blocks (up to 80%) which are inferred to have been deposited during the initial crater formation (Stoppa, 1996).

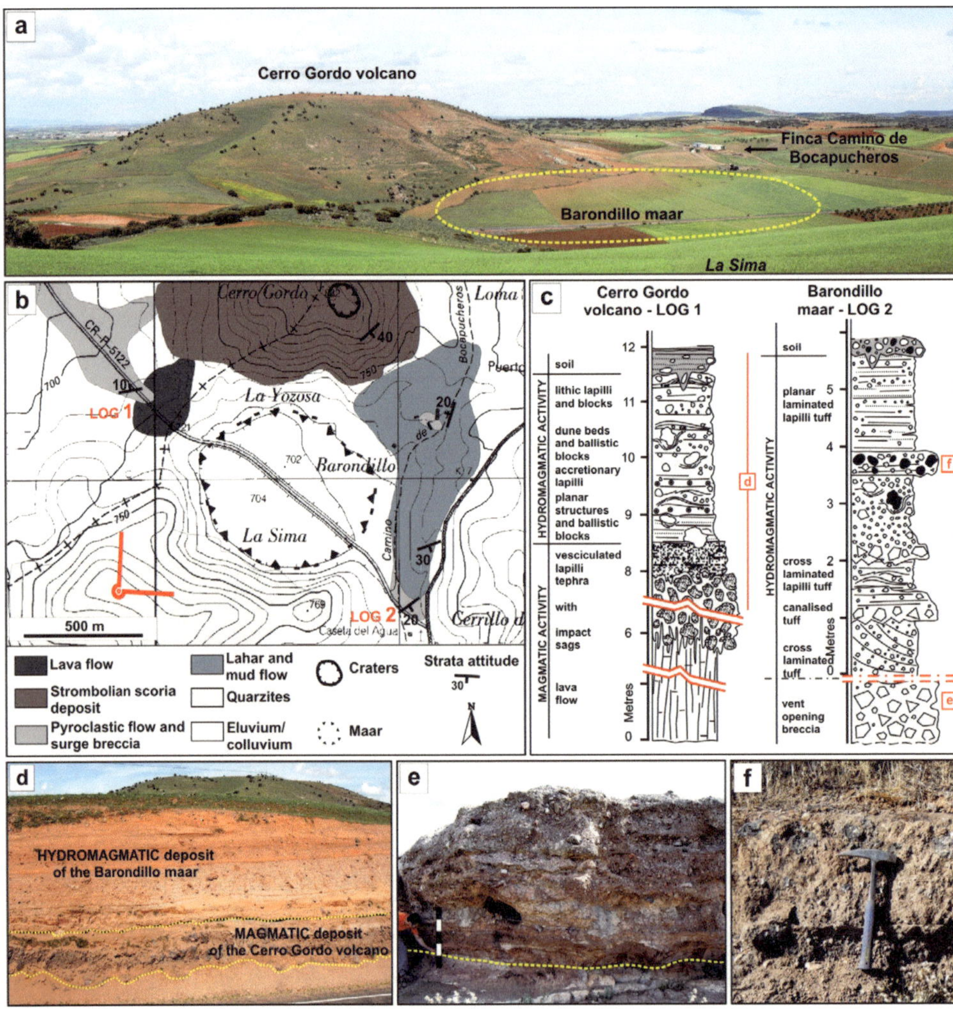

Fig. 4. a) General view of the Barrondillo maar and the adjacent volcano of Cerro Gordo; b) geological sketch map of the area, c) stratigraphy of Cerro Gordo and Barrondillo deposits (log position is indicated in the map and written in red), d) hydro-volcanic dune layers overlapping lava scoriae along the CR-P-5122 road, e) general view of vent opening breccia reported in the log of Barondillo maar and f) detail of peridotite nodules and scoria bomb-rich layer.

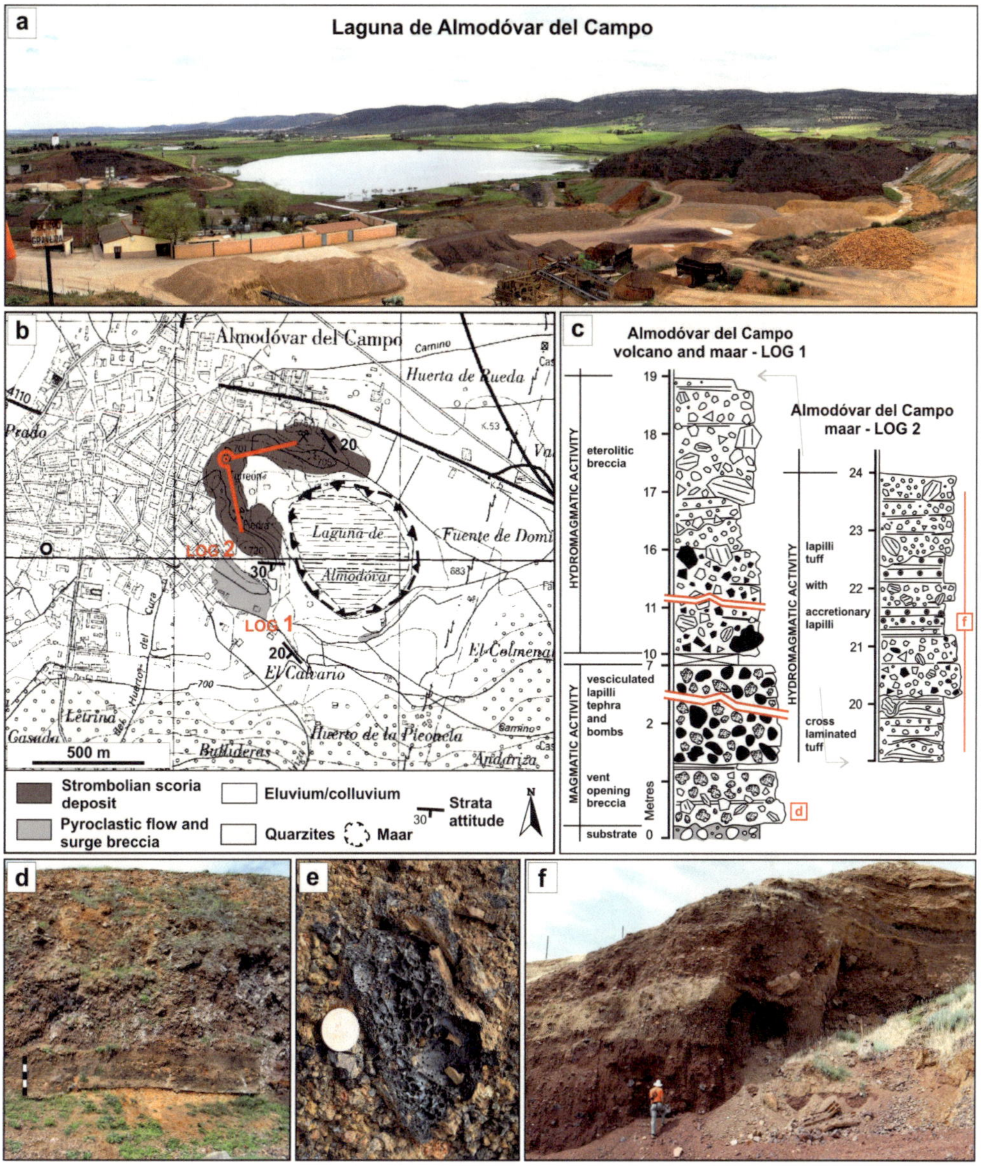

Fig. 5. Almodóvar del Campo: a) General view of Laguna de Almodóvar maar; b) geological sketch map of the area, c) stratigraphy of the volcanic sequence (logs indicated in inset b), d) Palaeosol and vent opening breccia, the bar scale = 1 metre, e) detail of vent opening breccia, showing a large mafic pumice, f) general view of the hydro-volcanic layers corresponding to the upper part of the sections of inset c, LOG 2.

Close to the top, there are about 2 m of laminated, carbonatitic tuffs containing melilitite lapilli concentric-shelled lapilli with a central kernel of amphibole and phlogopite xenocrysts or mantle nodules. Tuff layers are hardened by carbonate and show lapilli plastically moulded each other (Fig. 8). Large discrete peridotitic nodules are scattered as impacting blocks in the tuffs.

The Barondillo maar is adjacent to Cerro Gordo volcano located near La Sima volcano, currently the most significant CO_2 emission centre in CVF. This small maar, located at 700 m a.s.l., has a diameter of 80 m and is partially excavated in the quartzite of the Almagro massif (Figs. 1, 4a,b). Road-cut exposures along CR-P-5122 road, towards Valenzuela de

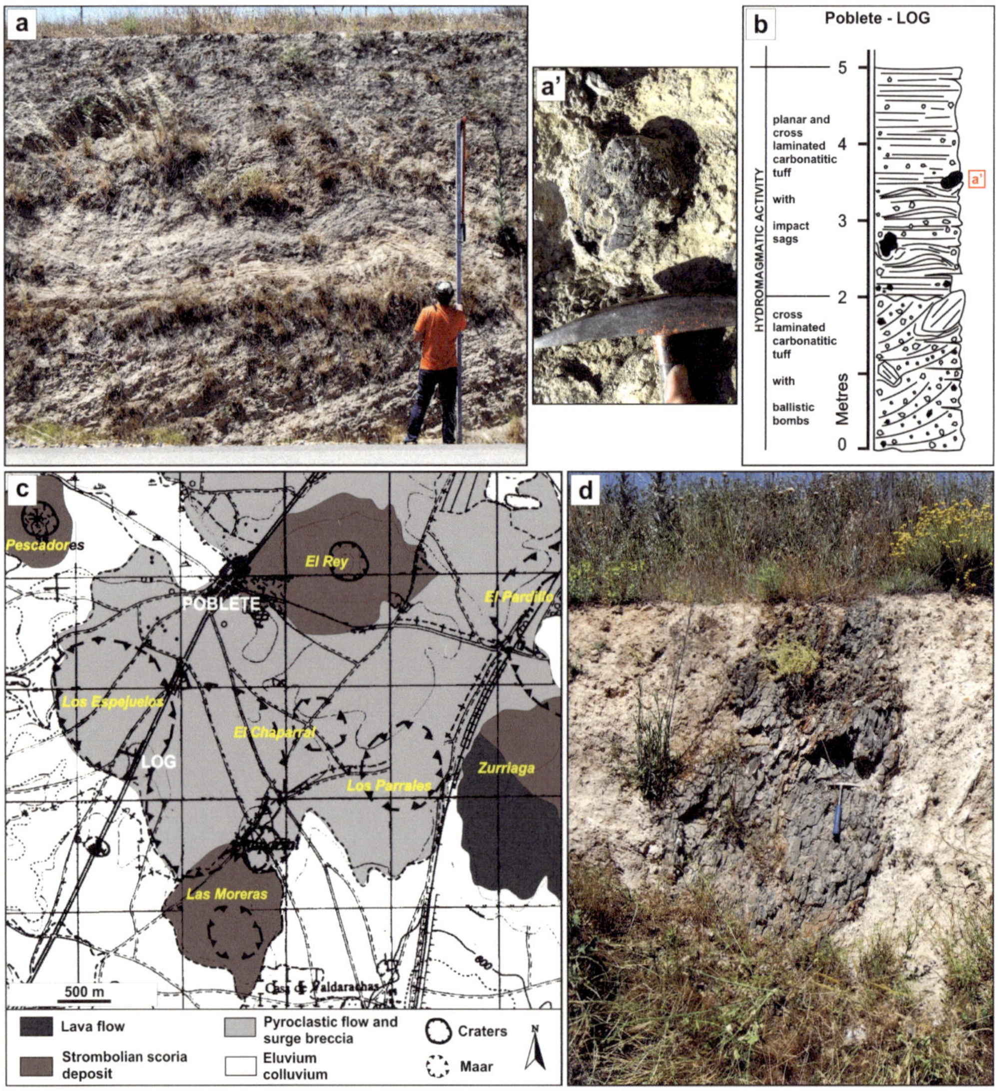

Fig. 6. Pyroclastic deposits in the Poblete area: a) carbonatite dune layers on a road cut, a') ballistic block composed of an amphibole mega-crystal, b) stratigraphy at the road cut (see map for reference), c) sketch map of the area, d) large country-rock ejecta in carbonatitic ash tuff. The names of other volcanic centres are in yellow in the sketch map.

Calatrava and near Mina de San Carlos, offer a good view of volcanic deposits (LOG 1, Fig. 4c,d). South and SW of Ciudad Real, pyroclastic units have unique carbonatitic features. They have been substantially neglected sso far because not clearly related to a volcanic centre. These pyroclastic deposits are located 3 km south of the village of Poblete and close to the Los Espejuelos maar. The scoria cones of Cabezo Segura and Volcán de la Zurriaga are to the SE and SW; the maars of Hoya del Pardillo, El Chaparral and Hoya del Mortero are 2 km to the E, the small edifice of Volcán de Cabezo de Pescadores is 2 km to the NW, while a cluster of others maars and the scoria cones are located north (Fig. 6c). The exposures show lapilli tuffs and dune layers with juvenile mafic lapilli immersed in a carbonate matrix (Figs. 6a,a'). Metre-sized ballistic block of country-rocks impacted this soft substrate and produced notable impact sags (Fig. 6d). Sedimentation interference between impact blocks and carbonate layers emplacement indicate a rapid accumulation in a turbulent volcanic regime (surge). Centimetre-sized clinopyroxene crystals are scattered in the tuff. The size of the blocks suggests that the sequence is very proximal to the emission centre (Los Espejuelos?). More distal outcrops show lapilli tuffs organised in co-sets of layers with the ash matrix being composed of carbonate but large ballistic blocks are absent and only mafic bombs are visible (Fig. 6a').

5.2 Scoria cones and lava flows

The Volcán Morrón de Villamayor (Fig. 1) is associated with lava flows of olivine leucite melilitite (Humphreys et al., 2010). The vent area shows two necks having columnar jointing; the largest neck is located north and represents the highest point of the volcano at 840 m a.s.l.. The south flank of the pyroclastic edifice is covered by spatter bombs indicating a lava fountain activity. Two lava flows originated from this area and, after a short travel, flooded the area where the Cantera del Morrón quarry is located. The two lava flows have different modal composition, mostly due to abundant country rocks lithics in the lower flow and olivine in the upper flow. Columnar jointing cutting through the two lava flows indicates that they form a single cooling-unit. The pyroclastic rocks and lava flows cover about 2 km^2.

The Volcán Cabeza Segura II has an olivine melilitite-carbonatite and nephelinite composition (Stoppa et al., 2011). It consists of a 100 m-thick accumulation of massive or layered agglomerate that become a lapilli tuffs towards the top (Fig. 7b). Distinctive layers of inverse graded, ovoid concentric-shelled lapilli and bombs (Carracedo Sánchez et al., 2009), intercalated with undulated beds of carbonatitic lapilli tuff, occur in the middle of the volcanic pile (Stoppa et al., 2011). Towards the top there are cross-laminated, carbonate-rich beds, 20-30 cm thick (Fig. 7b,e,f). They contain melilititic lapilli with internal concentric layers of carbonatitic melilitite (Fig. 8b). Lava a few metres wide and about one metre thick lava sheets are repeted in the volcanic sequence, while a thicker tabular lava flow close the eruptive cycle. The upper part of the sequence surrounding the lava is a layered, reddened scoriaceous lapilli fall deposit with a thin palaeosol on the top.

The Cerro San Marcos (Fig. 1) is an isolated scoria cone of olivine melilitite composition. The volcano is located about 2 km south of Torralba de Calatrava. The volcanic deposit is an inverse-graded agglomerate of spherical bombs and lapilli (Fig. 8a,b) with a few cm thick ash levels. Some bombs are nucleated by large lithics. On top of the sequence, ribbon-bombs and pumiceous lapilli prevail over other pyroclasts and lithics. Torralba lapilli have distinct concentric lava shells (Fig 8a,b).

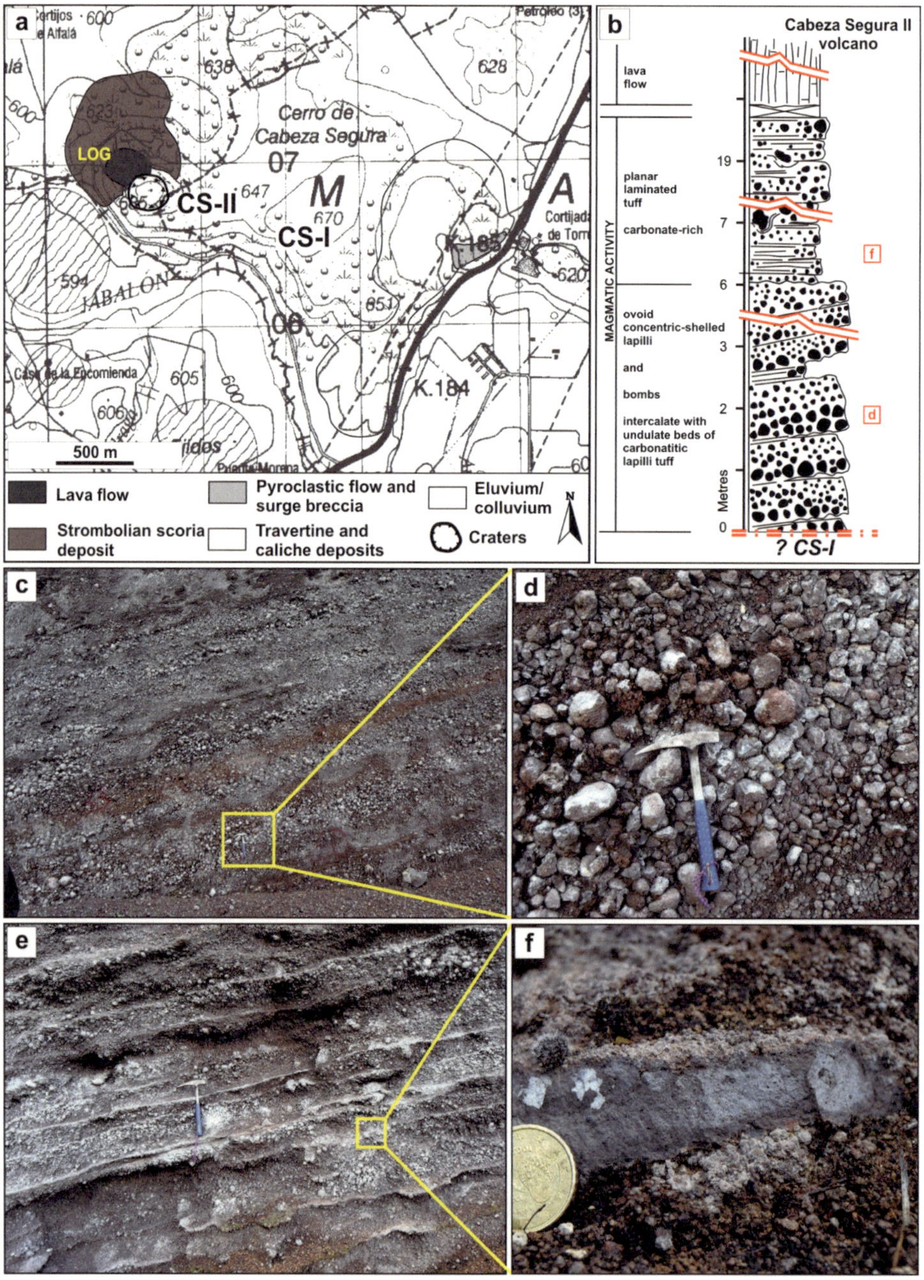

Fig. 7. Cabezo Segura II: a) geological sketch map of the area and b) stratigraphic of the volcanic sequence, c) pyroclastic layers alternating inverse graded tephra and agglomerate, d) detail of spheroidal bombs and lapilli, e) agglomerate layers with carbonatitic ash-tuff showing cross lamination and undulated layers (dunes) and f) detail of carbonatite tuff.

6. Features of CVF pyroclastites

6.1 Lapilli

Concentric-shelled lapilli are juvenile pyroclasts specifically associated with maar/diatreme deposits as discussed in a number of study to which we remand for details (Junqueira-Brod, 1999; Mitchell, 1997; Stoppa & Wolley, 1997). These peculiar lapilli have been described from recent African and other European provinces (Bailey, 1989; Bednarz & Schmincke, 1990; Hay, 1978; Lloyd, 1985; Lloyd et al., 2002; Riley et al., 1996; Stachel et al., 1994; Stoppa et al., 2003). Stoppa & Principe 1998) described similar lapilli in maar deposit at Monte Vulture Italy and defined them "concentric-shelled lapilli". Concentric-shelled lapilli are associated with ultra-alkaline rocks such as carbonatites and melilitites found in intracratonic settings and, to our knowledge, they have not been found in any other volcanic rock association or tectonic settings (Callicoat et al., 2008; Cloos, 1941; Ferguson et al., 1973; Gurney et al., 1991; Keller, 1981; Mitchell, 1997; Stoppa & Lupini, 1993; Stoppa, 1996). They have been interpreted to have formed as sub-volcanic, fluid, "spinning droplets" in a conduit by Junqueira-Brod et al. (1999). A similar model has been discussed by Stoppa et al., at CVF (2011).

Fig. 8. a) Agglomerate of spheroidal lapilli from Cerro San Marcos-Torralba volcano; b) sectioned concentric-shelled lapilli from CVF (Stoppa et al., 2011), c) Laguna Blanca tuff: juvenile lapilli and angular crustal lithics suspended in a carbonate matrix, d) carbonatite ash-tuff showing teardrops lapilli.

Concentric-shelled lapilli invariably cored by mantle material, xenocryst or peridotite nodule, are typical of the CVF pyroclastic rocks. This structure has been observed also in bombs which have a larger mantle nodule in the core (Carracedo Sánchez et al., 2009). They occur in maar deposits such as Almodóvar del Campo, la Nava, La Hoya de la Cervera, Laguna Blanca as well as scoria cone such as Cabeza Segura II, Cerro Gordo and Cerro San Marcos (Fig. 8a,b; Plate 2a,b in Stoppa et al., 2011). The lapilli shells are formed by densely packed, welded, glassy spherules of melilitite containing microphenocrysts of olivine, clinopyroxene, melilite, nepheline, haüyne, opaques, plus glass. In some spherules, melilite laths are abundant and concentrically arranged around a core which is a mafic xenocrysts (Fig. 9a,b). Concentric-shelled lapilli at the CVF contain up to 50% of primary carbonate in the form of coalescent globules having menisci necks and amoeboid shape which testify segregation under liquid condition (Fig. 9c,d) (Rosatelli et al., 2007). Primary features of igneous carbonate are largely targeted by geochemical studies and have been the specific object of several papers (e.g., Rosatelli et al., 2010; Stoppa et al., 2005). Detailed geochemical studies are outside the aim of this paper and the textural criteria, well known from previous papers, are used to asses a primary igneous origin of carbonates. Primary calcite composition in the CVF lapilli is given in Stoppa et al. (2011).

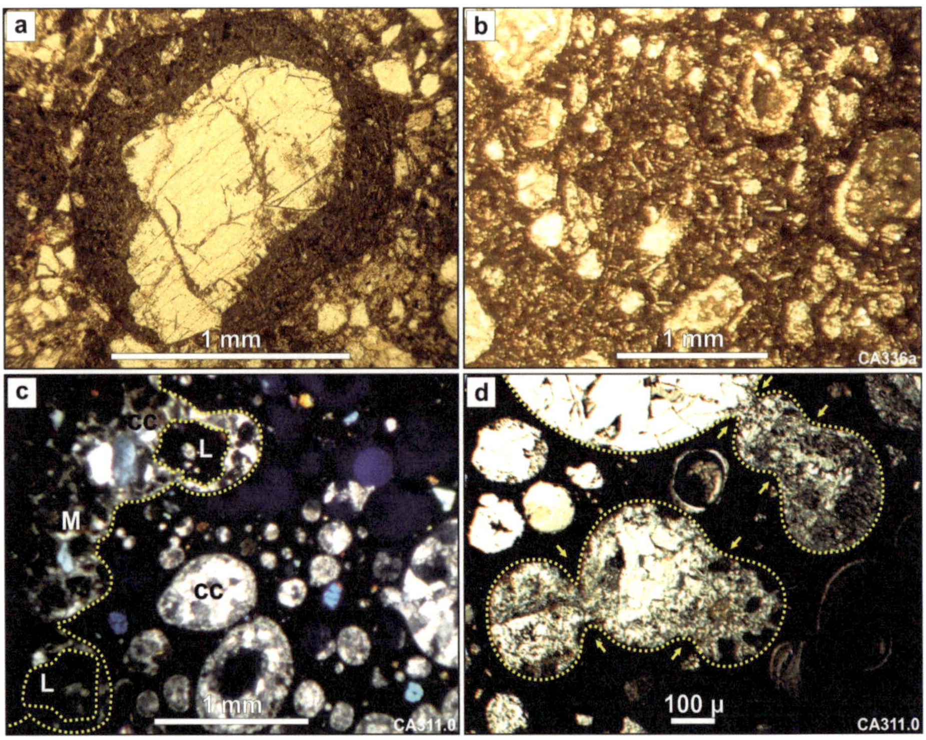

Fig. 9. a) Hoya de la Cervera melilitite cored ash-sized lapillus in thin section, b) Cerro Gordo melilitite ash-sized lapillus showing concentric internal structure, c) and d) Laguna Blanca tuff in thin section, melilitite lapillus-carbonatitic matrix contact and coalescent calcite globules preserving menisci (indicated by yellow arrows). Key symbols: CC - carbonatic matrix; L - lapillus; M - matrix.

6.2 Matrix

At CVF the matrix of concentric-shelled lapilli tuff is composed of particles having distinctive morphology and arrangements which are shown in Fig. 9. We consider as matrix the <2 mm clastic component of tuffs. At CVF, ash-sized particles are very often composed of turbid, micro or cryptocristalline calcite. Ash-sized particles structure, have often with two or more concentric shells, shells, resembling that of the lapilli irrespectively of the size (Fig. 8c,d). Their sub-spherical shape and "teardrops" reveals that they quenched from a melt under surface-tension condition (Keller, 1981). In spite of the strong agglutination and compaction the faint contours of the spherules can be easily distinguished (Fig. 9a,b). Carbonate spherules can be in contact or moulded with melilitite spherules.

The interstices among larger particles, which are in general melilitites, are filled by progressively smaller, densely packed spherules which are in general carbonatitic. Intragranular spaces are filled by amoeboid turbid cryptocrystalline carbonate (micrite). Similar textural occurrences at CVF and elsewhere were proven to be primary igneous or re-crystallised primary carbonate, by geochemistry (Hay, 1978; Stoppa & Lupini, 1993; Stoppa et al., 2005; Stoppa et al., 2011). Calcite globules in the melilitites spherules are broken towards their host spherule rim, indicating that their content was poured out and incorporated in the matrix, which is in fact composed of smaller melilititic fragments and carbonate having very similar composition with respect to those in the spherules (Fig. 9g). Patches of mosaic textured limpid carbonate and vugs with sparitic calcite are also present and represent secondary calcite and cement.

6.3 Dry and wet pyroclastic deposits

Dry-magmatic deposits at CVF are composed of juvenile, spherical or ovoidal, concentric-shelled lapilli and bomb and/or high vesiculated fragments, variable in size. They generally form agglomerate or lapilli layers devoided in fine-ash component. Presence of abundant lithics gives to many deposits a brecciated aspect (vent opening facies). These “breccia” layers are interbedded by hardned carbonate-rich lapilli tuffs. The lapilli-tuffs have compacted matrices formed by isotropic or nucleated spherules of carbonatites and melilitite which can maintain a spherical shape or mould each other (Fig. 9). This matrix structure is conducive of dry high temperature deposition at least able to produce welding and agglutination of smaller spherules in the matrix. Deposits made by these juvenile pyroclasts are found at CVF both in the maars and the scoria cone. Layer attitude, volcano sedimentary structures and texture indicate they have been deposited mainly by pyroclastic surge and ballistic fall related to moderate to strong vulcanian to strombolian activity.

Hydro-volcanic deposits in the CVF are dominated by lava blocks interbedded with vesiculated fine-ash matrix, where vesicles are probably produced by expansion of water vapour in a muddy matrix, and other cold-emplaced, wet pyroclastic deposits. Microscopic features reveal abundance of sharp-cut glass fragments, specific alteration and zeolite cement related to interaction with a magma/water interaction. In addition, condensed water is reflected by mud-flows formation and fall-out deposits composed of accretionary lapilli that often plastically mould each other. Dominant emplacement mechanism is surge, wet pyroclastic flow and mud flow possibly linked to moderate collapse of hydrovulcanic strombolian column.

7. Discussion

7.1 Sub-surface vs surface fragmentation mechanism

The CVF provides the opportunity to highlight and to contrast different mechanisms of magma fragmentation associated to maar-related pyroclasts including deep-seated gas exolution, sub-surface magma vesiculation and superficial hydrovolcanic phenomena. Juvenile pyroclasts at CVF varies from highly vesiculated scorias (pumices) to mostly unvesiculated concentric-shelled bombs, lapilli and ash sized spherules. These juvenile pyroclasts have morphology and structure indicating they reached the final shape when still plastic and hot. This process has to occur at magmatic temperature as it involves co-eruptive carbonatite and melilitite, the latter composition having high liquidus temperature (Brey and Green, 1977). A high degree of vesiculation indicate high content of juvenile volatiles and possible dry magmatic fragmentation in sub-surface condition (e.g., Mattsson & Tripoli, 2011) but unvesiculated concentric-shelled lapilli fragments require different origin.

Despite the fact that crustal debris far outweighs the proportion of mantle fragments in the CVF pyroclastic rocks, the concentric-shelled pyroclasts have only mafic crystal or peridotite fragment as kernels (Fig. 9a). Clearly, at the CVF the concentric lapilli were formed before crustal fragment incorporation in the rising magma. They formed before that other process may modify eruptive style, as they are found in both maar and scoria cones. Lack of vesicularity need not necessarily indicate that their magma was low in initial juvenile volatile component, but rather that the juvenile gases have been concentrated as the fluidising medium in the subvolcanic genetic environments after magma fragmentation (Stoppa et al., 2011 and references therein).

The shell structure suggests that the lapilli do not form by separation of discrete lumps of melt. They are obviously produced by agglutination of very small spherules of melt (Stoppa et al., 2011) - a mechanism which favours the CO_2 exsolution due to the very large surface of the spherules with respects the lapilli (Junqueira-Brod et al., 1999; Stoppa et al., 2003). Concentric-shelled lapilli found into maars are important because they represent the interface between the erupting magma and the volatile component. This is germane to the concept of sub volcanic fluidisation of tuffisite which implies pristine exolution of large amount of juvenile gases (see discussion in Lloyd & Stoppa, 2003).

The CVF concentric-shelled lapilli and spherules are completely different from particles produced by Zimanowski et al. (1997) experiment because experimental spherules lack concentric structure and a core. Instead, Kelvin-Helmholtz instability is produced by liquids having different densities moving at various speeds. The Kelvin-Helmholtz instability occurs when shear is present within a continuous fluid, or when there is sufficient velocity difference across the interface between two fluids. This can start lapilli formation in a conduit where carbonatite and silicate liquids may be physically separated according their immiscibility predicted by experiments of Kjarsgaard & Hamilton (1989) and formed according the "spin-lapilli" model proposed by Junqueira-Brod et al. (1999). Textural study of tuffisites and related extrusive deposits suggest that the immiscible carbonate fraction can be incorporated in the lapilli or form the external carbonatite layers of them and the rest sprayed as ash-sized fragments and spherules (Stoppa et al., 2003). Lapilli in tuffisite and related extrusive deposits are typically shelled by a smooth cover of very fine-grained carbonate (Carracedo Sánchez et al., 2009; Stoppa & Woolley, 1997). This may explain why concentric shelled lapilli are restricted to rocks containing igneous carbonate and spinning lapilli are found also deep-inside diatremes. Notably, concentric-lapilli and dense welded matrix composed of melt spherules are not found in natural hydrovolcanic tuffs.

Concentric-shelled lapilli and accretionary lapilli formation require totally different genetic conditions. Accretionary lapilli have low density (1.2 gr/cm^3, Schumacher & Schmincke, 1991) and can be sustained in suspension by vigorous convective cells in the eruptive column. Concentric lapilli and bombs, have much higher density and formation in the same condition of accretionary lapilli is unrealistic. Stoppa et al. (2011) report terminal velocity (Vt) curves built using measured density and size for accretionary lapilli and concentric-shelled lapilli from Calatrava and Monte Vulture Italy. They argue that concentric shelled lapilli and bombs cannot be generated in the convective region of a volcanic plume as are accretionary lapilli and require Vt similar to that required to transport mantle nodules. Owing their deep origin it is clear that concentric- shelled lapilli and mantle nodules are not carried to the surface by hydrovolcanism.

7.2 Role of H_2O

There is a general agreement in the previous Spanish literature that CVF eruption style was influenced by heterogeneous ground permeability producing interaction of a low-viscosity, high-temperature melt feeding system with a spatially restricted aquifer, i.e. hydrovolcanism (Sheridan & Wohletz, 1983). If groundwater-flow rates are insufficient to maintain persistent hydro-volcanic eruption, activity evolves towards strombolian activity or hawaiian fountains and lava flows (Peinado et al., 2009). We note, that in CVF this may be restricted to high hydraulic conductivity facies in the basin sedimentary infilling as there are no remarks of "impure coolants" such as wet sediments in the deposits. In addition, hydro-volcanic activity often ends the eruptions (e.g., Almodóvar del Campo, Fig. 5c-h) suggesting that water entered the conduit late owing the dropping of pressure in the conduit itself. In these rocks, hydrous phases such as phlogopite and amphibole indicate minor juvenile H_2O as well. It was argued that ultramafic melts (e.g., kimberlite magma) may contain high amount of juvenile H_2O rather than CO_2. In the latter case, it is important to note that the presence of juvenile water, inferred for kimberlitic magma, which also contain spin-lapilli and abundant mantle nodules but not concentric-shelled lapilli, does not affect CO_2-triggered diatresis and cannot be confused with hydrovolcanic feature because it does not imply a contribution of external water.

7.3 Role of CO_2

There is general consensus that carbonatite and associated silicate melts (foidite melilitite) are near-primary melts generated in the mantle (e.g., Bailey, 1993). Carbon dioxide is the major volatile phase in carbonatites and melilitites where concentric-shelled lapilli have been reported (Junqueria-Brod et al., 1999; Lloyd & Stoppa, 2003). Low viscosity, high temperature, ultramafic carbonate-rich magma needs rapid ascent to erupt both silicate liquid and mantle xenoliths from the asthenospheric depths (Humphreys et al., 2010). A conduit flow mechanism is sufficient; diapiric ascent and melt percolation are too slow (Anderson, 1979; Spera, 1987). Initial carbonatitic percolating melts start out with high CO_2 and those that follow an eruptive adiabat will exsolve gases and pass through Olafsson & Eggler's (1983) carbonate-out boundary, leading to massive CO_2 exsolution (Bailey, 1985). At the same time, the melt temperature approaches the liquidus where major gas foaming (CO_2) occurs. In the CO_2 diatresis model of Bailey (1985), foaming at mantle depth in near conjunction with the carbonate-out boundary has three effects: (i) production of minute spherules when melt is fragmented; (ii) very rapid quenching of these spherules, (iii) concentration of the juvenile gases largely as the fluidising medium. If a firm link between deep-seated CO_2, diatresis and tuffisite formation is established, no other model is required to explain the formation of these maars. In fact, the diatresis phenomena responsible for diatrema formation is well able to excavate the maar itself and to fragmentate the country rocks. This hypotheses confirmed by several studies on maar associated to ultra-alkaline melts poses the problem to classify this peculiar volcanic activity which may be considered a form of strombolian to sub-plinian activity modified by physical proprieties of CO_2 instead of H_2O vapour/gas. High density of CO_2 probably prevents the formation of high buoyant volcanic column and favours later expansion of surges. Diatreme formation is also important in the formation of a maar crater which is considered here the superficial expression of the growing of a diatreme beneath it. We suggest the term "diatremic eruption" for this kind of CO_2 related maar formation.

8. Conclusions

1. At CVF, hydrovolcanic deposits are clearly distinguishable from magmatic deposits in terms of volcano sedimentary structures, texture, grain-size and lapilli key features. The accretionary lapilli found in hydro-volcanic deposit originate in the convective region of an eruptive column by wet ash-particles accretion and require passage from vapour to condensed water. The formation of concentric-shelled lapilli, which are typical of CVF maars and scoria cones as well as many other carbonatitic maars, cannot be realistic in a volcanic column for these high density, high temperature pyroclasts.
2. Concentric-shelled lapilli are ubiquitous; accretionary lapilli are limited to hydrovolcanic deposit. We deduce that hydrovolcanism at CVF was neither necessary, nor sufficient condition for concentric shelled lapilli. In fact, the concentric-shelled lapilli formation cannot be explained by hydrovolcanic process.
3. Highly vesiculated fragments in CVF maars indicate abundant presence of juvenile volatiles. On the other hand, agglutinated spherule of melilitite and carbonatite forming concentric shells around a mantle-rock kernel are considered as evidence of deep seated magma fragmentation in a diatreme. The coincidence of their sampling depth of the mantle kernel with the depth of CO_2 exolution (diatresis) strongly suggest that the magma propulsion and eruption was triggered and sustained by mantle CO_2 and not juvenile or external water.
4. We interpret the carbonate matrix of the CVF tuffs as the quench of an immiscible carbonatitic liquid separated by the eruptive carbonatitic melilitites magma. The matrix is composed of carbonatite spherules sprayed directly as a phisically separated magmatic liquid. This process is documented by carbonatite liquid blebs is the melilitite pyroclasts.
5. CO_2 diffused and climatic emissions are at present the most notable activity in CVF. We argue that due to the carbonatitic nature of CFV magma CO_2 was the dominant gas in the volcanic system. Diatresis is here preferred as boosting maar formation as also suggested by Mattsson & Tripoli (2011). Maar formation at CVF is interpreted as mainly

due to magmatic mechanisms. Passage from maar stage to strombolian/effusive stage is interpreted as the consequence of the dropping of volatiles concentrated on the top of the magmatic column in the conduit.

9. Acknowledgments

We thank Ken Bailey whose adamantine advocacy of carbonatites volcanism in Europe inspired our research. We are indebted with Mercedes Munoz, Eumenio Ancochea Soto of the Complutense University of Madrid and Elena Gonzáles Cárdenas of the Castilla-La Mancha University for the precious guidance during several field trips. We are greatly indebted with Karoly Nemeth, Felicity Lloyd and two anonymous referees for their precious comments.

10. References

Ancohea, E. & Ibarrola, E. (1982). Caracterizaciön geoquímica del vulcanismo de la regiön volcánica central española. *Boletin de la Real Sociedad Espanola de Historia Natural. Seccion Geologica*, Vol. 80, pp. 57-88.

Ancochea, E. (1982). Evolutiön espacial y temporal del volcanismo reciente de España Central, *Ph.D. Thesis*, Universidad Complutense de Madrid, Madrid.

Anderson, O.L. (1979). The role of fracture dynamics in kimberlite pipe formation, In: *Kimberlites, Diatremes and Diamonds: Their Geology, Petrology, and Geochemistry*, Boyd, F.R. & Meyer, H.O.A. (Eds.), pp. 344-353, American Geophysical Union, ISBN 9780875902128, Washington.

Bailey, D.K. & Collier, J.D. (2000). Carbonatite-melilitite association in the Italian collision zone and the Ugandan rifted craton: significant common factors. *Mineralogical Magazine*, Vol. 64, pp. 675-682.

Bailey, D.K. (1985). Fluids, melts, flowage, and styles of eruption in alkaline ultramafic magmatism. *Geological Society of South Africa*, Vol. 88, pp. 449-457.

Bailey, D.K. (1989). Carbonate melt from the mantle in the volcanoes of south-east Zambia. *Nature*, Vol. 388, pp. 415-418.

Bailey, D.K. (1993). Carbonate magmas. *Journal of the Geological Society London*, Vol. 150, pp. 637-651.

Bailey, K.; Garson, M.; Kearns, S. & Velasco, A.P. (2005). Carbonate volcanism in Calatrava, central Spain: a report on the initial findings. *Mineralogical Magazine*, Vol. 69, pp. 907-915.

Bailey, K.; Kearns, S.; Mergoil, J.; Daniel, J.M. & Paterson, B. (2006). Extensive dolomitic volcanism through the Limagne Basin, central France: a new form of carbonatite activity. *Mineralogical Magazine*, Vol. 70, pp. 231-236.

Balogh, K.; Németh, K.; Itaya, T.; Molnár, F.; Stewart, R.; Thanh, N.X.; Hyodo, H. & Daróczi, L. (2010). Loss of ^{40}Ar(rad) from leucite-bearing basanite at low temperature: implications on K/Ar dating. *Central European Journal of Geosciences*, Vol. 2, No. 3, pp. 385-398.

Becerra-Ramírez, R.; González Cárdenas, E.; Dóniz, J.; Gosálvez Rey, R.U. & Escobar, E. (2010). Análisis morfométrico de los volcanes de la cuenca media del río Jabalón.

Región Volcánica del Campo de Calatrava (Ciudad Real, España). In: Aportaciones recientes en volcanología 2005-2008, González, E.; Escobar, E.; Becerra, R.; Gosálvez, R.U. & Dóniz, J. (Eds.), pp. 111-115, Centro de Estudios Calatravos, UCLM, Ministerio de Ciencia y Tecnología, ISBN: 978-84-614-1025-5

Bednarz, U. & Schmincke, H.-U. (1990). Evolution of the Quaternary melilite-nephelinite Herchenberg volcano (East Eifel). *Bulletin of Volcanology*, Vol. 52, pp. 426-444.

Bergamín, J.F. & Carbo, A. (1986). Discusion de modelos para la corteza y manto superior en la zona sur del area centroiberica, basados en anomalias gravimetricas. *Estudios Geológicos*, Vol. 42, pp. 143-146.

Bonadonna, F.P. & Villa, I. (1986). Estudio geocrológico del volcanismo de las Higueruelas, In: *Actas de la I Reunión de Estudios Regionales de Castilla-La Mancha*, Vol. 3, pp. 249-253, Albacete, mayo 1984.

Brey, G. & Green, D.H. (1977). Systematic study of liquidus phase relations in olivine melilitite +H_2O +CO_2 at high pressures and petrogenesis of an olivine melilitite magma. *Contribution to Mineralogy and Petrology*, Vol. 61, pp. 141-162.

Callicoat, J.S.; Hamer, C. & Chesner, C.A. (2008). Pelletal lapilli in ultramafic diatremes, Avon Volcanic District, Missouri, *Proceedings of North-Central Section, GSA - 42nd Annual Meeting*, Evansville, Indiana, USA, 24-25 April, 2008.

Calvo, D.; Barrancos, J.; Padilla, G.; Brito, M.; Becerra, R.; González, E.; Gosálvez, R.; Escobar, E.; Melián, G.; Nolasco, D.; Padrón, E.; Marrero, R.; Hernández, P. & Pérez, N. (2010). Emisión difusa de CO_2 en el Campo de Calatrava, Ciudad Real. In: Aportaciones recientes en volcanología 2005-2008, González, E.; Escobar, E.; Becerra, R.; Gosálvez, R.U. & Dóniz, J. (Eds.), pp. 51-55, Centro de Estudios Calatravos, UCLM, Ministerio de Ciencia y Tecnología.

Carracedo Sánchez, M.; Sarrionandia, F.; Arostegui, J.; Larrondo, E. & Gil Ibarguchi, J.I. (2009). Development of spheroidal composite bombs by welding of juvenile spinning and isotropic droplets inside a mafic eruption column. *J. Volcanol. Geoth. Res.*, Vol. 186, pp. 265-279.

Cebriá, J.-M. & López-Ruiz, J. (1995). Alkali basalts and leucitites in an extensional intracontinental plate setting: The late Cenozoic Calatrava Volcanic Province (central Spain). *Lithos*, Vol. 35, pp. 27-46.

Cebriá, J.-M. & López-Ruiz, J. (1996). A refined method for trace element modelling of nonmodal batch partial melting processes: the Cenozoic continental volcanism of Calatrava, central Spain. *Geochim. Cosmochim. Ac.*, Vol. 60, pp. 1355-1366.

Cebriá, J.-M.; Martín-Escorza, C.; López-Ruiz, J.; Morán-Zenteno, D.J. & Martiny, B.M. (2011). Numerical recognition of alignments in monogenetic volcanic areas: Examples from the Michoacán-Guanajuato Volcanic Field in Mexico and Calatrava in Spain. *J. Volcanol. Geoth. Res.*, Vol. 201, pp. 73-82, doi:10.1016/j.jvolgeores.2010.07.016

Cloos, H. (1941). Bau und Tätigkeit von Tuffschloten. Untersuchungen an dem Schwäbischen Vulkan. *Geol. Rundsch.*, Vol. 32, pp. 709-800.

Crespo, A. & Lunar, R. (1997). Terrestrial hot-spring Co-rich Mn mineralisation in the Pliocene Quaternary Calatrava Region (Central Spain). Geological Society Special Publication, Vol. 119, pp. 253-264.

De Vicente, G.; Vegas, R.; Muñoz Martín, A.; Silva, P.G.; Andriessen, P.; Cloetingh, S.; González Casado, J.M.; Van Wees, J.D.; Álvarez, J.; Carbó, A. & Olaiz, A. (2007). Cenozoic thick-skinned deformation and topography evolution of the Spanish Central System. *Global Planet. Change*, Vol. 58, pp. 335-381.

Díaz, J. & Gallart, J. (2009). Crustal structure beneath the Iberian Peninsula and surrounding waters: a new compilation of deep seismic sounding results. *Phy. Earth Planet. In.*, Vol. 173, pp. 181-190.

EPTISA (2001). Estudio de caracterización geológica e hidrogeológica del área afectante al sondeo surgente de Granátula de Calatrava (Ciudad Real). Conclusiones. Inédito. Junta de Comunidades de Castilla-La Mancha, Ciudad Real, 2001.

Ferguson, J.; Danchin, R.V. & Nixon, P.H. (1973). Petrochemistry of kimberlite autoliths. In: *Lesotho Kimberlites*, Nixon, P.H. (Ed.), pp. 285-293, Lesotho National Development Corporation, Maseru, Lesotho.

Fernàndez, M.; Marzán, I. & Torne, M. (2004). Lithospheric transition from the Variscan Iberian Massif to the Jurassic oceanic crust of the Central Atlantic. *Tectonophysics*, Vol. 386, pp. 97-115.

Gonzáles Cárdenas, E.; Gosálvez Rey, R.U.; Becerra Ramírez, R. & Escobar Lahoz, E. (2007). Actividad eruptiva holocena en el Campo de Calatrava (Volcán Columba, Ciudad Real, España). In: *XII Reunión Nacional de Cuaternario*, Lario, J. & Silva, G. (Eds.), pp. 143-144, Ávila, España.

González Cárdenas, E. & Gosálvez Rey, R.U. (2004). Nuevas aportaciones al conocimiento del hidrovolcanismo en el Campo de Calatrava (España). In: *VIII Reunión Nacional de Geomorfología*, septiembre 2004, Toledo, España.

González Cárdenas, E.; Gosálvez, R.U.; Becerra, R. & Escobar, E. (2010). El trabajo reciente de los geógrafos en el volcanismo del Campo de Calatrava. In: Aportaciones recientes en volcanología 2005-2008, González, E.; Escobar, E.; Becerra, R.; Gosálvez, R.U. & Dóniz, J. (Eds.), pp. 91-95, Centro de Estudios Calatravos, UCLM, Ministerio de Ciencia y Tecnología.

Gosálvez, R.U.; Becerra, R.; González, E. & Escobar, E. (2010). Evolución de la emisión de CO_2 en La Sima. Campo de Calatrava (Ciudad Real, España). In: Aportaciones recientes en volcanología 2005-2008, González, E.; Escobar, E.; Becerra, R.; Gosálvez, R.U. & Dóniz, J. (Eds.), pp. 101-103, Centro de Estudios Calatravos, UCLM, Ministerio de Ciencia y Tecnología.

Gurney, J.J.; Moore, R.B.; Otter, M.L.; Kirkley, M.B.; Hops, J.J. & McCandless, T.E. (1991). Southern African kimberlites and their xenoliths. In: *Magmatism in Extensional Structural Settings*, Kampunzu, A.B. & Lubala, R.T. (Eds.), pp. 495-536, Springer, Heidelberg.

Hamilton, D.L.; Freestone, I.C.; Dawson, J.B. & Donaldson, C.H. (1979). Origin of carbonatites by liquid immiscibility. *Nature*, Vol. 279, pp. 52-54.

Hay, R.L. (1978). Melilite-carbonatite tuffs in the Laetolil Beds of Tanzania. *Contrib. Mineral. Petr.*, Vol. 67, pp. 357-367.

Hoernle, K.,Y. Zhang S., Graham D. (1995). Seismic and geochemical evidence for largescale mantle upwelling beneath the Eastern Atlantic and Western and Central Europe, *Nature*, Vol. 374, pp. 219-229.

Humphreys, E.R.; Bailey, K.; Hawkesworth, C.J.; Wall, F.; Najorka, J. & Rankin, A.H. (2010). Aragonite in olivine from Calatrava, Spain - Evidence for mantle carbonatite melts from >100 km depth. *Geology*, Vol. 38, pp. 911-914, doi:10.1130/G31199.1

Junqueira-Brod, T.C.; Brod, J.A.; Thompson, R.N. & Gibson, S.A. (1999). Spinning droplets - A conspicuous lapilli-size structure in kamafugitic diatremes of Southern Goiás, Brazil. *Revista Brasileira de Geociências*, Vol. 29, pp. 437-440.

Keller, J. (1981). Carbonatitic volcanism in the Kaiserstuhl alkaline complex: Evidence for high fluid carbonatitic melts at the earth's surface. *J. Volcanol. Geoth. Res.*, Vol. 9, pp. 423-431.

Kjarsgaard, B.A. & Hamilton, D.L. (1989). The genesis of carbonatites by immiscibility. In: *Carbonatites: Genesis and Evolution*, Bell, K. (Ed.), pp. 388-404, Unwin Hyman, London.

Le Bas M.J. (1981). Carbonatite magmas. *Mineral. Mag.*, Vol. 44, pp. 133-140.

Lloyd, F.E. & Stoppa, F. (2003). Pelletal lapilli in diatremes - some inspiration from the old masters. *Geolines*, Vol. 15, pp. 65-71.

Lloyd, F.E. (1985). Experimental melting and crystallisation of glassy olivine melilitites. *Contrib. Mineral. Petr.*, Vol. 90, pp. 236-243.

Lloyd, F.E.; Woolley, A.R.; Stoppa, F. & Eby, G.N. (2002). Phlogopite-biotite parageneses from the K-mafic-carbonatite effusive magmatic association of Katwe-Kikorongo, SW Uganda. *Miner. Petrol.*, Vol. 74, pp. 299-322.

López-Ruiz, J.; Cebriá, J.M. & Doblas, M. (2002). Cenozoic volcanism I: the Iberian Peninsula. In: *The Geology of Spain*, Gibbons, W. & Moreno, T. (Eds.), pp. 417-438, The Geological Society of London, London.

Martín-Serrano, A.; Vegas, J.; García-Cortés, A.; Galán, L.; Gallardo-Millán, J.L.; Martín-Alfageme, S.; Rubio, F.M.; Ibarra, P.I.; Granda, A.; Pérez-González, A. & García-Lobón, J.L. (2009). Morphotectonic setting of maar lakes in the Campo de Calatrava Volcanic Field (Central Spain, SW Europe). *Sediment. Geol.*, Vol. 222, pp. 52-63.

Mattsson, H.B. & Tripoli, B.A. (2011). Depositional characteristics and volcanic landforms in the Lake Natron-Engaruka monogenetic field, northern Tanzania. *J. Volcanol. Geoth. Res.*, Vol. 203, pp. 23-34, doi:10.1016/j.jvolgeores.2011.04.010

Mitchell, R.H. (1997). Kimberlites, Orangeites, Lamproites, Melilitites and Minettes: A Petrographic Atlas. Almaz Press Inc., Thunder Bay, Canada.

Olafsson, M. & Eggler, D.H. (1983). Phase relations of amphibole, amphibole-carbonate, and phlogopite-carbonate peridotite: petrologic constraints on the asthenosphere. *Earth Planet. Sc. Lett.*, Vol. 64, pp. 305-315.

Peinado, M.; García Rayego, J.L.; González Cárdenas, E. & Ruiz Pulpón, Á.R. (2009). Itinerarios geográficos y paisajes por la provincia de Ciudad Real. Guía de salidas de campo del XXI Congreso de Geógrafos Españoles. Imprenta Provincial, Ciudad Real.

Peña, L. (1934). Mapa geológico de España (1:50 000) Hoja nº 811, Moral de Calatrava. Instituto Geológico y Minero de España, Madrid.

Poblete Piedrabuena, M.A. (1997). Evolución y características geomorfológicas del sector central del Campo de Calatrava (Ciudad Real). In: *Elementos del Medio Natural en la provincia de Ciudad Real*, García Rayego, J.L. & González Cárdenas, E. (Eds.), pp. 131-159, UCLM, Cuenca.

Poblete Piedrabuena, M.A. (1992).. El empleo de los vocabolos maar, cráter de explosión y diatrema en morfología volcánica. *Ería*, Vol. 27, pp. 89-94, ISSN: 0211-0563

Portero García, J.M.; Ramírez Merino, J.I.; Ancochea Soto, E. & Pérez González, A. (1988). Mapa geológico de España (1:50 000) Hoja nº 784, Ciudad Real. Ministerio de Industria y Energia, Madrid.

Redondo, R. & Yélamos, J.G. (2005). Determination of CO_2 origin (natural or industrial) in sparkling bottled waters by $^{13}C/^{12}C$ isotope ratio analysis. *Food Chem.*, Vol. 92, pp. 507-514.

Riley, T.R.; Bailey, D.K. & Lloyd, F.E. (1996). Extrusive carbonatite from the quaternary Rockeskyll complex, West Eifel, Germany. *Can. Mineral.*, Vol. 34, pp. 389-401.

Rosatelli, G.; Wall, F. & Stoppa, F. (2007). Calcio-carbonatite melts and metasomatism in the mantle beneath Mt. Vulture (Southern Italy). *Lithos*, Vol. 99, pp. 229-248.

Rosatelli, G.; Wall, F.; Stoppa, F. & Brilli, M. (2010). Geochemical distinctions between igneous carbonate, calcite cements, and limestone xenoliths (Polino carbonatite, Italy): spatially resolved LAICPMS analyses. *Contrib. Mineral. Petr.*, Vol. 160, pp. 645-661.

Schumacher, R. & Schmincke, H.-U. (1991). Internal structure and occurrence of accretionary lapilli - a case study at Laacher See Volcano. *B. Volcanol.*, Vol. 53, pp. 612-634.

Sheridan, M.F. & Wohletz, K.H. (1983). Hydrovolcanism: basic considerations and review. *J. Volcanol. Geoth. Res.*, Vol. 17, pp. 1-29.

Spera, F.J. (1987). Dynamics of translithospheric migration of metasomatic fluid and alkaline magma. In: *Mantle metasomatism*, Menzies, M.A. & Hawkesworth, C.J. (Eds.), pp. 1-20, Academic Press, London.

Stachel, T.; Lorenz, V. & Stanistreet, I.G. (1994). Gross Brukkaros (Namibia) - an enigmatic crater-fill reinterpreted as due to Cretaceous caldera evolution. *B. Volcanol.*, Vol. 56, pp. 386-397.

Stoppa, F. & Lupini, L. (1993). Mineralogy and petrology of the Polino monticellite calciocarbonatite (Central Italy). *Miner. Petrol.*, Vol. 49, pp. 213-231.

Stoppa, F. & Principe, C. (1998). Eruption style and petrology of a new carbonatitic suite from the Mt. Vulture (Southern Italy): The Monticchio Lakes Formation. *J. Volcanol. Geoth. Res.*, Vol. 80, pp. 137-153.

Stoppa, F. & Woolley, A.R. (1997). The Italian carbonatites: field occurrence, petrology and regional significance. *Miner. Petrol.*, Vol. 59, pp. 43-67.

Stoppa, F. (1996). The San Venanzo maar and tuff ring, Umbria, Italy: eruptive behaviour of a carbonatite-melilitite volcano. *B. Volcanol.*, Vol. 57, pp. 563-577.

Stoppa, F.; Lloyd, F.E. & Rosatelli, G. (2003). CO_2 as the propellant of carbonatite-kamafugite cognate pairs and the eruption of diatremic tuffisite. *Period. Mineral.*, Vol. 72, pp. 205-222.

Stoppa, F.; Lloyd, F.E.; Tranquilli, A. & Schiazza, M. (2011). Comment on: Development of spheroid "composite" bombs by welding of juvenile spinning and isotropic droplets inside a mafic "eruption" column by Carracedo Sánchez et al. (2009). *J. Volcanol. Geoth. Res.*, Vol. 204, pp. 107-116, doi: 10.1016/j.jvolgeores.2010.11.017

Stoppa, F.; Rosatelli, G.; Wall, F. & Jeffries, T. (2005). Geochemistry of carbonatite-silicate pairs in nature: a case history from Central Italy. *Lithos*, Vol. 85, pp. 26-47.

Villaseca, C.; Ancochea, E.; Orejana, D. & Jeffries, T.E. (2010). Composition and evolution of the lithospheric mantle in central Spain: inferences from peridotite xenoliths from the Cenozoic Calatrava volcanic field. *Geol. Soc.*, London, Vol. 337, pp. 125-151, doi:10.1144/SP337.7

White, J.D.L. & Ross, P.-S. (2011). Maar-diatreme volcanoes: A review. *J. Volcanol. Geoth. Res.*, Vol. 201, pp. 1-29, doi:10.1016/j.jvolgeores.2011.01.010

Woolley, A.R. & Church, A.A. (2005). Extrusive carbonatites: a brief review. *Lithos*, Vol. 85, pp. 1-14.

Yélamos, J.G. & Villarroya Gil, F.I. (1991). Variacíon de la pizometría y el caudal en cuatro explotaciones de aguas subterráneas en el acuífero del terciario detritico de Madrid. *Boletín geológico y minero*, Vol. 102, pp. 857-874, ISSN: 0366-0176

Yu, X.; Mo, X.; Liao, Z.; Zhao, X. & Su, Q. (2003). Geochemistry of kamafugites and carbonatites from West Qinling area (China). *Period. Mineral.*, Vol. 1, pp. 161-179.

Zimanowski, B.; Büttner, R.; Lorenz, V. & Häfele, H.-G. (1997). Fragmentation of basaltic melt in the course of explosive volcanism. *J. Geophys. Res.*, Vol. 102, pp. 803-814.

2

An Overview of the Monogenetic Volcanic Fields of the Western Pannonian Basin: Their Field Characteristics and Outlook for Future Research from a Global Perspective

Károly Németh
Massey University
New Zealand

1. Introduction

Miocene to Pleistocene basaltic volcanic fields are common in the Pannonian Basin in Central Europe (Figs 1 & 2). Included in these fields are the here described monogenetic volcanic fields of western Hungary. These volcanic fields provide excellent exposures to explore the volcanic facies architecture of monogenetic volcanoes that formed during a period of intra-continental volcanism that lasted over 6 million years (Fig. 2). Over this 6 millions of years eruptive history (Wijbrans et al., 2007), volcanic fields such as the Bakony-Balaton Highland (BBHVF) or the Little Hungarian Plain Volcanic Field (LHPVF) formed (Fig. 1) as typical low magma-flux, time-predicted fields that were largely tectonically-controlled rather than magmatically-controlled (Martin & Németh, 2004; Kereszturi et al., 2011). The preserved volcanic eruptive products of the BBHVF, including pyroclastic, effusive and intrusive rocks, have been estimated to be about 3 km^3, significantly larger than those erupted through the LHPVF (Martin & Németh, 2004; Kereszturi et al., 2011). Considering the potential erosion of distal air fall tephras and the common juvenile pyroclast-poor nature of the majority of the preserved pyroclastic rocks, a recalculation of eruptive volumes to dense rock equivalent (DRE) values would likely yield a total erupted volume of less than 5 km^3 for the western Hungarian Miocene to Pleistocene volcanic fields. Here we provide a short review of the current research on these monogenetic volcanic fields in western Hungary with an aim to characterise their pyroclastic successions and infer the eruptive environment where they erupted and accumulated. Furthermore, we define key research subjects for future study on these fields on the basis of our current knowledge. Such future research directions for the western Hungarian monogenetic volcanic fields could significantly contribute to our understanding of the volcanic evolution, eruption styles, and preservation potential of monogenetic volcanic fields in general. A "sister" volcanic field approach is also proposed to link these volcanic fields to other, similar volcanic fields worldwide (Németh et al., 2010).

2. Monogenetic volcanic fields of the Western Pannonian Basin (WPB)

The volcanic fields of the Western Pannonian Basin (WPB) are erosional remnants forming buttes and mesas (Fig. 3A) that are commonly composed of gently inward dipping primary pyroclastic rocks (Fig. 3B) and capping lavas (Martin & Németh, 2004). The centre parts of the volcanic buttes are composed of tuff breccias commonly rich in accidental lithic fragments from the underlying basement rocks (Fig. 3C). The majority of the preserved pyroclastic rocks are lapilli tuffs that are rich in volcanic glass shards (Fig. 3D) and accidental lithic fragments (Martin & Németh, 2004; Németh, 2010b). These volcanic fields are commonly referred to as monogenetic volcanic fields (White, 1991a; Valentine & Gregg, 2008; Manville et al., 2009; Németh, 2010a), attesting to the small-volume of the individual volcanoes that make up the field. The small volume of the individual volcanoes and the generally simple volcanic architecture of the volcanoes are the key features that define these volcanoes as monogenetic. In spite of the small volume nature of the preserved and inferred original volcanic landforms of these volcanic fields, there are volcanic complexes that were clearly erupted in various eruptive episodes, commonly through laterally shifted vents that produced nested volcanic complexes (Auer et al., 2007; Kereszturi & Németh, 2011). These volcanic fields can also be classified as typical phreatomagmatic volcanic fields (Németh, 2010a) on the basis of the overwhelming evidence of magma - water interaction driven explosive eruptions, at least in the initial stage of the eruptive history of the majority of the volcanoes of western Hungary. The phreatomagmatic explosive eruption style has been interpreted due to the abundance of preserved pyroclastic rock units in volcanic glass shards with macro- and micro-textural features characteristic of sudden chilling of the rising basaltic melt upon contact with external water, as demonstrated by comparison of experimental volcanology results (Büttner et al., 2002; Büttner et al., 2006) with natural glass shards (Dellino & LaVolpe, 1996; Büttner et al., 1999; Dellino & Liotino, 2002). Textural features, such as the low vesicularity and angular and rugged shape, evident in volcanic glass shards from the western Hungarian volcanic fields are generally accepted to support magma and water explosive interaction in other locations (Heiken & Wohletz, 1986; Büttner et al., 1999; Dellino, 2000; Morrissey et al., 2000; Dellino & Kyriakopoulos, 2003) (Fig. 4). The volcanic rocks of the western Hungarian Mio/Pleistocene volcanic fields are preserved in a very diverse type of volcanic landforms such as erosional remnants of maar-diatremes, tuff rings, scoria cones, lava shields and lava fields (Németh & Martin, 1999).

The original volcanic landforms of the western Hungarian volcanic fields have been reconstructed on the basis of the 3D facies architecture of the preserved pyroclastic and coherent lava rock units, the volcanic stratigraphy and the associated volcanic facies relationship with syn-eruptive country rock units. This method has been widely used in older, erosion-advanced volcanic fields such as Hopi Butte in Arizona (White, 1989; White, 1990; White, 1991b; Vazquez & Ort, 2006), Chubut in Argentina (Németh et al., 2007), Waipiata in New Zealand (Németh & White, 2003), Western Snake River Plain in Idaho (Godchaux et al., 1992; Godchaux & Bonnichsen, 2002; Brand & White, 2007) or the east Oregon volcanic fields (Heiken, 1971; Brand & Clarke, 2009), among many known fields.

In addition, the micro- and macro textural analysis of the juvenile particles of the preserved pyroclastic rocks of the WPB volcanic erosion remnants, and the component analysis of the same rocks, demonstrated clearly the abundance of country rock fragments from the known basement and Neogene basin-filling sediments which indicated a significant excavation of country rocks in the course of the eruptions (Martin & Németh, 2004). The abundance of country rocks in the pyroclastic successions is also a sign that magma fragmentation must have taken place in those strata and the released kinetic energy fragmented and excavated the rocks in situ (Lorenz & Kurszlaukis, 2007). Such a process can take place when hot magma and ground-water interact explosively below the surface and the generated shock wave fragments the wall rock (Lorenz, 1986; Wohletz, 1986; Zimanowski et al., 1986; Morrissey et al., 2000; White & Ross, 2011), allowing accidental lithic-dominated debris to exit the vent (Lorenz, 1986; Lorenz & Kurszlaukis, 2007; White & Ross, 2011). The result of this is a significant volume of excavated country rocks, the formation of mass deficit that eventually leads to a gradual collapse, and the formation of volcanic debris-filled volcanic conduit, or diatreme (White & Ross, 2011). The mechanism of the formation of a diatreme is far from well-known, and there is still argument about whether it is magmatic gas (Stoppa, 1996; Stoppa & Principe, 1997; Sparks et al., 2006; Walters et al., 2006; Suiting & Schmincke, 2009; 2010) or magma and water explosive interaction (Lorenz, 1973; 1986; Zimanowski et al., 1986; Wohletz & Heiken, 1992; Mastrolorenzo, 1994; Zimanowski et al., 1995; Zimanowski et al., 1997; Calvari & Tanner, 2011) that drives the energy release that fragments the country rocks. However, there is agreement that the resulting subsurface pipe is a volcanic and non-volcanic debris dominated zone with collapsed blocks of wall rock and complex arrays of juvenile particle enriched sub-vertical regions (Lorenz & Kurszlaukis, 2007; White & Ross, 2011).

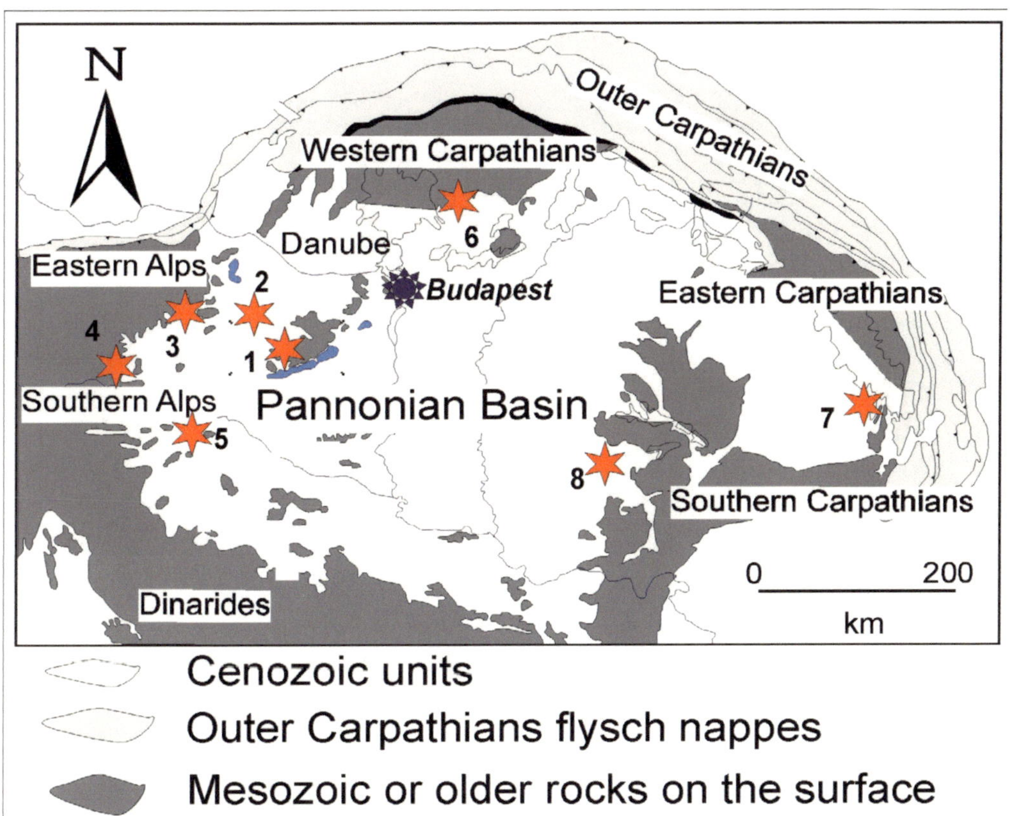

Fig. 1. Mio/Pleistocene monogenetic volcanic fields of the Pannonian Basin and their relationship with major stratigraphic units of the Carpathian – Pannon region. 1 – Bakony-Balaton Highland Volcanic Field; 2 - Little Hungarian Plain Volcanic Field; 3 – Burgenland; 4 - Styria Basin; 5 - Northern Slovenian Volcanic Field; 6 – Nógrád – Gemer Volcanic Field; 7- Persanyi Mts; and 8 - Bánát.

The Mio/Pleistocene eroded volcanoes visible in western Hungary today are inferred to represent the preserved part of the original volcanic edifices with only the crater to upper conduit-filling deposits of the former volcanoes exposed (Németh & Martin, 1999; Németh et al., 2003). This provides the opportunity to understand the 3D architecture of the proximal volcanic facies of such small-volume intra-continental volcanoes. Due to the eroded state of the Western Pannonian Basin (WPB) volcanic fields in general, these sites are not suitable for the systematic volcanic facies analysis commonly performed on young volcanoes (Vazquez & Ort, 2006) where the aim is to identify proximal to distal facies variations and understand pyroclast transportation by pyroclastic fall and density currents. Nearly each of the known individual volcanoes of the WPB volcanic fields have at least in their base pyroclastic units a record of an initial stage magma-water interaction triggered explosive eruption that formed basal phreatomagmatic tephra ring deposits around the active vents (Martin & Németh, 2004). However, gradual eruption style changes have been recognized in sites where proximal crater rim deposits are preserved, indicating a complex interplay between internal and external governing parameters on mafic explosive volcanic eruptions of western Hungary (Martin & Németh, 2004).

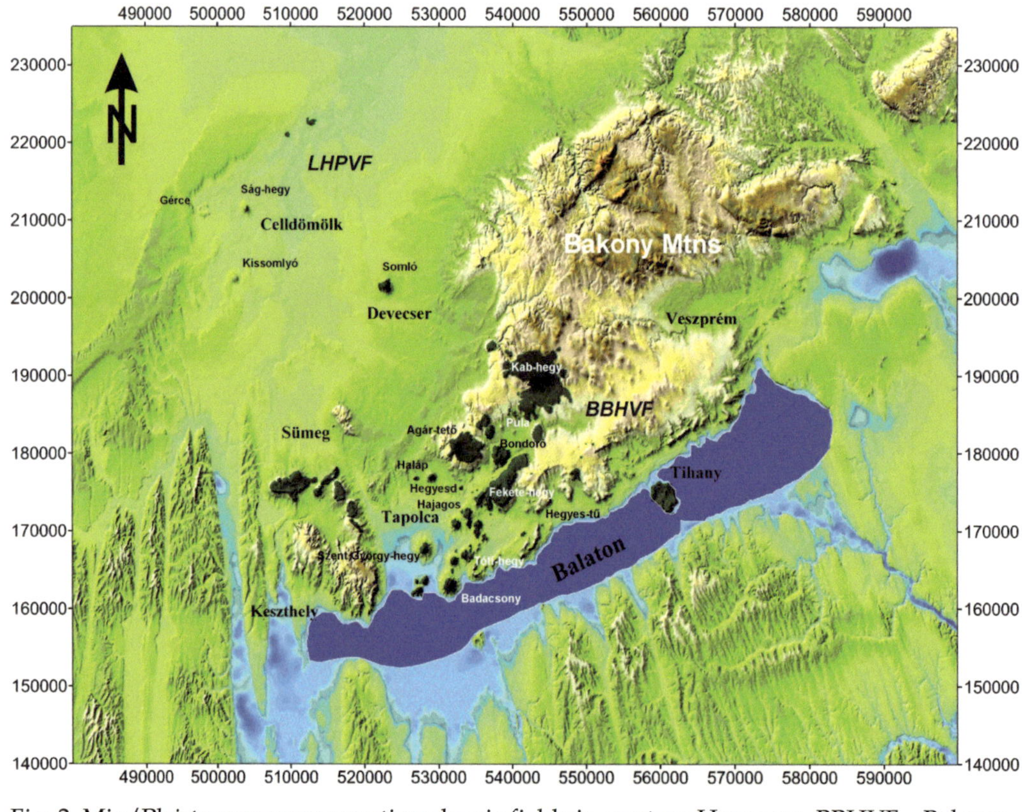

Fig. 2. Mio/Pleistocene monogenetic volcanic fields in western Hungary: BBHVF – Bakony-Balaton Highland Volcanic Field, and LHPVF - Little Hungarian Plain Volcanic Field. Distribution of volcanic rocks on the surface is marked by dark green. Hungarian Grid Reference shown in the margin with 10 km rectangular spacing.

The WPB is of particular interest in volcanic research, due to the strong correlation recently recognized between gradually changing environmental elements over millions of years and volcanic eruptions that were produced by changing eruption styles from phreatomagmatic to magmatic explosive over about 6 million years of evolution (Kereszturi et al., 2011). In addition to the long-term eruption style changes, abrupt to gradual changes in eruption styles have been recognized in a short time-scale comparable to the lifetime of an individual monogenetic volcano. The past two decades the volcanology research in the WPB confirmed the overwhelming dominance of magma-water interaction driven explosive eruption styles during the eruption history of nearly each individual volcano which is the basis of the definition of these volcanic fields as externally dominated (Martin & Németh, 2004; Németh et al., 2010; Kereszturi et al., 2011). In addition, these volcanic fields also show a marked link between the syn-volcanic country rock hydrology, the surface water abundance and the resulting volcanic eruption styles (Martin & Németh, 2004).

Fig. 3. Volcanic features preserved in typical monogenetic basaltic volcanoes of the WPB. A) Volcanic buttes of the western margin of the Kál basin of the Bakony- Balaton Highland Volcanic Field. Badacsony has an extensive lava cap, a former lava lake and associated scoria cone, while Szigliget is dominantly a pyroclastic rock dominated erosional remnant; B) Well-bedded juvenile ash- and lapilli-rich pyroclastic succession of the Gérce tuff ring in the Little Hungarian plain Volcanic Field with interbedded tuff breccia horizon (arrow) abundant in lapilli-sized Neogene sediments as country rocks; C) Peperitic domain from the Hajagos-hegy maar-diatreme from the BBHVF (lt -lapilli tuff host, fl - fluidized zone, b - basanite fragments from coherent intrusive body; D) Glassy pyroclasts from the Hegyesd diatreme core zone (BBHVF). Sideromelane glass shards (s) are blocky, moderately vesicular and their microlite content varies greatly.

The volcanic fields studied in detail in the WPB are among those that could be looked as type localities for the characterization of low-land volcanic fields that were erupted through a combined aquifer that has laterally changeable thickness and hydrological characteristics similar to those described in low-lying alluvial plains (Németh et al., 2010; Ross et al., 2011). A combined aquifer is defined to be a country rock pile beneath the volcano (especially the upper 500 metres below surface) that consists of a layer-cake-like strata of rocks with very great diversity of hydraulic conductivity, porosity, permeability and tortuosity, as defined by the state of diagenesis, grain size, bedding, and abundance of fractures and fissures in the rocks. A porous-media aquifer is defined as a rock-sediment pile with moderate hydraulic conductivity, storage capacity, and a typical but complex relationship between water-recharge and water-withdraw (i.e. the system needs time to recharge after withdrawal).

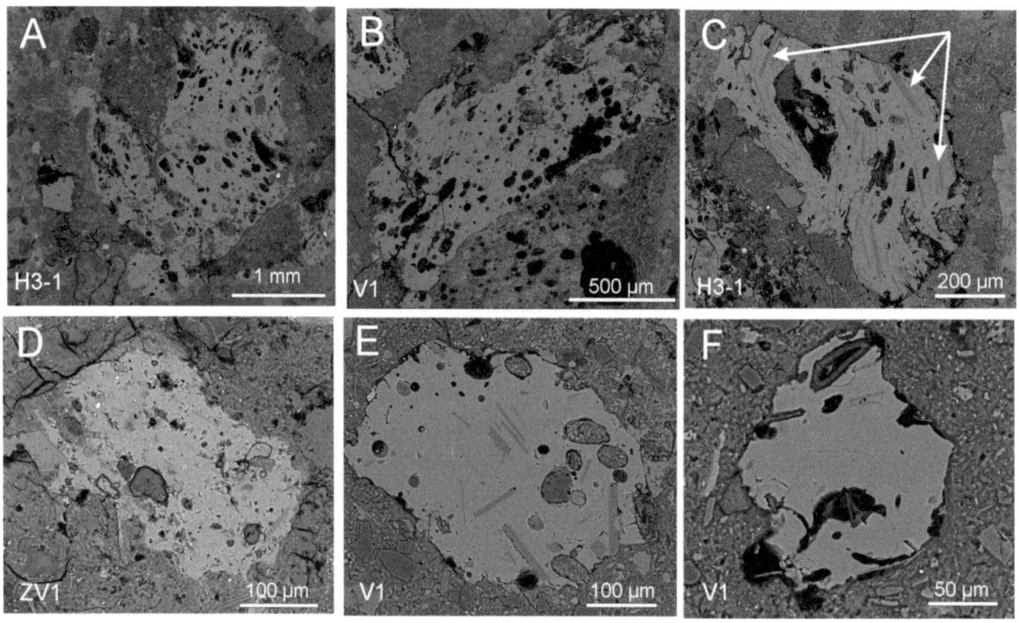

Fig. 4. Volcanic glass shards on back-scattered electron-microscopy (BSE) images from the BBHVF exhibit blocky shapes and complex particle outlines typical for brittle fragmented, and therefore fast cooled (chilled) melt upon contact with coolant (water or water-saturated sediment). The upper row (A, B & C) shows glass shards of coarse ash that are glassy in texture, bulky in shape, but carry textural features characteristic of chilling in the time the melt was still deforming in a ductile fashion. The coarser particles are considered to represent pyroclasts that were not directly derived from the interaction zone of magma and external water, but fragmented by the release of explosion energy from the main body of quickly cooling magma around the interaction zone (non-interactive particles). The lower row (D, E, & F) are fine ash particles that are more angular, blocky, with large patches of glassy areas with no microlite or vesicles indicating sudden chilling of the magma and its brittle fragmentation. These fine ash particles are considered to represent pyrolcasts directly derived from the interaction zone between hot magma and external water. Note on "C" the trachytic texture defined by microlites (arrows).

A fracture-controlled aquifer is dominated by rocks that store water in fractures (cavities) that can supply an infinite volume of water upon withdrawal if water-filled zones were encountered, but can behave as a complete aquitard in areas of no fractures or cavities. While these type of aquifers are end-members, in nature some sort of combination of these basic types form the zone that magma encounters in the upper few hundreds of metres of its to the surface. We can express the type of aquifers beneath a volcanic field to define the dominant behaviour type, such as soft-substrate versus hard-substrate aquifers (Lorenz, 2003; Sohn & Park, 2005; Auer et al., 2007; Németh et al., 2010; Ross et al., 2011). For a global comparison, the WPB's volcanic fields are compared with other localities that are erupted through an aquifer defined as a combined substrate type (e.g. soft substrate covered hard substrate) which highlights the rheology, and therefore the hydrology, of the country rocks the magma encounters (Németh et al., 2010). The WPB's volcanic fields are relatively well-described from a physical volcanology point of view. However, new globally significant research has recently identified the following as the critical parameters that strongly influencing the basic characteristics of the resulting volcanic fields: the interplay between the external and internal forcing of the eruption styles of small-volume mafic volcanoes; the influence of long term environmental changes on the variations of the dominant eruption styles in the evolution of the volcanic field; and the long term fluctuation of magmatic flux and output rates (Valentine & Perry, 2006; Valentine & Keating, 2007; Valentine & Perry, 2007; Keating et al., 2008; Brenna et al., 2010; Genareau et al., 2010; Valentine & Hirano, 2010; Brenna et al., 2011). An application of these new results for WPB volcanism could lead to a better understanding of the eruption history of the individual monogenetic volcanoes of WPB and could allow them to be comparedto similar volcanoes worldwide. In addition, new research is needed to understand the magmatic evolution over shorter time-scales that may produce complex monogenetic volcanoes closely resembling polygenetic volcanoes. Overall the current knowledge on the volcanic field evolution of the WPB is substantial enough to be able to provide a good volcanic reconstruction, applying the "sister volcanic field" approach to understand the overall volcanic field and individual volcano eruption history (Németh et al., 2010).

3. Geological setting

Volcanic fields in the western Pannonian Basin are Late Miocene to Pleistocene alkaline basaltic intracontinental fields (Szabó et al., 1992; Balogh & Németh, 2005; Wijbrans et al., 2007; Lexa et al., 2010). They consist of erosional remnants of maars, tuff rings, scoria cones, lava flows and lava fields (Martin & Németh, 2004). Four individual fields have been separated in the WPB on the basis of their location (Fig. 1): Bakony-Balaton Highland Volcanic Field (BBHVF) in Hungary (Fig. 5), Little Hungarian Plain Volcanic Field (LHPVF) in Hungary, Burgenland - Styria Volcanic Fields (BSVF) in Austria, and Northern Slovenian Volcanic Field (NSVF) in Slovenia. Research on the volcanic history of the BBHVF and LHPVF has concluded their extensively phreatomagmatic origin (Martin & Németh, 2004). In contrast we know very little about the volcanic eruption mechanism, eruptive environment and style in the case of the Austrian and Slovenian volcanic fields. Preliminary research, however, indicates their similarity to those volcanic fields located in western Hungary (Martin & Németh, 2004; Kralj, 2011). Time and space distribution of monogenetic volcanism in the western Pannonian Basin seems to show a random pattern (Pécskay et al., 1995). The earliest known alkaline basaltic rocks are located in the NSVF marking an onset of volcanism about 10-13 million years ago (Pécskay et al., 1995; Lexa et al., 2010). The date of the onset of volcanism in the BBHVF is well established at about 8 million years ago

(Balogh & Németh, 2005; Wijbrans et al., 2007; Balogh et al., 2010). The main phase of volcanism in the BBHVF falls in the 4 to 3 million years ago time period; while in the LHPVF the peak of activity took place slightly earlier, about 5 million years ago (Wijbrans et al., 2007). The volcanism ceased in the western Pannonian Basin in the Pleistocene around 2.3 Ma (Kereszturi et al., 2010; Kereszturi et al., 2011).

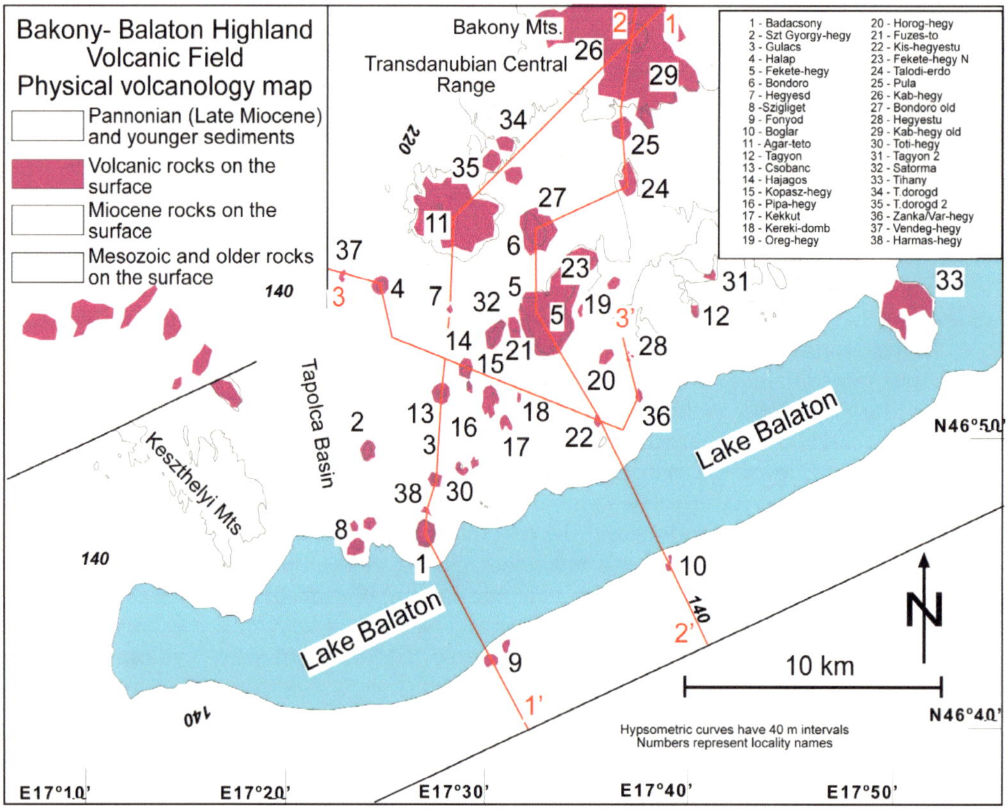

Fig. 5. Simplified geological map of the Bakony- Balaton Highland Volcanic Field marking Miocene to Pleistocene volcanic rocks on the surface. Numbers correspond to identified volcanic centres (some cases with multiple vents reconstructed). Cross sections along section lines (1-1′; 2-2′, 3-3′) are shown on Fig. 9.

4. Eruptive environment of the WPB

At the onset of the eruption of monogenetic volcanoes in the WPB in western Hungary (Fig. 2), magma began to interact with a moderate amount of groundwater in the water-saturated Neogene fluvio-lacustrine sand, silt and gravel beds (Martin & Németh, 2004). As the eruptions continued, the craters grew both vertically and laterally and the repeated phreatomagmatic blasts fragmented the deeper fractured hard rock substrate around the explosion locus, commonly giving the karst water (or any fracture controlled aquifer-store) direct access to the rising hot basaltic magma (Németh et al., 2001).

Evidence to support magma and water interaction between rising basalt magma and ground water are recorded in the lapilli tuff and tuff units abundant in glassy pyroclasts (Németh,

2010b). The micro-texture of volcanic glass shards are dominantly angular, bulky, low in vesicularity and rich in surface features typical of brittle fragmentation. This points to the fast cooling rate of these pyroclasts as they formed (Fig. 4). Microlite-rich glassy pyroclasts with micro-vesicles indicate active degassing and crystallisation of the magma upon contact with external water that freeze and lock these textures (Fig. 4).

The complex shape of pyroclasts identified from fine grained, accidental lithic fragment-rich rock units indicate complex shape parameters and fractal values typical for bulky particles with complex boundaries (Németh, 2010b), as compared to glass particles studied elsewhere (Dellino & LaVolpe, 1996; Büttner et al., 1999; Zimanowski et al., 2003). These rock types are exclusively rich in accidental rock fragments derived from the underlying pre-volcanic strata. In areas where the basement rocks of Mesozoic carbonate, Paleozoic sandstones or schist rocks are covered by thick Neogene siliciclastic semi-consolidated deposits, the preserved pyroclastic rocks contain abundant volumes of rock fragments derived from this cover bed (Fig. 6A). Pyroclastic rocks identified from regions located in areas with thin Neogene siliciclastic cover are abundant in accidental lithic rocks fragments derived from various basement rocks (Fig. 6B). On the basis of the abundance of country rocks in the majority of the preserved pyroclastic rocks of the volcanic fields in western Hungary, it can be inferred that the eruptions excavated a significant portion of the underlying pre-volcanic substrates and incorporated the material in the accumulating pyroclastic debris. The abundant volume of excavated country rocks attests to the formation of a volcanic depression, commonly referred to as a maar-diatreme volcano (Lorenz, 1986; White & Ross, 2011). While original maar volcanic landforms are rarely preserved in the WPB, the textural characteristics of the preserved pyroclastic rocks allow us to reconstruct their shape, size and volcanic facies architecture.

Fig. 6. Pyroclastic rocks of the monogenetic volcanic erosion remnants of the WPB are rich in accidental lithic fragments. Pyroclastic rocks of the Ság-hegy (A) in the LHPVF are dominated by various glassy pyroclasts (g) and abundant fragments from the Neogene siliciclastic underlying sedimentary successions such as mud aggregates (m), abundant quartz (arrows) or just mud in the matrix (yellowish homogeneous background). The view on "A" is about 2 mm across. In areas where the Neogene siliciclastic sedimentary cover was thin during the volcanism such as in Pula in the BBHVF (B), the accidental lithic fragments are dominated by clasts derived from the basement, such as Mesozoic limestone and dolomite fragments (arrows). Fragments from the Neogene sedimentary successions are commonly milled and hydrothermally altered (circle). Lens cap is about 5 cm across.

The appearance of maar volcanoes and their deposits in the Western Pannonian Basin are inferred to be strongly dependent on the paleo-hydrological conditions of the near-surface porous media, as well as the deep fracture-controlled aquifer (Martin & Németh, 2004). The seasonal variability of water-saturation of the karstic systems, as well as the climatic influence on surface and/or near sub-surface water, are considered to be potential controlling parameters of the style of explosive volcanism that took place over the evolution of an individual volcanic field (Németh et al., 2001).
Shallow but broad maar volcanoes are inferred to have been formed due to phreatomagmatic explosions of mixing magma with water-saturated siliciclastic sediments in areas where thick Neogene silicilcastic units build up the immediate pre-volcanic strata, such as in the LHPVF (Martin & Németh, 2005). Such volcanoes have often formed late magmatic infill in their maar basins, such as scoria cones and lava lakes. Today these volcanoes are preserved as lensoid shaped volcanic successions, usually capped by solidified lava lakes forming low aspect ratio mounds (Martin & Németh, 2005). The pyroclastic successions of this type of phreatomagmatic volcano are rich in sand, silt and mud from the Neogene siliciclastic basin filling sediments (Martin & Németh, 2005). Deep seated xenoliths are rare. In areas where the Neogene sedimentary cover was thin, deep maar crater formation has been inferred on the basis of the present day steep and abrupt 3D architecture of phreatomagmatic pyroclastic rock facies. The abundance of sand and silt in the matrix of lapilli tuff and tuff breaccia units with a high proportion of angular accidental lithic fragments from deep-seated hard rock units suggests that these volcanoes must have had deeper fragmentation sites that allowed excavation country rocks from deeper regions. The presence of abundant deep seated xenoliths in such volcanic erosional remnants suggests that water must have been available in those zones in fractures. This type of maar volcanoes is interpreted to develop in areas, where relatively thin Neogene fluvio-lacustrine units rested on the Mesozoic or Paleozoic fracture-controlled, e.g. karst water-bearing, aquifer (Németh et al., 2001). Nemeth et al. posed the idea that the seasonal variation of karst water aquifers and their fracture and cavity controlled hydrogeological nature (e.g. fast recharge rate, zero or unlimited hydraulic conductivity across such rock units) would vary the available water that magma could meet so that magma could pass a karst water aquifer at a time when it is nearly dry or completely filled with water. This could result strikingly different volcanic landforms forming in "spring" (maximum water capacity - phreatomagmatic) and "summer" (minimum water capacity - magmatic) (Németh et al., 2001).

5. Pyroclastic architecture of typical monogenetic volcano of the WPB

In the western Hungarian monogenetic volcanic fields, each of the identified volcanic eruptive centres represents a proximal zone of a former volcano. Nearly all of the known volcanoes had at least a short period of phreatomagmatic activity in their vent opening stage. This is recorded in the preserved fine grained, accretionary lapilli bearing (Figs 7A & B), massive to mega-ripple bedded pyroclastic rocks (Fig. 7C) formed by pyroclasts deposited by pyroclastic density currents and phreatomagmatic falls associated with intermittent initial vent breccias. These deposits are known from proximal pyroclastic successions that are commonly deposited on inward dipping inner crater walls and are represented by large blocks that collapsed in the growing volcanic crater. Erosional talus commonly covers key outcrops around the preserved volcanic buttes; however, steep cliff faces can expose pyroclastic rock facies which are typically accumulated beneath the syn-eruptive paleosurface and considered to be rock types forming diatremes. Such rocks are typically chaotic in texture, abundant in accidental lithic rock fragments from fine to coarse grained sizes. Volcaniclastic deposits accumulated in the maar craters are also known (e.g. Pula) that are typically deposited from volcaniclastic debris flows that transported volcanic debris from the tephra rings surrounding the maar basin (Németh et al., 2008). In few places (e.g. in Pula, Gérce), thick laminated rhythmic lacustrine deposits are preserved in maar craters, occasionally disturbed by ash falls from distal volcanic eruptions (Fig. 7D) or contorted the accumulated deposits by paleoseismicity (Németh et al., 2008).

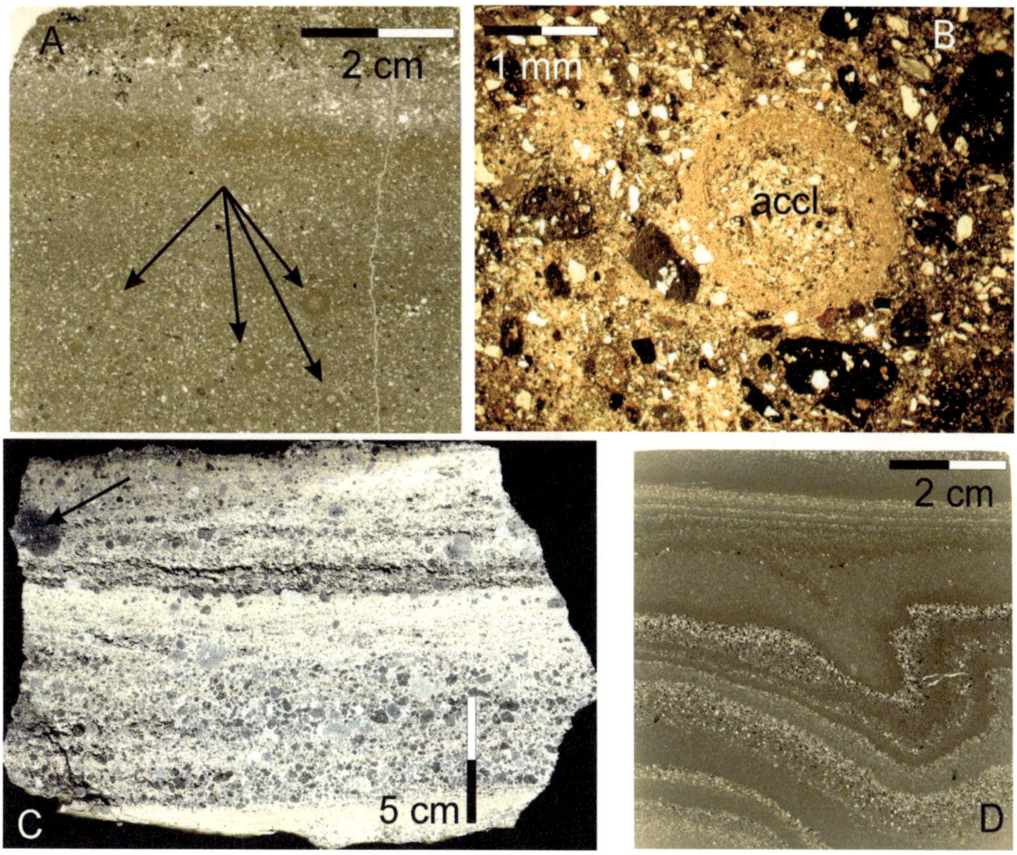

Fig. 7. Accretionary lapilli-bearing fine ash beds from the BBHVF. Accretionary lapilli are rim type (from Tihany on "A", marked by arrows and Szentbékkálla on "B" signed as "accl"). Typical lapilli tuff and tuff succession in proximal setting form pyroclastic units such as the sample shown from Kissomlyó ("C"). Fine ash accumulation in maar lakes has been recorded from the Pula maar lake deposits (dark grains).

Exposed diatreme-filling rocks in the erosion remnants in the western Pannonian Basin are rich in sedimentary grains, as well as mineral phases, from Neogene shallow marine to fluvio-lacustrine sedimentary units. These units are not preserved anymore, but their existence suggests a near intact sedimentary cover over the basement in syn-volcanic time (Németh et al., 2003). The general abundance of such clasts in the pyroclastic rocks also indicates the importance of soft substrate environment for phreatomagmatic volcanoes. Such volcanoes are commonly interpreted to form "champagne-glass" shaped maar/diatremes that are suspected to underlie the eruptive centres of the LHPVF (Lorenz, 2003). However, recent studies showed evidence that steep-walled diatremes can equally form in hard as well as soft-substrate (White & Ross, 2011). The link between substrate type and the resulting maar-diatreme volcano is so far not well understood (White & Ross, 2011). The WPB volcanic erosion remnants are abundant in rock textures which indicate interaction between magmatic bodies confined between the limit of crater walls and tephra rings and water-saturated sediments to form a great variety of peperite textures (Martin & Németh, 2007). The identification of these intra-crater peperites, accompanied with lava domes and shallow intrusions, indicates that maar/tuff ring volcanoes were likely to have been quickly flooded by ground and/or surface water, suggesting that they were excavating their craters into the region close to the level of the syn-eruptive ground-water table.

Fig. 8. Typical butte of the Tapolca Basin in the Bakony- Balaton Highland Volcanic Field with a thick lava cap sitting over pyroclastic rocks abundant in accidental lithic fragments and angular volcanic glass shards typical for pyroclastic rocks with phreatomagmatic origin.

The Neogene alkaline basaltic volcanic erosional remnants of the western Pannonian Basin are exposed from former subsurface to surface levels of the maar-diatreme volcanoes (Fig. 8). Today the lower parts of the exhumed diatremes are commonly covered by Quaternary talus flanks (Fig. 8). The outcrop availability strongly controls the identification of the facies relationships. The deepest levels of exposures are located in the western and the southern part of the area. The level of exposure of the diatreme facies reflects the ability to remove the host sediments. In the western part of the BBHVF for instance, the Neogene sedimentary cover was easily eroded and, once penetrated, allowed the diatremes to be exhumed (Fig. 9). The most strongly eroded regions are those where no subsequent lava caps sheltered the volcaniclastic sequences. Probably the eruptive centres of Balatonboglár (Boglár Volcano), Kereki-domb, Vár-hegy of Zánka, Hármas-hegy and Véndek-hegy (Fig. 9) represent the deepest exposed level of the phreatomagmatic eruptive centers (Martin & Németh, 2004) and are inferred to be exposed zones of lower diatremes, a term used for the Hopi Buttes' diatreme field by (White, 1991b). Interestingly, the diatremes and plugs located in the eastern part of the BBHVF are among the oldest volcanic features known from the western Hungarian monogenetic volcanic fields, and represent the eruptive products of the onset of the volcanism about 8 million years ago (Balogh et al., 2010).

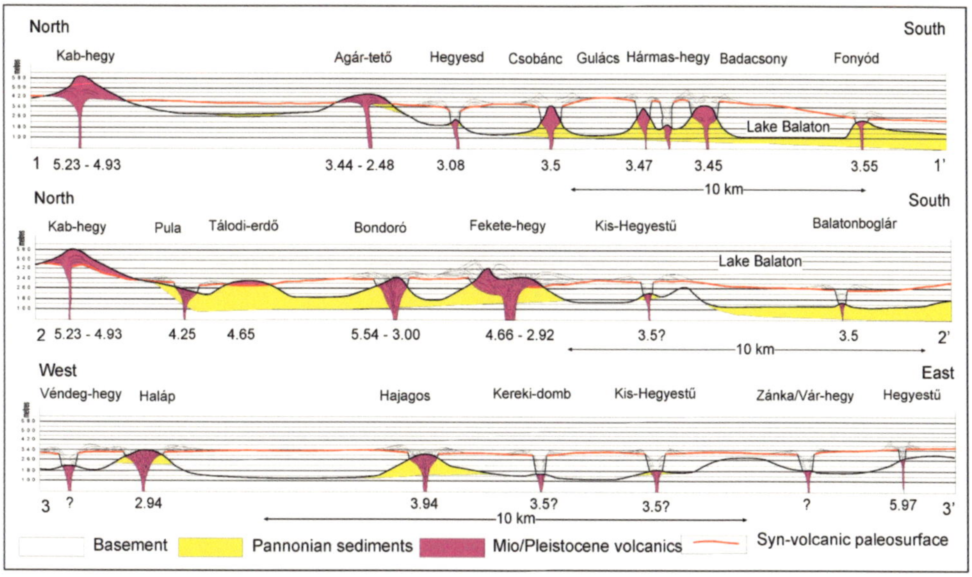

Fig. 9. Cross-section across the BBHVF with erosional remnants of monogenetic volcanoes as reconstructed by Németh et al 2003. Cross section lines are shown on Fig. 5. Numbers below each cross section represent age ranges of the volcanic erosion remnants determined from numerous K-Ar radiometric age datings (Balogh et al., 1986; Balogh & Pécskay, 2001).

Apart from this deep level of exposure, there are no exposed irregular shaped, fragmented wall rock rich dykes (Lorenz & Kurszlaukis, 2007), like those that are widely reported from other monogenetic maar-diatreme volcanic fields, including Hopi Buttes (White, 1991b). Such levels of exposures and preserved outcrops are more common in the northern part of the Pannonian Basin, in southern Slovakia and northern Hungary (Fig. 1) (Lexa et al., 2010). The best example of an individual plugs as an exposed base of a volcanic lava filled crater is at Hegyes-tű (Figs 2 & 5) where a small remnant of vent filling mixture of volcanic and siliciclastic debris is preserved and intruded by the plug, indicating that explosive fragmentation preceded the formation of the basanite plug (Balogh et al., 2010).

Upper diatremes (White & Ross, 2011) represent scoria cones and associated lava plugs built on the basal phreatomagmatic volcano. However, they do not necessarily represent volcanic landforms grown over the syn-volcanic landscape. After erosion, a significant part of the

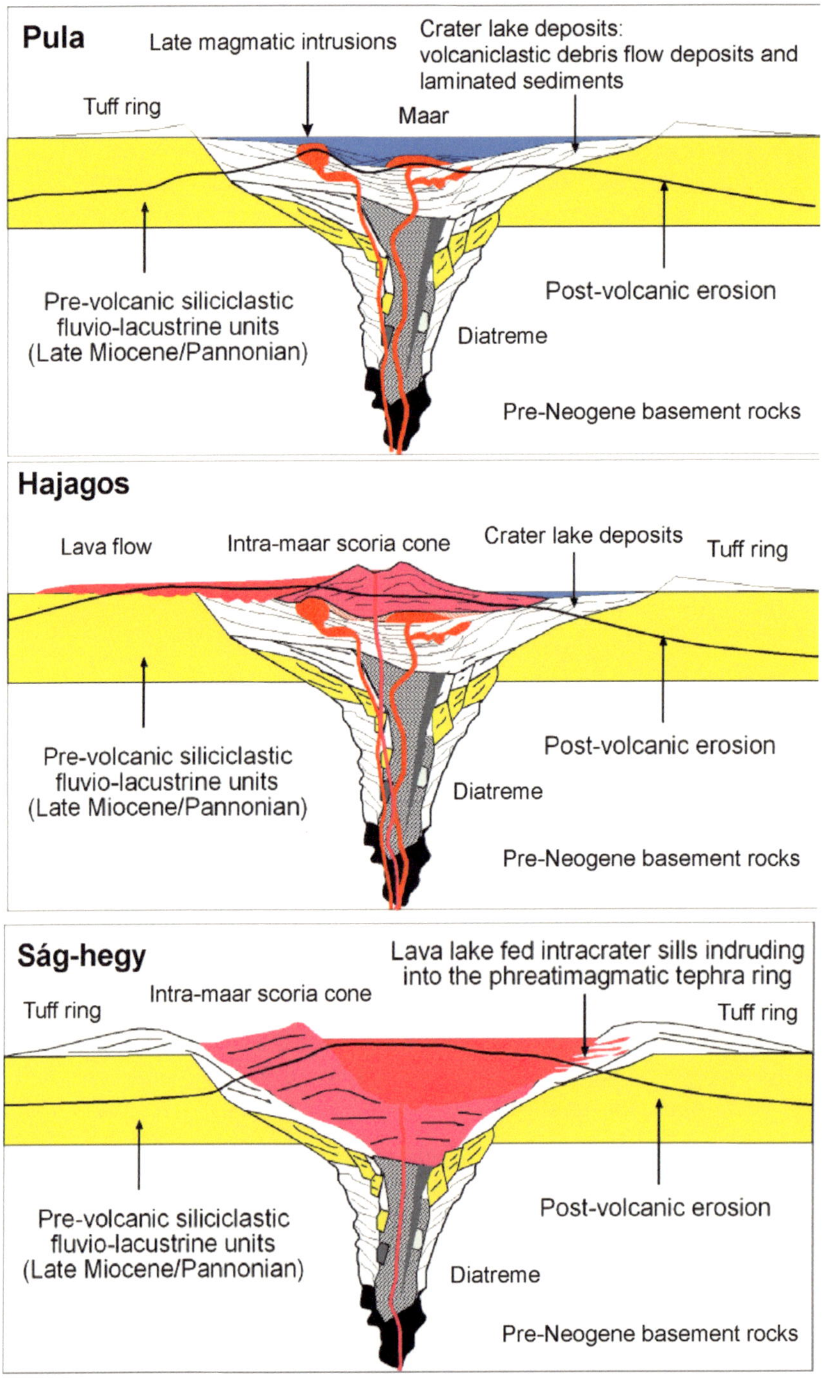

Fig. 10. Theoretical cross-sections of complex maar volcanoes reconstructed on the basis of preserved pyroclastic rocks, their texture and distribution patterns. A combination of crater lake sedimentation and/or magmatic crater filling successions are reconstructed on the basis of preserved 3D facies architecture of volcanic rocks and their textural characteristics.

volcanic edifice could be eliminated, and in the eroded phreatomagmatic volcanic field the identification of such remnants should be undertaken with precaution to establish the syn-volcanic paleo-surface, to estimate the erosion. Surface volcanic edifices are preserved in areas of low erosion, and include volcanoes produced by both phreatomagmatic and magmatic eruptions. These vents are characterized by magmatic fragmentation triggered explosive eruptions. Vent remnants of these volcanoes are concentrated in the northern and central part of the BBHVF, such as Kab-hegy, Agár-tető, Haláp and Hegyesd (Fig. 2 & 5).

The preserved phreatomagmatic pyroclastic successions and their distributional pattern suggests that the original maar diameters of the western Pannonian region ranges from few hundreds of meters up to 5 km in diameter (Fekete-hegy - 5 km; Tihany - 4 km; Bondoró - 2.5 km; Badacsony - 2.5 km), however, the largest centres probably represent maar volcanic complexes with inter-connected large basins similar to those maars known from South-Australia and Victoria (Jones et al., 2001). The average maar basins are inferred to have been 1-1.5 km wide originally, which is within the range of most maars worldwide (Lorenz, 1986; Ross et al., 2011).

The volcanoes of the western Pannonian region have been reconstructed to be hybrids of phreatomagmatic and magmatic volcanic edifices and formed by initial maar or tuff ring forming events especially those erupted in an area with thick Neogene siliciclastic sedimentary cover (Fig. 10). The gradual exhaustion of water source to fuel the magma/water interaction led to "drier" phreatomagmatic, then pure magmatic, fragmentation of the uprising melt, often building large scoria cones inside the phreatomagmatic volcanoes, as witnessed at Vulkaneifel in Germany (Houghton & Schmincke, 1986) or Auckland Volcanic Field in New Zealand (Houghton et al., 1999). The typical types of such volcanoes are located in the southwestern site of the BBHVF in the Tapolca Basin (Badacsony, Szent György-hegy, Hajagos-hegy, Fekete-hegy).

6. "Dry" volcanoes

In the central part of the BBHVF erosion remnants of scoria cones and shield volcanoes give evidence for a smaller impact of the ground and surface water in control of the volcanic eruptions. The age distribution of erosional remnants of scoria cones suggest a peak in their formation about 3 million years ago, which coincides well with a dryer period of the environmental history of the region, suggesting a potential link between the large scale climatic changes and eruption style variations over long time periods in this region (Kereszturi et al., 2011).

Erosion remnants of scoria cones are commonly strongly modified after erosion, and their original volcanic landforms can be hardly recognized (Fig. 11A). In spite of the general assumption of the fast erosion of scoria cones, there are remarkable well-preserved scoria cones known from the central part of the BBHVF (Kereszturi et al., 2011). These scoria cone remnants are about 3 - 2.3 million years old and still have retained their original crater morphology and some part of their constructional edifice (Kereszturi et al., 2011). Many of the late magmatic capping units over basal tuff rings are abundant in welded lava spatter and or lava spindle bombs commonly cored with mantle origin xenoliths (Fig. 11B).

Fig. 11. One of the youngest, relatively intact scoria cone of Agár-tető of the BBHVF with a still recognisable cone morphology and breached crater (A) and a collection of spindle bombs (commonly filled with mantle-derived xenoliths such as peridotite lherzolite) from capping pyroclastic units of one of the best preserved scoria cone, Kopácsi-hegy, just west of the Fekete-hegy maar volcanic complex in the Bakony- Balaton Highland Volcanic Field (B).

7. Discussion on future research

Intense research over the past 10 years allowed characterisation of the WPB volcanic fields as phreatomagmatic monogenetic volcanic fields. Alongside this recognition, substantial research has been done to describe in detail the eruption scenarios these volcanoes may have provided during their activity. In spite of the huge step forward in knowledge of Miocene to Pleistocene volcanism in the Pannonian Basin, there are major questions that can be formulated and need to be answered in the near future. The recent volcanic research highlighted the need to formulate our research effort along 4 major lines of enquiry (Fig. 12): to understand 1) how monogenetic these monogenetic volcanoes are; 2) what the relative role is of the external and internal forces that may have controlled the formation of individual volcanoes; 3) how the long term environmental changes may effect the overall manifestation of volcanism over the nearly 6 million years of volcanic field evolution; and 4) what the syn-eruptive landscape and the volcanic landform looked like, and how these can be connected with the preserved pyroclastic rock units.

7.1 The "monogenetic enigma"

Current research in the Western Pannonian Basin's phreatomagmatic volcanoes has documented clear field and textural evidence of volcanism that accumulated multiple pyroclastic units separated by volcaniclastic successions indicating some break in the eruption. Locations (on Figs 2 & 5), such as Fekete-hegy (Auer et al., 2007), Bondoró (Kereszturi et al., 2010) and Tihany (Németh et al., 2001), are prime suspects to demonstrate a multiphase nature of the eruption of these volcanoes commonly accompanied by vent shifting in the course of their eruption. These volcanoes are complex phreatomagmatic to magmatic volcanic edifices, inferred to have been erupted over a long time, leaving behind pyroclastic successions separated by well-marked discordance horizons. Newly initiated research intends to identify any signatures of the chemical zoning or polymagmatic nature of these volcanoes, similar to those recently identified in pyroclastic units of Jeju Island in

Korea. A combined effort of sedimentology and geochemical research from complex volcanic erosion remnants can provide answers to identify small chemical changes that may be related to magma plumbing systems, volcanic conduit complexity and volcanic conduit dynamics and their changes over short time (Fig. 13A). Such chemical signatures are likely to be reflected in the type and texture of pyroclastic rocks preserved. Such research has not been completed yet from the WPB, and potentially could provide significant new views on monogenetic volcanism, that could be linked to research in New Zealand, Korea, Argentina or the western USA.

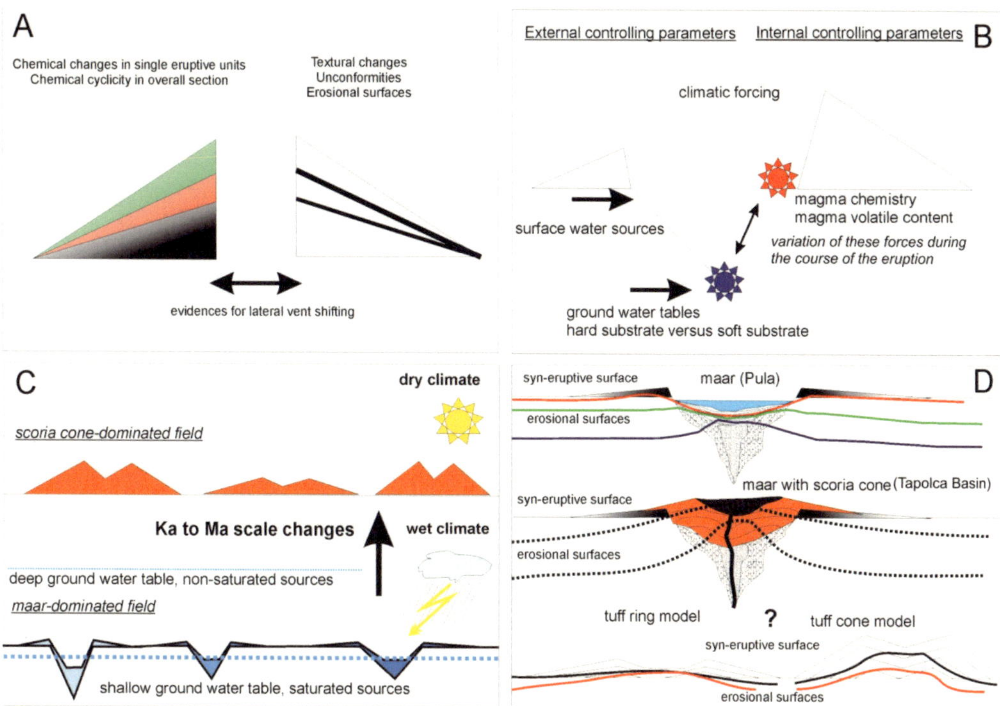

Fig. 12. Graphic expressions of basic research questions and orientations for future work on the WPB's volcanic fields: A) Question on the complexity of the eruption history of a single volcano of WPB such as 1) simple monogenetic versus complex monogenetic volcanism, 2) monomagmatic versus polymagmatic activity, 3) distinct eruptive episodes versus eruption cycles, and 4) recognition of potential lateral vent migration. B) Role of internal versus external controlling parameters - or the balance between phreatomagmatic and magmatic fragmentation styles. Recognition of shallow and deep water sources to fuel phreatomagmatism and the hydrogeology of such water sources. Definition of the role of the magma composition, source and volatile content to define the magma potential for the style of magmatic fragmentation. C) Recognition of how long term environmental changes may have influenced the eruption styles of the volcanism of the volcanic fields such as dry climate (scoria cone-dominated) versus wet climate (phreatomagmatism-dominated). D)To reconstruct primary volcanic landforms (negative landforms - e.g. maars - versus positive landforms - e.g. tuff rings, tuff cones and scoria cones) to establish landscape evolution.

Fig. 13. A) Explosion breccia horizons (arrows) from Suwolbong tuff ring in Jeju Island, Korea represents not only textural changes but also chemical breaks in the small-volume volcanic edifice's history; B) Joya Honda maar in San Luis Potosi, Mexico with its nearly 300 metres deep maar crater is a fine example to question how external and internal forces control the final result of monogenetic volcanism. The crater rim tuff ring-forming deposit is about 50 m thick in the view above dashed line; C) Maar volcanoes can form in areas that are located currently in arid climate due to the availability of ground-water in their recent past (such as Mekegölü in central Turkey), and collect aeolian deposits. Climatic changes can influence the eruptive styles a monogenetic volcanic field can be dominated by; D) Recognition of constructional monogenetic edifices and their facies architecture can help us to delineate the erosional history of the area where the monogenetic volcano erupted. This tuff cone in the figure in a small off shore island of Chagui-do (Jeju, South Korea) exposes its inner, crater filling deeply inward dipping crater facies that is strikingly similar to upper diatreme facies of a maar volcano.

7.2 External versus internal control on eruption styles

It has been recently recognized in few volcanic fields, that the volcanic landform and its eruption styles are controlled by the changes of internal and external parameters (Valentine & Gregg, 2008). Internal parameters that considered to be controlling the magma fragmentation and therefore the eruption style can be defined as those that reflect the type of source of the magma, its degassing, vesiculation, crystallisation and the style of its road to the surface. The magma flux and output rates are the key parameters that seem to play an important role in volcanic eruptions (Houghton et al., 1999; Houghton & Gonnermann, 2008; Valentine & Gregg, 2008). However, volcanic fields that erupted through broad areas with variable country rocks and underlying sedimentary sequences with diverse water saturation, permeability and hydraulic conductivity levels (Fig. 13B). These external parameters strongly affect the style of volcanism such as magmatic effusive, explosive versus phreatomagmatic.

In a simple way, small magma supply (rate and flux) can create a situation where the external parameters overrun the system providing the development of phreatomagmatic volcanoes with substantial volume. On the other hand, if the magma supply is continuous, the external water sources can be exhausted quickly and the magmatic (internal) controlling parameters take over the control of the eruption style (Fig. 13C). The resulting pyroclastic successions, therefore, will provide evidence for gradual volcanic facies changes in a single eruption sequence of a single volcano.

The WPB seems to be a perfect site which has volcanoes clearly dominated by internal parameters and also those that were run over by the external parameters. As a result, the WPB is a perfect and complex amalgamation of complex volcanoes.

7.3 Understanding long-term environmental evolution and its influence on eruption styles

It is a logical assumption if external factors can play a significant role in the style of eruptions, and the overall landform evolution of a monogenetic volcano, such parameters can change over a long time (e.g. thousands to millions of years). As a result one can expect that climatic changes maybe reflected in the overall attitude of a volcanic field in certain time periods (e.g. phreatomagmatic versus magmatic dominated fields vary over time) (Fig 13C).

Recent research attempted to understand the climatic evolution of the WPB, and confirmed the potential of such an approach as it seems that some changes could be related to volcanic field wide environmental changes (Kereszturi et al., 2011). While this idea is new, and not too easy to test, the first results are promising, and it certainly merits further investigation.

7.4 Landscape evolution models

Landscape evolution models can be separated into two major approaches: 1) understanding and reconstructing the original volcanic landforms and 2) on the basis of the type and style of the volcanic eruptions, reconstruct the syn-eruptive volcanic landscapes and model the long-term erosion of the volcanic field.

Currently it is accepted that the majority of the volcanic erosion remnants of the WPB are maar-diatreme volcanoes, many of them preserving exposed diatreme facies forming butte-like hills. While this model seems to be consistent with volcanic textures recognized in the preserved volcanic successions, there are large numbers of sites where the original volcanic landforms need to be reconstructed in a far more detailed manner. It is a critical to know whether a certain volcanic erosion remnant represents "something" which was beneath or on the syn-eruptive surface (Fig. 13D). To answer this fundamental question is especially important in areas such as the Tapolca Basin (Figs 2 & 5), where clear evidence indicates that the eruptions were in their late stage when scoria and lava spatter cones were built that are constructional landforms.

To reconstruct precisely the volcanic landforms will provide vital information on the syn-eruptive environment and its potential landforms. A concentrated research effort, systematically targeting these questions, would provide a more united and more precise syn-eruptive landscape model to the WPB than the currently existing ones.

8. Conclusion

The general features of the volcanic fields of the Western Pannonian Basin are very similar to other eroded volcanic fields which erupted into wet environments such as Fort Rock Christmas Valley, Oregon (Heiken, 1971), Snake River Plain, Idaho (Brand and White, 2007; Godchaux et al., 1992; Németh and White, 2009), Hopi Buttes, Arizona (Vazquez and Ort, 2006; White, 1989; White, 1990; White, 1991), and Saar-Nahe (Germany) (Lorenz and Haneke, 2004). It seems that the WPB phreatomagmatic volcanoes evolved at a time when climatic and environmental changes were dramatic, and that is likely reflected in the overall eruptive styles of the newly formed volcanoes. It is generally accepted that the WPB phreatomagmatic volcanoes are eroded maar-diatreme type volcanoes, many with a complex history commonly forming nested volcanic complexes. While this idea is generally plausible, it is time to pursue future research to refine our understanding of the WPB phreatomagmatism, and utilise this knowledge to contribute to our general model on monogenetic volcanism, landscape evolution and eruption style changes over time. It is suggested that cooperative work with an intention to identify volcanic field analogies on a global scale can help to develop more accurate models to understand the Mio-Pleistocene basaltic volcanism in the Pannonian Basin.

9. Acknowledgment

A travel grant awarded by the Massey University Leave & Ancillary Appointments Committee (LAAC10/37) to K. Németh allowed him to participate in the International Maar Workshop held in Tapolca, Hungary in 2010 to discuss the current state of volcanic research in the WPB. The results presented here are part of research projects supported by the Foundation for Research, Science and Technology International Investment Opportunities Fund Project MAUX0808 "Facing the challenge of Auckland volcanism", and Massey University Research Fund (RM13227).

10. References

Auer, A.; Martin, U. & Németh, K. (2007). The Fekete-Hegy (Balaton Highland Hungary) "Soft-Substrate" And "Hard-Substrate" Maar Volcanoes in an Aligned Volcanic Complex - Implications for Vent Geometry, Subsurface Stratigraphy and the Palaeoenvironmental Setting. *Journal of Volcanology and Geothermal Research*, Vol.159, No.1-3, (Jan), pp. 225-245, 0377-0273

Balogh, K.; Árva-Sós, E.; Pécskay, Z. & Ravasz-Baranyai, L. (1986). K/Ar Dating of Post-Sarmatian Alkali Basaltic Rocks in Hungary. *Acta Mineralogica et Petrographica, Szeged*, Vol.28, pp. 75-94, HU ISSN 0365 8006

Balogh, K. & Németh, K. (2005). Evidence for the Neogene Small-Volume Intracontinental. Volcanism in Western Hungary: K/Ar Geochronology of the Tihany Maar Volcanic Complex. *Geologica Carpathica*, Vol.56, No.1, (Feb), pp. 91-99, ISSN 1335-0552

Balogh, K.; Németh, K.; Itaya, T.; Molnár, F.; Stewart, R.; Thanh, N.X.; Hyodo, H. & Daróczi, L. (2010). Loss of 40ar(Rad) from Leucite-Bearing Basanite at Low Temperature: Implications on K/Ar Dating. . *Central European Journal of Geosciences*, Vol.2, No.3, (September 2010), pp. 385-398, ISSN: 2081-9900 (print version) - ISSN: 1896-1517 (electronic version)

Balogh, K. & Pécskay, Z. (2001). K/Ar and Ar/Ar Geochronological Studies in the Pannonian-Carpathians-Dinarides (Pancardi) Region. *Acta Geologica Academiae Scientiarum Hungaricae*, Vol.44, No.2-3, pp. 281-301, ISSN 1788-2281 (Print) ISSN 1789-3348 (Online)

Brand, B.D. & Clarke, A.B. (2009). The Architecture, Eruptive History, and Evolution of the Table Rock Complex, Oregon: From a Surtseyan to an Energetic Maar Eruption. *Journal of Volcanology and Geothermal Research*, Vol.180, No.2-4, (Mar), pp. 203-224, 0377-0273

Brand, B.D. & White, C.M. (2007). Origin and Stratigraphy of Phreatomagmatic Deposits at the Pleistocene Sinker Butte Volcano, Western Snake River Plain, Idaho. *Journal of Volcanology and Geothermal Research*, Vol.160, No.3-4, (Feb 15), pp. 319-339, ISSN: 0377-0273

Brenna, M.; Cronin, S.J.; Németh, K.; Smith, I.E.M. & Sohn, Y.K. (2011). The Influence of Magma Plumbing Complexity on Monogenetic Eruptions, Jeju Island, Korea. *Terra Nova*, Vol.23, No.2, (April 2011), pp. 70-75, ISSN 1365-3121

Brenna, M.; Cronin, S.J.; Smith, I.E.M.; Sohn, Y.K. & Németh, K. (2010). Mechanisms Driving Polymagmatic Activity at a Monogenetic Volcano, Udo, Jeju Island, South Korea. *Contributions to Mineralogy and Petrology* Vol.160, (December 2010), pp. 931-950, 0010-7999

Büttner, R.; Dellino, P.; La Volpe, L.; Lorenz, V. & Zimanowski, B. (2002). Thermohydraulic Explosions in Phreatomagmatic Eruptions as Evidenced by the Comparison between Pyroclasts and Products from Molten Fuel Coolant Interaction Experiments. *Journal of Geophysical Research-Solid Earth*, Vol.107, No.B11, pp. art. no.-2277, ISSN: 1934-8843

Büttner, R.; Dellino, P.; Raue, H.; Sonder, I. & Zimanowski, B. (2006). Stress-Induced Brittle Fragmentation of Magmatic Melts: Theory and Experiments. *Journal of Geophysical Research-Solid Earth*, Vol.111, No.B8, (Aug 24), pp. ISSN: 1934-8843

Büttner, R.; Dellino, P. & Zimanowski, B. (1999). Identifying Magma-Water Interaction from the Surface Features of Ash Particles. *Nature*, Vol.401, No.6754, (14 October 1999), pp. 688-690, ISSN: 0028-0836

Calvari, S. & Tanner, L.H. (2011). The Miocene Costa Giardini Diatreme, Iblean Mountains, Southern Italy: Model for Maar-Diatreme Formation on a Submerged Carbonate Platform. *Bulletin of Volcanology*, Vol.73, No.5 (June), pp. 557-576, pp. ISSN 0258-8900 (Print) 1432-0819 (Online)

Dellino, P. (2000). Phreatomagmatic Deposits: Fragmentation, Transportation and Deposition Mechanisms. *Terra Nostra*, Vol.6, pp. 99-105, ISSN 0946-8978

Dellino, P. & Kyriakopoulos, K. (2003). Phreatomagmatic Ash from the Ongoing Eruption of Etna Reaching the Greek Island of Cefalonia. *Journal of Volcanology and Geothermal Research*, Vol.126, No.3-4, (Aug 20), pp. 341-345,

Dellino, P. & LaVolpe, L. (1996). Image Processing Analysis in Reconstructing Fragmentation and Transportation Mechanisms of Pyroclastic Deposits. The Case of Monte Pilato-Rocche Rosse Eruptions, Lipari (Aeolian Islands, Italy). *Journal of Volcanology and Geothermal Research*, Vol.71, No.1, (Apr), pp. 13-29, 0377-0273

Dellino, P. & Liotino, G. (2002). The Fractal and Multifractal Dimension of Volcanic Ash Particles Contour: A Test Study on the Utility and Volcanological Relevance. *Journal of Volcanology and Geothermal Research*, Vol.113, No.1-2, pp. 1-18, ISSN: 0377-0273

Genareau, K.; Valentine, G.A.; Moore, G. & Hervig, R.L. (2010). Mechanisms for Transition in Eruptive Style at a Monogenetic Scoria Cone Revealed by Microtextural Analyses (Lathrop Wells Volcano, Nevada, USA). *Bulletin of Volcanology*, Vol.72, No.5, (Jul), pp. 593-607, 0258-8900

Godchaux, M. & Bonnichsen, B. (2002). Syneruptive Magma-Water and Posteruptive Lava-Water Interactions in the Western Snake River Plain, Idaho, During the Past 12 Million Years., In: *Tectonic and Magmatic Evolution of the Snake River Plain Volcanic Province*, B. Bonnichsen; C.M. White & M. McCurry, (Ed.), 387-435, Idaho Geological Survey, University of Idaho, ISBN-10: 1557650292, Moscow, Idaho

Godchaux, M.M.; Bonnichsen, B. & Jenks, M.D. (1992). Types of Phreatomagmatic Volcanos in the Western Snake River Plain, Idaho, USA. *Journal of Volcanology and Geothermal Research*, Vol.52, No.1-3, (Sep), pp. 1-25, ISSN: 0377-0273

Heiken, G.H. (1971). Tuff Rings: Examples from the Fort Rock-Christmas Lake Valley Basin, South-Central Oregon. *Journal of Geophysical Research*, Vol.76, No.23, pp. 5615-5626, ISSN: 0148-0227

Heiken, G.H. & Wohletz, K.H. (1986). *Volcanic Ash*, University of California Press, ISBN 0520052412 (0-520-05241-2), Berkeley

Houghton, B.F. & Gonnermann, H.M. (2008). Basaltic Explosive Volcanism: Constraints from Deposits and Models. *Chemie der Erde-Geochemistry*, Vol.68, No.2, pp. 117-140, 0009-2819

Houghton, B.F. & Schmincke, H.U. (1986). Mixed Deposits of Simultaneous Strombolian and Phreatomagmatic Volcanism; Rothenberg Volcano, East Eifel Volcanic Field. *Journal of Volcanology and Geothermal Research*, Vol.30, No.1-2, pp. 117-130, ISSN 0377-0273

Houghton, B.F.; Wilson, C.J.N. & Smith, I.E.M. (1999). Shallow-Seated Controls on Styles of Explosive Basaltic Volcanism: A Case Study from New Zealand. *Journal of Volcanology and Geothermal Research*, Vol.91, No.1, pp. 97-120, ISSN 0377-0273

Jones, R.N.; McMahon, T. & Bowler, J.M. (2001). Modelling Historical Lake Levels and Recent Climate Change at Three Closed Lakes, Western Victoria, Australia (C.1840-1990). *Journal of Hydrology*, Vol.246, No.1-4, pp. 159-180, ISSN: 0022-1694

Keating, G.N.; Valentine, G.A.; Krier, D.J. & Perry, F.V. (2008). Shallow Plumbing Systems for Small-Volume Basaltic Volcanoes. *Bulletin of Volcanology*, Vol.70, No.5, (Mar), pp. 563-582, 0258-8900

Kereszturi, G.; Csillag, G.; Németh, K.; Sebe, K.; Kadosa, B. & Jáger, V. (2010). Volcanic Architecture, Eruption Mechanism and Landform Evolution of a Plio/Pleistocene Intracontinental Basaltic Polycyclic Monogenetic Volcano from the Bakony-Balaton Highland Volcanic Field, Hungary. *Central European Journal of Geosciences*, Vol.2, No.3, (September 2010), pp. 362-384, ISSN: 2081-9900 (print version) - ISSN: 1896-1517 (electronic version)

Kereszturi, G. & Németh, K. (2011). Shallow-Seated Controls on the Evolution of the Pleistocene Kopasz-Hegy Volcanic Complex; a Monogenetic Volcanic Chain in the

Western Pannonian Basin, Hungary. *Geologica Carpathica*, Vol.[accepted], pp. ISSN: 1335-0552 (print version); ISSN: 1336-8052 (electronic version)

Kereszturi, G.; Németh, K.; Csillag, G.; Balogh, K. & Kovács, J. (2011). The Role of External Environmental Factors in Changing Eruption Styles of Monogenetic Volcanoes in a Mio/Pleistocene Continental Volcanic Field in Western Hungary *Journal of Volcanology and Geothermal Research*, Vol.201, No.1-4, (15 April 2011), pp. 227-240, 0377-0273

Kralj, P. (2011). Eruptive and Sedimentary Evolution of the Pliocene Grad Volcanic Field, North-East Slovenia. *Journal of Volcanology and Geothermal Research*, Vol.201, No.1-4, (15 April 2011), pp. 272-284, ISSN 0377-0273

Lexa, J.; Seghedi, I.; Németh, K.; Szakács, A.; Konečný, V.; Pécskay, Z.; Fülöp, A. & Kovacs, M. (2010). Neogene-Quaternary Volcanic Forms in the Carpathian-Pannonian Region: A Review. *Central European Journal of Geosciences*, Vol.2, No.3, pp. 207-270, ISSN: 2081-9900 (print version) - ISSN: 1896-1517 (electronic version)

Lorenz, V. (1973). On the Formation of Maars. *Bulletin of Volcanology*, Vol.37, No.2, pp. 183-204, ISSN 0258-8900 (Print) 1432-0819 (Online)

Lorenz, V. (1986). On the Growth of Maars and Diatremes and Its Relevance to the Formation of Tuff Rings. *Bulletin of Volcanology*, Vol.48, pp. 265-274, ISSN 0258-8900 (Print) 1432-0819 (Online)

Lorenz, V. (2003). Maar-Diatreme Volcanoes, Their Formation, and Their Setting in Hard-Rock or Soft-Rock Environments. *Geolines - Journal of the Geological Institute of AS Czech Republic*, Vol.15, pp. 72-83, ISSN: 1210–9606

Lorenz, V. & Kurszlaukis, S. (2007). Root Zone Processes in the Phreatomagmatic Pipe Emplacement Model and Consequences for the Evolution of Maar-Diatreme Volcanoes. *Journal of Volcanology and Geothermal Research*, Vol.159, No.1-3, (Jan 1), pp. 4-32, ISSN: 0377-0273

Manville, V.; Németh, K. & Kano, K. (2009). Source to Sink: A Review of Three Decades of Progress in the Understanding of Volcaniclastic Processes, Deposits, and Hazards. *Sedimentary Geology*, Vol.220, No.3-4, (Oct), pp. 136-161, ISSN 0037-0738

Martin, U. & Németh, K. (2004). *Mio/Pliocene Phreatomagmatic Volcanism in the Western Pannonian Basin.*, Geological Institute of Hungary, 963-671-238-7, Budapest, Hungary ISBN 963-671-238-7

Martin, U. & Németh, K. (2005). Eruptive and Depositional History of a Pliocene Tuff Ring That Developed in a Fluvio-Lacustrine Basin: Kissomlyó Volcano (Western Hungary). *Journal of Volcanology and Geothermal Research*, Vol.147, No.3-4, (Oct), pp. 342-356, ISSN 0377-0273

Martin, U. & Németh, K. (2007). Blocky Versus Fluidal Peperite Textures Developed in Volcanic Conduits, Vents and Crater Lakes of Phreatomagmatic Volcanoes in Mio/Pliocene Volcanic Fields of Western Hungary. *Journal of Volcanology and Geothermal Research*, Vol.159, No.1-3, (Jan), pp. 164-178, ISSN 0377-0273

Mastrolorenzo, G. (1994). Averno Tuff Ring in Campi-Flegrei (South Italy). *Bulletin of Volcanology*, Vol.56, No.6-7, pp. 561-572, ISSN 0258-8900 (Print) 1432-0819 (Online)

Morrissey, M.M.; Zimanowski, B.; Wohletz, K. & Büttner, R. (2000). Phreatomagmatic Fragmentation, In: *Encyclopedia of Volcanoes*, H. Sigurdsson; B.F. Houghton; S.R. McNutt; H. Rymer & J. Stix, (Ed.), 431-446, Academic Press, ISBN: 978-0126431407, New York

Németh, K. (2010a). Monogenetic Volcanic Fields: Origin, Sedimentary Record, and Relationship with Polygenetic Volcanism In: *What Is a Volcano? Gsa Special Papers Volume 470*, E. Cañón-Tapia & A. Szakács, (Ed.), 43-67, Geological Society of America, ISBN: 9780813724706 Boulder, Colorado

Németh, K. (2010b). Volcanic Glass Textures, Shape Characteristics and Compositions from Phreatomagmatic Rock Units of the Western Hungarian Monogenetic Volcanic Fields and Their Implication to Magma Fragmentation. *Central European Journal of Geosciences*, Vol.2, No.3, (September 2010), pp. 399-419, ISSN: 2081-9900 (print version) - ISSN: 1896-1517 (electronic version)

Németh, K.; Cronin, S.J.; Haller, M.J.; Brenna, M. & Csillag, G. (2010). Modern Analogues for Miocene to Pleistocene Alkali Basaltic Phreatomagmatic Fields in the Pannonian Basin: "Soft-Substrate" To "Combined" Aquifer Controlled Phreatomagmatism in Intraplate Volcanic Fields. *Central European Journal of Geosciences*, Vol.2, No.3, (September 2010), pp. 339-361, ISSN: 2081-9900 (print version) - ISSN: 1896-1517 (electronic version)

Németh, K.; Goth, K.; Martin, U.; Csillag, G. & Suhr, P. (2008). Reconstructing Paleoenvironment, Eruption Mechanism and Paleomorphology of the Pliocene Pula Maar, (Hungary). *Journal of Volcanology and Geothermal Research*, Vol.177, No.2, (Oct), pp. 441-456, 0377-0273

Németh, K. & Martin, U. (1999). Late Miocene Paleo-Geomorphology of the Bakony-Balaton Highland Volcanic Field (Hungary) Using Physical Volcanology Data. *Zeitschrift für Geomorphologie*, Vol.43, No.4, (December 1999), pp. 417-438, ISSN 0372-8854

Németh, K.; Martin, U. & Csillag, G. (2003). Calculation of Erosion Rates Based on Remnants of Monogenetic Alkaline Basaltic Volcanoes in the Bakony–Balaton Highland Volcanic Field (Western Hungary) of Mio/Pliocene Age. *Geolines - Journal of the Geological Institute of AS Czech Republic*, Vol.15, pp. 93-97, ISSN: 1210–9606

Németh, K.; Martin, U.; Haller, M.J. & Alric, V.L. (2007). Cenozoic Diatreme Field in Chubut (Argentina) as Evidence of Phreatomagmatic Volcanism Accompanied with Extensive Patagonian Plateau Basalt Volcanism? *Episodes*, Vol.30, No.3, (Sep), pp. 217-223, ISSN 0705-3797

Németh, K.; Martin, U. & Harangi, S. (2001). Miocene Phreatomagmatic Volcanism at Tihany (Pannonian Basin, Hungary). *Journal of Volcanology and Geothermal Research*, Vol.111, No.1-4, (November 2001), pp. 111-135, ISSN 0377-0273

Németh, K. & White, J.D.L. (2003). Reconstructing Eruption Processes of a Miocene Monogenetic Volcanic Field from Vent Remnants: Waipiata Volcanic Field, South Island, New Zealand. *Journal of Volcanology and Geothermal Research*, Vol.124, No.1-2, (May), pp. 1-21, ISSN 0377-0273

Pécskay, Z.; Lexa, J.; Szakács, A.; Balogh, K.; Seghedi, I.; Konečný, V.; Kovács, M.; Márton, E.; Kaliciak, M.; Széky-Fux, V.; Póka, T.; Gyarmati, P.; Edelstein, O.; Rosu, E. & Zec, B. (1995). Space and Time Distribution of Neogene-Quaternary Volcanism in the Carpatho-Pannonian Region. *Acta Vulcanologica*, Vol.7, No.2, pp. 15-28, ISSN: 1121-9114

Ross, P.-S.; Delpit, S.; Haller, M.J.; Németh, K. & Corbella, H. (2011). Influence of the Substrate on Maar–Diatreme Volcanoes — an Example of a Mixed Setting from the Pali Aike Volcanic Field, Argentina *Journal of Volcanology and Geothermal Research*, Vol.2001, No.1-4, (15 April 2011), pp. ISSN 0377-0273

Sohn, Y.K. & Park, K.H. (2005). Composite Tuff Ring/Cone Complexes in Jeju Island, Korea: Possible Consequences of Substrate Collapse and Vent Migration. *Journal of Volcanology and Geothermal Research*, Vol.141, No.1-2, (Mar 1), pp. ISSN 0377-0273

Sparks, R.S.J.; Baker, L.; Brown, R.J.; Field, M.; Schumacher, J.; Stripp, G. & Walters, A. (2006). Dynamical Constraints on Kimberlite Volcanism. *Journal Of Volcanology And Geothermal Research*, Vol.155, No.1-2, (Jul), pp. 18-48, ISSN 0377-0273

Stoppa, F. (1996). The San Venanzo Maar and Tuff Ring, Umbria, Italy: Eruptive Behaviour of a Carbonatite-Melilitite Volcano. *Bulletin of Volcanology*, Vol.57, No.7, pp. 563-577, ISSN 0258-8900 (Print) 1432-0819 (Online)

Stoppa, F. & Principe, C. (1997). Eruption Style and Petrology of a New Carbonatitic Suite from the Mt. Vulture Southern Italy: The Monticchio Lakes Formation. *Journal of Volcanology and Geothermal Research*, Vol.78, No.3-4, (Sep), pp. 251-265, ISSN: 0377-0273

Suiting, I. & Schmincke, H.U. (2009). Internal Vs. External Forcing in Shallow Marine Diatreme Formation: A Case Study from the Iblean Mountains (Se-Sicily, Central Mediterranean). *Journal of Volcanology and Geothermal Research*, Vol.186, No.3-4, (OCT 10 2009), pp. 361-378, ISSN: 0377-0273

Suiting, I. & Schmincke, H.U. (2010). Iblean Diatremes 2: Shallow Marine Volcanism in the Central Mediterranean at the Onset of the Messinian Salinity Crisis (Iblean Mountains, Se-Sicily)-a Multidisciplinary Approach. *International Journal of Earth Sciences*, Vol.99, No.8, (DEC 2010), pp. 1917-1940, ISSN: 1437-3254

Szabó, C.; Harangi, S. & Csontos, L. (1992). Review of Neogene and Quaternary Volcanism of the Carpathian Pannonian Region. *Tectonophysics*, Vol.208, No.1-3, (Jul 30), pp. 243-256, ISSN 0040-1951

Valentine, G.A. & Gregg, T.K.P. (2008). Continental Basaltic Volcanoes - Processes and Problems. *Journal of Volcanology and Geothermal Research*, Vol.177, No.4, (Nov), pp. 857-873, ISSN 0377-0273

Valentine, G.A. & Hirano, N. (2010). Mechanisms of Low-Flux Intraplate Volcanic Fields-Basin and Range (North America) and Northwest Pacific Ocean. *Geology*, Vol.38, No.1, (Jan), pp. 55-58, ISSN 0091-7613

Valentine, G.A. & Keating, G.N. (2007). Eruptive Styles and Inferences About Plumbing Systems at Hidden Cone and Little Black Peak Scoria Cone Volcanoes (Nevada, USA). *Bulletin of Volcanology*, Vol.70, No.1, (Sep), pp. 105-113, ISSN 0258-8900

Valentine, G.A. & Perry, F.V. (2006). Decreasing Magmatic Footprints of Individual Volcanoes in a Waning Basaltic Field. *Geophysical Research Letters*, Vol.33, No.14, (Jul), pp. ISSN 0094-8276

Valentine, G.A. & Perry, F.V. (2007). Tectonically Controlled, Time-Predictable Basaltic Volcanism from a Lithospheric Mantle Source (Central Basin and Range Province, USA). *Earth and Planetary Science Letters*, Vol.261, No.1-2, (Sep), pp. 201-216, ISSN 0012-821X

Vazquez, J.A. & Ort, M.H. (2006). Facies Variation of Eruption Units Produced by the Passage of Single Pyroclastic Surge Currents, Hopi Buttes Volcanic Field, USA. *Journal of Volcanology and Geothermal Research*, Vol.154, No.3-4, (Jun 15), pp. 222-236, ISSN 0377-0273

Walters, A.L.; Phillips, J.C.; Brown, R.J.; Field, M.; Gernon, T.; Stripp, G. & Sparks, R.S.J. (2006). The Role of Fluidisation in the Formation of Volcaniclastic Kimberlite: Grain

Size Observations and Experimental Investigation. *Journal of Volcanology and Geothermal Research*, Vol.155, No.1-2, (Jul 1), pp. 119-137, ISSN: 0377-0273

White, J.D.L. (1989). Basic Elements of Maar-Crater Deposits in the Hopi Buttes Volcanic Field, Northeastern Arizona, USA. *Journal of Geology*, Vol.97, pp. 117-125, ISSN:0022-1376

White, J.D.L. (1990). Depositional Architecture of a Maar-Pitted Playa - Sedimentation in the Hopi Buttes Volcanic Field, Northeastern Arizona, USA. *Sedimentary Geology*, Vol.67, No.1-2, pp. 55-84, ISSN: 0037-0738

White, J.D.L. (1991a). The Depositional Record of Small, Monogenetic Volcanoes within Terrestrical Basins, In: *Sedimentation in Volcanic Settings*, R.V. Fisher & G.A. Smith, (Ed.), 155-171, Society for Sedimentary Geology, ISBN-10: 0918985897; ISBN-13: 978-0918985897, Tulsa (Oklahoma)

White, J.D.L. (1991b). Maar-Diatreme Phreatomagmatism at Hopi Buttes, Navajo Nation (Arizona), USA. *Bulletin of Volcanology*, Vol.53, pp. 239-258, ISSN 0258-8900 (Print) 1432-0819 (Online)

White, J.D.L. & Ross, P.-S. (2011). Maar-Diatreme Volcanoes: A Review. *Journal of Volcanology and Geothermal Research*, Vol.201, No.1-4, (15 April 2011), pp. 1-29, ISSN 0377-0273

Wijbrans, J.; Németh, K.; Martin, U. & Balogh, K. (2007). Ar-40/Ar-39 Geochronology of Neogene Phreatomagmatic Volcanism in the Western Pannonian Basin, Hungary. *Journal of Volcanology and Geothermal Research*, Vol.164, No.4, (Aug), pp. 193-204, ISSN 0377-0273

Wohletz, K. & Heiken, G. (1992). *Volcanology and Geothermal Energy*, University of California Press, ISBN-10: 0520079140; ISBN-13: 978-0520079144, Berkeley

Wohletz, K.H. (1986). Explosive Magma-Water Interactions: Thermodynamics, Explosion Mechanisms, and Field Studies. *Bulletin of Volcanology*, Vol.48, pp. 245 - 264, ISSN 0258-8900 (Print) 1432-0819 (Online)

Zimanowski, B.; Büttner, R.; Lorenz, V. & Hafele, H.G. (1997). Fragmentation of Basaltic Melt in the Course of Explosive Volcanism. *Journal of Geophysical Research-Solid Earth*, Vol.102, No.B1, pp. 803-814, ISSN: 0148-0227

Zimanowski, B.; Fröhlich, G. & Lorenz, V. (1995). Experiments on Steam Explosion by Interaction of Water with Silicate Melts. *Nuclear Engineering and Design*, Vol.155, No.1-2, pp. 335-343, ISSN: 0029-5493

Zimanowski, B.; Lorenz, V. & Fröhlich, G. (1986). Experiments on Phreatomagmatic Explosions with Silicate and Carbonatitic Melts. *Journal of Volcanology and Geothermal Research*, Vol.30, pp. 149-153, ISSN: 0377-0273

Zimanowski, B.; Wohletz, K.; Dellino, P. & Büttner, R. (2003). The Volcanic Ash Problem. *Journal of Volcanology and Geothermal Research*, Vol.122, No.1-2, pp. 1-5, ISSN: 0377-0273

3

Quaternary Volcanism Along the Volcanic Front in Northeast Japan

Koji Umeda[1] and Masao Ban[2]
[1]*Geological Isolation Research and Development Directorate, Japan Atomic Energy Agency*
[2]*Department of Earth and Environmental Sciences, Yamagata University*
Japan

1. Introduction

Northeast Japan parallels a subduction zone where the Pacific plate converges against the North American plate. The axial part of Northeast Japan is composed of an uplifted mountain range called the Ou Backbone Range, along which a number of Quaternary volcanoes are distributed. The eastern margin of these volcanoes defines part of the Quaternary volcanic front of Northeast Japan (Fig. 1). The chemical composition of the volcanic rocks indicates a strong across variation in the alkali content and other incompatible elements, which are lower along the volcanic front and gradually increase rearward (Nakagawa et al., 1988; Yoshida, 2001). The Sr isotope compositions also indicate across-arc variation; the fore-arc volcanoes have higher $^{87}Sr/^{86}Sr$ ratios (0.704-0.705) than the rear-arc volcanoes (around 0.703) (Notsu, 1983; Kumura & Yoshida, 2006). Such variations can be ascribed to heterogeneous subcontinental lithosphere and/or additional of components from the subducted slab (e.g., Sakuyama & Nesbitt, 1986; Tatsumi & Eggins, 1995). This trench-parallel chemical zonation in Northeast Japan has been established since ca. 12 Ma (Yoshida, 2001).

The late Miocene to Quaternary evolution of the volcanic arc of Northeast Japan has been accompanied by some remarkable features. These include (1) Late Miocene to Pliocene caldera-forming volcanism phase, under a direction of maximum compression oblique to the arc and (2) Quaternary andesite stratovolcano-forming volcanism phase, under orthogonal convergence settings (Acocella et al., 2008). A compressive stress regime under orthogonal convergence is unfavourable to facilitate caldera-forming volcanism requiring the formation of a large magma reservoir at shallow depth (Yoshida, 2001). The predominance of stratovolcanoes is reconciled with compressional tectonic settings in the present-day subduction system. Nevertheless, it remains obscure as to when the andesite stratovolcano-forming volcanism has been established under Quaternary orthogonal convergence in Northeast Japan.

In general, characteristics of volcanism such as distribution of volcanoes, type of eruptions, magma discharge rate are closely associated with tectonics surrounding the volcanoes. It is very important to examine the relationship between them for better understanding magmatism in various tectonic settings. In this chapter, the temporal changes in the distribution, type and magma discharge rate of the volcanoes near the volcanic front (i.e., Nasu Volcanic Zone) during the last 2.0 m. y. were clarified based on the age and volume data of the Quaternary volcanoes in Northeast Japan presented by Martin et al. (2004). In addition, we examine the relationship between variations in Quaternary volcanism and tectonics specifically with regard to faulting and uplifting.

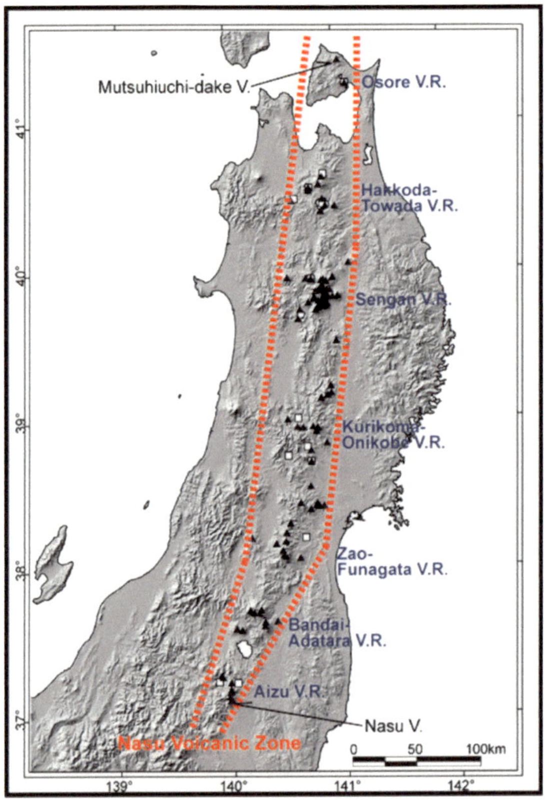

Fig. 1. Distribution of volcanic centers in Northeast Japan since 2.0 Ma. Solid triangle and open square represent stratovolcano and large-scale caldera volcanoes, respectively.

2. Volcanism during the last 2.0 million years

Thirty four Quaternary volcanoes have been recognized along the volcanic front between Mutsuhiuchi-dake volcano and Nasu volcano (Ono et al., 1981), and their eruptive volumes were calculated (Aramaki & Ui 1978). To refine the sequence of volcanism during the last 2.0 million years, Umeda et al. (1999) subdivided individual volcanoes into as small a unit as possible, and estimated their active periods from radioactive age and stratigraphic data, and calculated their eruptive volumes. Recently, Martin et al. (2004) revised the database of Umeda et al. (1999) using the "Catalog of Quaternary volcanoes in Japan" of Committee for Catalogue of Quaternary Volcanoes in Japan eds. (1999) and other new radiometric age data for each volcano along the volcanic front in Northeast Japan. Martin et al. (2004) refers to "Volcanic Event" that is defined as multiple eruptions from the same conduit occurring over several tens to hundreds of thousands of years (Table 1). By defining the highest points of the individual stratovolcanoes or the geometrical centers of the calderas as volcanic centers (the main vents) for each volcanic event, the locations, magma volume and eruption styles were evaluated to clarify the temporal change in volcanism.

Volcano Complex	Volcanic event	Location		Age (Ma)			Volume
		Latitude	Longitude	Oldest	Approxi-mate	Younge-st	(km³, DRE)
Mutsuhiuchi-dake	Older Mutsuhiuchi-dake	41.437	141.057		ca.0.73		5.9
Mutsuhiuchi-dake	Younger Mutsuhiuchi-dake	41.437	141.057	0.45		0.2	3.6
Osorezan	Kamabuse-yama	41.277	141.123		ca.0.8		11.4
Hakkoda	Hakkoda P.F.1st.	40.667	140.897		0.65		17.8
Hakkoda	South-Hakkoda	40.600	140.850	0.65	–	0.4	52.4
Hakkoda	Hakkoda P.F.2nd.	40.667	140.897		0.4		17.3
Hakkoda	North-Hakkoda	40.650	140.883	0.16	–	0	30.4
Okiura	Aoni F. Aonigawa P.F.	40.573	140.763		ca.1.7		17.6
Okiura	Aoni F. Other P.F.	40.573	140.763	1.7	–	0.9	3.7
Okiura	Okogawasawa lava	40.579	140.759	0.9	-	0.65	0.9
Okiura	Okiura dacite	40.557	140.755	0.9	-	0.7	2.1
Ikarigaseki	Nijikai Tuff	40.500	140.625		ca.2.0		20.2
Ikarigaseki	Ajarayama	40.490	140.600	1.91	–	1.89	2.1
Towada	Herai-dake	40.450	141.000				5.1
Towada	Ohanabe-yama	40.500	140.883	0.4	–	0.05	8.9
Towada	Hakka	40.417	140.867				1.4
Towada	Towada Okuse	40.468	140.888		0.055		4.8
Towada	Towada Ofudo	40.468	140.888		0.025		22.1
Towada	Towada Hachinohe	40.468	140.888		0.013		26.9
Towada	Post-caldera cones	40.457	140.913	0.013	–	0	14.4
Nanashigure	Nanashigure	40.068	141.112	1.06	–	0.72	55.5
Moriyoshi	Moriyoshi	39.973	140.547	1.07	–	0.78	18.1
Bunamori	Bunamori	39.967	140.717		1.2		0.1
Akita-Yakeyama	Akita-Yakeyama	39.963	140.763	0.5	–	0	9.9
Nishimori/ Maemori	NIshimori/Maemori	39.973	140.962	0.5	–	0.3	2.6
Hachimantai/ Chausu	Hachimantai	39.953	140.857	1	–	0.7	5.5
Hachimantai/ Chausu	Chausu-dake	39.948	140.902	0.85	–	0.75	13.7
Hachimantai/ Chausu	Fukenoyu	39.953	140.857		ca.0.7		0.2
Hachimantai/ Chausu	Gentamri	39.956	140.878				0.2
Yasemori/ Magarisaki-yama	Magarisaki-yama	39.878	140.803	1.9	–	1.52	0.3
Yasemori/ Magarisaki-yama	Yasemori	39.883	140.828		1.8		0.9
Kensomori/ Morobidake	Kensomori	39.897	140.871		ca.0.8		0.8
Kensomori/ Morobidake	Morobi-dake	39.919	140.862	1	–	0.8	2.5

Volcano Complex	Volcanic event	Location		Age (Ma)			Volume (km^3, DRE)
		Latitude	Longitude	Oldest	Approximate	Youngest	
Kensomori/ Morobidake	1470m Mt. lava	39.909	140.872				0.1
Kensomori/ Morobidake	Mokko-dake	39.953	140.857		ca.1.0		0.5
Tamagawa Welded Tuff	Tamagawa Welded Tuffs R4	39.963	140.763		ca.2.0		83.2
Tamagawa Welded Tuff	Tamagawa Welded Tuffs D	39.963	140.763		ca.1.0		32.0
Nakakura/ Shimokura	Obuka-dake	39.878	140.883	0.8	–	0.7	2.9
Nakakura/ Shimokura	Shimokura-yama	39.889	140.933				0.4
Nakakura/ Shimokura	Nakakura-yama	39.888	140.910				0.4
Matsukawa	Matsukawa andesite	39.850	140.900	2.6	–	1.29	11.6
Iwate/ Amihari	Iwate	39.847	141.004	0.2	–	0	25.1
Iwate/ Amihari	Amihari	39.842	140.958	0.3	–	0.1	10.6
Iwate/ Amihari	Omatsukura-yama	39.841	140.919	0.7	–	0.6	3.3
Iwate/ Amihari	Kurikigahara	39.849	140.882				0.2
Iwate/ Amihari	Mitsuishi-yama	39.848	140.900		0.46		0.6
Shizukuishi/ Takakura	Marumori	39.775	140.877	0.4	–	0.3	2.4
Shizukuishi/ Takakura	Shizukuishi-Takakura-yama	39.783	140.893	0.5	–	0.4	5.2
Shizukuishi/ Takakura	Older Kotakakura-yama	39.800	140.900		1.4		2.7
Shizukuishi/ Takakura	North Mikado-yama	39.800	140.875				0.3
Shizukuishi/ Takakura	Kotakakura-yama	39.797	140.907	0.6	–	0.5	1.8
Shizukuishi/ Takakura	Mikado-yama	39.788	140.870		ca.0.3		0.2
Shizukuishi/ Takakura	Tairagakura-yama	39.808	140.878		ca.0.3		0.1
Nyuto/ Zarumori	Tashirotai	39.812	140.827	0.3	–	0.2	0.6
Nyuto/ Zarumori	Sasamori-yama	39.770	140.820	0.23	–	0.1	0.4
Nyuto/ Zarumori	Yunomori-yama	39.772	140.827		ca.0.3		0.5
Nyuto/ Zarumori	Zarumori-yama	39.788	140.850		0.56		0.9
Nyuto/ Zarumori	Nyutozan	39.802	140.843	0.58	–	0.5	5.0

Volcano Complex	Volcanic event	Location		Age (Ma)			Volume
		Latitude	Longitude	Oldest	Approxi-mate	Younge-st	(km³, DRE)
Nyuto/ Zarumori	Nyuto-kita	39.817	140.855		ca.0.4		0.1
Akita-Komagatake	Akita-Komagatake	39.754	140.802	0.1	–	0	2.9
Kayo	Kayo	39.803	140.735	2.2	–	1.17	5.9
Kayo	KoJiromori	39.828	140.787		0.94		0.3
Kayo	Akita-Ojiromori	39.839	140.788	1.7	1.7	1.7	0.3
Innai/ Takahachi	Takahachi-yama	39.755	140.655	1.7	1.7	1.7	0.0
Innai/ Takahachi	Innai	39.692	140.638	2	–	1.6	0.5
Kuzumaru	Aonokimori andesites	39.543	140.983		2.06		0.3
Yakeishi	Yakeishidake	39.161	140.832	0.7	–	0.6	9.5
Yakeishi	Komagatake	39.193	140.924		ca.1.0		7.6
Yakeishi	Kyozukayama	39.178	140.892	0.6	–	0.4	5.7
Yakeishi	Usagimoriyama	39.239	140.924	0.07	–	0.04	2.3
Kobinai	Kobinai	39.018	140.523	1	–	0.57	2.3
Takamatsu/ Kabutoyama	Kabutoyama Welded Tuff	39.025	140.618		1.16		3.2
Takamatsu/ Kabutoyama	Kiji-yama Welded Tuffs	39.025	140.618		0.30		5.1
Takamatsu	Takamatsu	38.965	140.610	0.3	–	0.27	3.8
Takamatsu	Futsutsuki-dake	38.961	140.661		ca.0.3		0.8
Kurikoma	Tsurugi-dake	38.963	140.792	0.1	–	0	0.2
Kurikoma	Magusa-dake	38.968	140.751	0.32	–	0.1	1.5
Kurikoma	Kurikoma	38.963	140.792	0.4	–	0.1	0.9
Kurikoma	South volcanoes	38.852	140.875		ca.0.5		0.3
Kurikoma	Older Higashi Kurikoma	38.934	140.779		ca.0.5		2.2
Kurikoma	Younger Higashi Kurikoma	38.934	140.779	0.4	–	0.1	0.7
Mukaimachi	Mukaimachi	38.770	140.520		ca.0.8		12.0
Onikobe	Shimoyamasato tuff	38.830	140.695	0.21	0.21	0.21	1.0
Onikobe	Onikobe Centeral cones	38.805	140.727		ca.0.2		1.1
Onikobe	Ikezuki tuff	38.830	140.695	0.3	–	0.2	17.3
Naruko	Naruko Central cones	38.730	140.727		ca.0.045		0.1
Naruko	Yanagizawa tuff	38.730	140.727		ca.0.045		4.8
Naruko	Nizaka tuff	38.730	140.727		ca.0.073		4.8
Funagata	Izumigatake	38.408	140.712	1.45	–	1.14	2.3
Funagata	Funagatayama	38.453	140.623	0.85	–	0.56	19.0
Yakuraisan	Yakuraisan	38.563	140.717	1.65	–	1.04	0.2
Nanatsumori	Nanatsumori lava	38.430	140.835	2.3	–	2	0.5
Nanatsumori	Miyatoko Tuffs	38.428	140.793		ca.2.5		6.1
Nanatsumori	Akakuzure-yama lava	38.433	140.768	1.6	–	1.5	1.5
Nanatsumori	Kamikadajin lava	38.447	140.772	1.6	–	1.5	0.8
Shirataka	Shirataka	38.220	140.177	1	–	0.8	3.8

Volcano Complex	Volcanic event	Location		Age (Ma)			Volume (km³, DRE)
		Latitude	Longitude	Oldest	Approxi-mate	Younge-st	
Adachi	Adachi	38.218	140.662		ca.0.08		0.9
Gantosan	Gantosan	38.195	140.480	0.4	–	0.3	4.6
Kamuro-dake	Kamuro-dake	38.253	140.488		ca.1.67		5.7
Daito-dake	Daito-dake	38.316	140.527				5.7
Ryuzan	Ryuzan	38.181	140.397	1.1	–	0.9	4.6
Zao	Central Zao 1st.	38.133	140.453	1.46	–	0.79	0.8
Zao	Central Zao 2nd.	38.133	140.453	0.32	–	0.12	15.2
Zao	Central Zao 3rd.	38.133	140.453	0.03	–	0	0.0
Zao	Sugigamine	38.103	140.462		1		9.9
Zao	Fubosan/byobudake	38.093	140.478	0.31	–	0.17	15.2
Aoso-yama	Gairinzan	38.082	140.610	0.7	–	0.4	6.1
Aoso-yama	Central Cone	38.082	140.610	0.4	–	0.38	3.0
Azuma	Azuma Kitei lava	37.733	140.247	1.3	–	1	24.7
Azuma	Higashi Azumasan	37.710	140.233	0.7	–	0	22.8
Azuma	Nishi Azumasan	37.730	140.150	0.6	–	0.4	7.2
Azuma	Naka Azumasan	37.713	140.188	0.4	–	0.3	4.6
Nishikarasu-gawa andesite	Nishikarasugawa andesite	37.650	140.283		ca.1.5		1.9
Adatara	Adatara Stage 1	37.625	140.280	0.55	–	0.44	0.3
Adatara	Adatara Stage 2	37.625	140.280		ca.0.35		0.4
Adatara	Adatara Stage 3a	37.625	140.280		ca.0.20		2.0
Adatara	Adatara Stage 3b	37.625	140.280	0.12	–	0.0024	0.3
Sasamori-yama	Sasamari-yama andesite	37.655	140.391	2.5	–	2	0.4
Bandai	Pre-Bandai	37.598	140.075		ca.0.7		0.1
Bandai	Bandai	37.598	140.075	0.3	–	0	14.0
Nekoma	Old Nekoma	37.608	140.030	1	–	0.7	11.4
Nekoma	New Nekoma	37.608	140.030	0.5	–	0.4	0.9
Kasshi/ Oshiromori	Kasshi	37.184	139.973				0.1
Kasshi/ Oshiromori	Oshiromori	37.199	139.970				0.7
Kasshi/ Oshiromori	Matami-yama	37.292	139.886				0.3
Kasshi/ Oshiromori	Naka-yama	37.282	139.899				0.0
Shirakawa	Kumado P.F.	37.242	140.032		1.31		19.2
Shirakawa	Tokaichi A.F. tuffs	37.242	140.032	1.31	–	1.24	12.0
Shirakawa	Ashino P.F.	37.242	140.032		1.2		19.2
Shirakawa	Nn3 P.F.	37.242	140.032	1.2	–	1.17	0.0
Shirakawa	Kinshoji A.F. tuffs	37.242	140.032	1.2	–	1.18	9.0
Shirakawa	Nishigo P.F.	37.252	139.869		1.11		28.8
Shirakawa	Tenei P.F.	37.242	140.032		1.06		7.7
Nasu	Futamata-yama	37.244	139.971		0.14		3.2

Volcano Complex	Volcanic event	Location		Age (Ma)			Volume (km³, DRE)
		Latitude	Longitude	Oldest	Approxi-mate	Younge-st	
Nasu	Kasshiasahi-dake	37.177	139.963	0.6	–	0.4	12.3
Nasu	Sanbonyari-dake	37.147	139.965	0.4	–	0.25	5.5
Nasu	Minami-gassan	37.123	139.967	0.2	–	0.05	8.7
Nasu	Asahi-dake	37.134	139.971	0.2	–	0.05	4.6
Nasu	Chausu-dake	37.122	139.966	0.04	–	0	0.3
Hakkoda	South-Hakkoda	41.437	141.057		ca.0.73		5.9
Hakkoda	Hakkoda P.F.2nd.	41.437	141.057	0.45		0.2	3.6
Hakkoda	North-Hakkoda	41.277	141.123		ca.0.8		11.4
Okiura	Aoni F. Aonigawa P.F.	40.667	140.897		0.65		17.8
Okiura	Aoni F. Other P.F.	40.600	140.850	0.65	–	0.4	52.4
Okiura	Okogawa-sawa lava	40.667	140.897		0.4		17.3
Okiura	Okiura dacite	40.650	140.883	0.16	–	0	30.4
Ikarigaseki	Nijikai Tuff	40.573	140.763		ca.1.7		17.6
Ikarigaseki	Ajarayama	40.573	140.763	1.7	–	0.9	3.7
Towada	Herai-dake	40.579	140.759	0.9	-	0.65	0.9
Towada	Ohanabe-yama	40.557	140.755	0.9	-	0.7	2.1
Towada	Hakka	40.500	140.625		ca.2.0		20.2
Towada	Towada Okuse	40.490	140.600	1.91	–	1.89	2.1
Towada	Towada Ofudo	40.450	141.000				5.1
Towada	Towada Hachinohe	40.500	140.883	0.4	–	0.05	8.9
Towada	Post-caldera cones	40.417	140.867				1.4
Nanashigure	Nanashi-gure	40.468	140.888		0.055		4.8
Moriyoshi	Moriyoshi	40.468	140.888		0.025		22.1
Bunamori	Bunamori	40.468	140.888		0.013		26.9
Akita-Yakeyama	Akita-Yakeyama	40.457	140.913	0.013	–	0	14.4
Nishimori/ Maemori	NIshimori/Maemori	40.068	141.112	1.06	–	0.72	55.5
Hachimantai/ Chausu	Hachimantai	39.973	140.547	1.07	–	0.78	18.1
Hachimantai/ Chausu	Chausu-dake	39.967	140.717		1.2		0.1
Hachimantai/ Chausu	Fukenoyu	39.963	140.763	0.5	–	0	9.9
Hachimantai/ Chausu	Gentamri	39.973	140.962	0.5	–	0.3	2.6
Yasemori/ Magarisaki-yama	Magarisaki-yama	39.953	140.857	1	–	0.7	5.5

Table 1. List of volcanoes along the volcanic front in Northeast Japan (after Martin et al., 2004)

2.1 Number of volcanoes and volcanic regions

The volcanoes are clustered near the volcanic front. Seven volcanic regions (V.R.) can be identified as along the arc and named named as follow: the Osore V.R., the Hakkoda-Towada V.R., the Sengan V.R., the Kurikoma-Onikobe V.R., Zao-Funagata V.R., the Bandai-

Adatara V.R. and the Aizu V.R. Each volcanic region is consists of a number of small- to medium-sized stratovolcanoes, typically measuring less than 10 km^3 in magmatic eruption (DRE). This feature of volcano clustering was first pointed out by Umeda et al. (1999) and re-pointed out by Tamura et al. (2002). However, several volcanic centers have produced large-sized stratovolcanoes or large-scale felsic pyroclastic flows, attaining as much as tens of cubic kilometers in DRE volume. Along the volcanic front in Northeast Japan, 139 volcanic events are recognized (Fig. 1). 113 are stratovolcano-forming events, and the rest are caldera-forming.

2.2 Temporal change in magma discharge rate and eruption style

In order to elucidate temporal variations in the long-term magma discharge rate all over the NE Japan arc, the magma volume erupted every 100 kilo years (long-term discharge rate of magma) was calculated for each volcano during the last 2.0 million years In the case of South-Hakkoda eruptive episode between 0.65 and 0.40 Ma (million years ago) belonging to Hakkoda volcano (Table 1), the erupted volume was estimated to be 52.4 km^3 magma from which 10.5 km^3, 21.0 km^3 and 21.0 km^3 magmas can be allocated to the periods of 0.7 to 0.6 Ma, 0.6 to 0.5 Ma and 0.5 to 0.4 Ma, respectively.

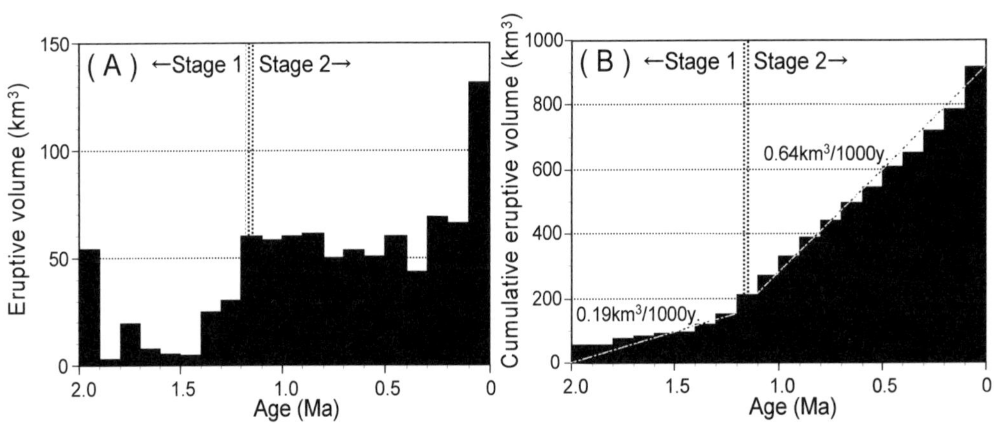

Fig. 2. Temporal changes in eruptive volume (A) and cumulative eruptive volume (B) per 100 ky along the volcanic front in Northeast Japan.

To identify when andesite stratovolcano-forming volcanism was initiated all over the volcanic arc, instead of felsic caldera-forming volcanism, the temporal change in the amount of magma discharged from all the volcanoes is shown in Fig. 2. The figure shows that erupted magma volume increased after 1.2 Ma and more than 50 km^3 of magma per 100,000 years were steadily erupted along the volcanic front.

On the one hand, the temporal change in the amount of erupted magma associated with stratovolcanoes and calderas is shown in Fig. 3. Felsic caldera-forming volcanism with large-scale pyroclastic flows is occurred intermittently since 2.0 Ma, and possible hiatuses of several one hundred thousand years exist during the Quaternary. Andesite stratovolcano-forming volcanism is recognized in the early Quaternary time, and note that it intensified after 1.1 Ma. Thus, Quaternary volcanism along the volcanic front changed in erupted magma volume and eruption style around 1.2 to 1.1 Ma. It can be characterized in two stages: stage 1 (before ca. 1.2 Ma), dominated by felsic caldera-forming volcanism; stage 2 (ca. 1.2 Ma onwards), was characterized by the predominance of andesite stratovolcano-forming volcanism, and marked by a significant increase in erupted magma volume.

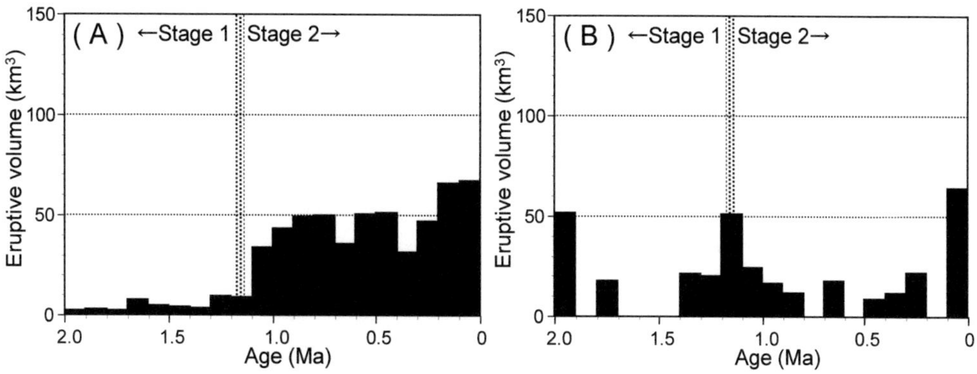

Fig. 3. Temporal changes in eruptive volume for each type of volcanism. (A) is stratovolcano-building volcanism associated with stratovolcanoes, and (B) is felsic caldera-forming volcanism.

2.3 Temporal change in distribution of volcanic centers

The distribution of volcanic centers identified in the two stages discussed above is shown in Fig. 4. Stage 1 volcanism is only recognized in the Hakkoda-Towada V.R., the Sengan V.R., the Zao-Funagata V.R., the Bandai-Adatara V.R. and the Aizu V.R. In contrast, additional volcanic centers in stage 2 emarged in the Osore V.R. and the Kurikoma-Onikobe V.R. in the frontal arc. All seven volcanic regions and the volcanic front in Northeast Japan have only been established since ca.1.2 Ma. Moreover, the distribution of volcanic centers indicates that the northern part of the volcanic front has shifted about 10 to 20 km toward the trench side around 1.2 Ma.

3. Overview of cenozoic tectonism, Northeast Japan

So far, a great deal of effort has been made to obtain information about the Cenozoic tectonism in Northeast Japan (e.g., Sato, 1994; Acocella et al., 2008) and the published results are summarized as follows. The Cenozoic tectonic sequence is directory associated with the separation of the present-day Northeast Japan arc from the Asian continental margin due to the subduction of the Pacific plate and the opening of the Japan Sea rifted. Main rifting started at ~ 23 Ma, and from 21 to 18 Ma, was accompanied by significant counterclockwise rotation of the Northeast Japan arc (Jolivet et al., 1994). Owing to the cessation of the opening of the Japan Sea, the extensional stress field changed at about 13 Ma. In the Middle Miocene to the Pliocene, the tectonics is characterized by very weak crustal deformation under the moderate regional stress field related to the convergence of the Pacific plate (Sato, 1994). The maximum horizontal stress oriented in the NE or ENE direction was manifested during this period. This is one of the reasons why the SW migration of the Kuril sliver due to the oblique convergence along the Kurile arc results from a NE or ENE trending maximum compression (e.g., Otsuki., 1990).

The tectonic shortening became apparent in an E-W direction of compression around the Pliocene to Quaternary boundary, which may be associated with the increase in the motion of the Pacific plate between 5 and 2 Ma (Cox and Engebretson, 1985; Pollitz, 1986). In contrast, the crustal shortening of Northeast Japan might be triggered by the eastward motion of the Amur plate including in the Eurasian plate in the Quaternary (Taira, 2001). This is the reason why the Amur plate is considered to have initiated an incipient subduction on the eastern margin of the Japan Sea (Nakamura, 1983, Tamaki & Honza, 1985). A compressional stress field during the Quaternary is responsible for the development of two narrow uplift zones oriented in the N-S direction, in the Northeast Japan arc: the Ou Backbone Range (fore-arc) and the Dewa Hills (rear-arc). They appear to be an active pop-up structure bounded by opposite-facing reverse faults accommodating < 5 mm/y. of E-W shortening across the range (Hasegawa et al., 2005). Based on the subsurface geology and deformation of river terraces, the initiation time of reverse faulting was estimated at several sites in the Northeast Japan. These results suggest that reverse faulting started in the rear-arc side between 3.4 and 2.4 Ma (Awata and Kakimi, 1985), and in the fore-arc side between 0.9 and 0.5 Ma (Otsuki et al., 1977), corresponding to the onset time of uplift of the Dewa Hills and the Ou Backbone Range. The compressional regime have reactivated normal faults related to the extensional back-arc rifting until 18 Ma as reverse faults and accommodate much of the ongoing shortening across the arc (e.g., Sato, 1994).

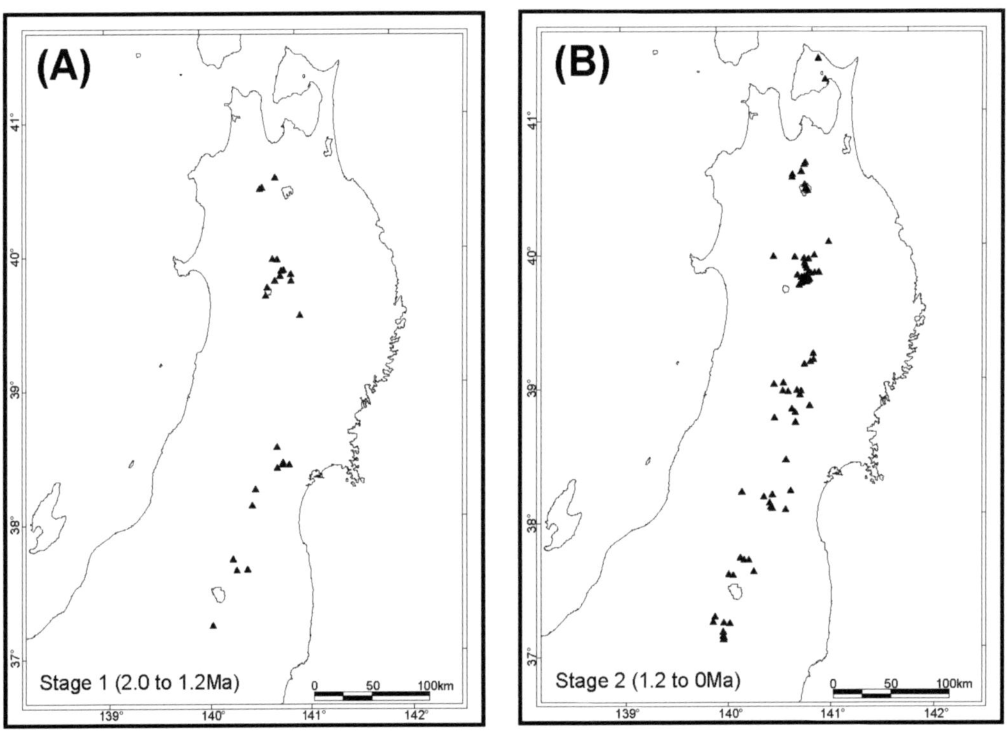

Fig. 4. Distribution of volcanic centers for stage 1 (2.0 -1.2 Ma) and stage 2 (1.2 - 0 Ma).

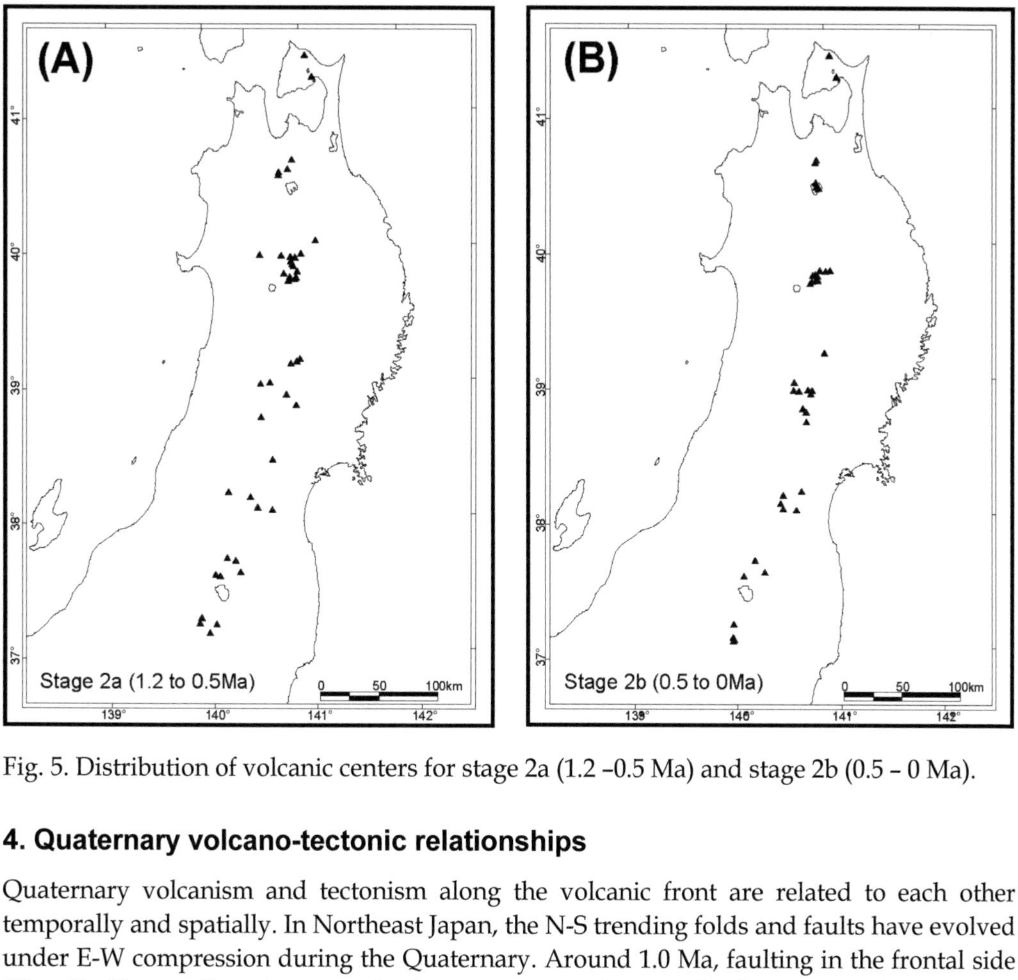

Fig. 5. Distribution of volcanic centers for stage 2a (1.2 -0.5 Ma) and stage 2b (0.5 - 0 Ma).

4. Quaternary volcano-tectonic relationships

Quaternary volcanism and tectonism along the volcanic front are related to each other temporally and spatially. In Northeast Japan, the N-S trending folds and faults have evolved under E-W compression during the Quaternary. Around 1.0 Ma, faulting in the frontal side (Ou Backbone Range), caused the concentrated crustal shortening there. Some contemporaneous changes occurred in volcanism as well; Around 1.2 to 1.1 Ma, felsic caldera-forming volcanism changed to andesite straovolcano-bulding volcanism. Moreover, the total erupted magma volumes along the volcanic front have notably increased since ca. 1.1 Ma. At the same time, magma underwent a systematic change in chemical composition. A significant volume of medium-K andesite has been erupted along the Ou Backbone Range since 1.0 Ma to 0.7 Ma, together with subordinate low-K andesite (Ban et al., 1992). Thus some synchronization between volcanism and tectonism is apparent.

To examine the spatial connections between volcanism and tectonism, the distribution of volcanic centers is compared to those of active faults, amplitudes of uplift and subsidence. Faulting along the volcanic front was initiated around 1.0 Ma and has been intensely activated all over the Ou Backbone Range since 0.5 Ma. The uplift of the mountain range might be accelerated due to resulting in reactivation of more faults. Based on these results, the distribution of volcanic centers along the volcanic front before and after 0.5 Ma, in stage 2, is shown in Fig. 5. The figure indicates that the volcanically active areas became localized near the volcanic front (shifted to the eastern margin of the Ou Backbone Range) after 0.5

Ma, and the alignment of volcanic centers exhibits a weak N-S trend in each volcanic region. Thus, volcanism in stage 2 can be divided into two sub-stages: stage 2a (1.2 to 0.5 Ma), marked by volcanism extended over a wide area and stage 2b (0.5 to 0 Ma), dominated by volcanic centers localized near the volcanic front.

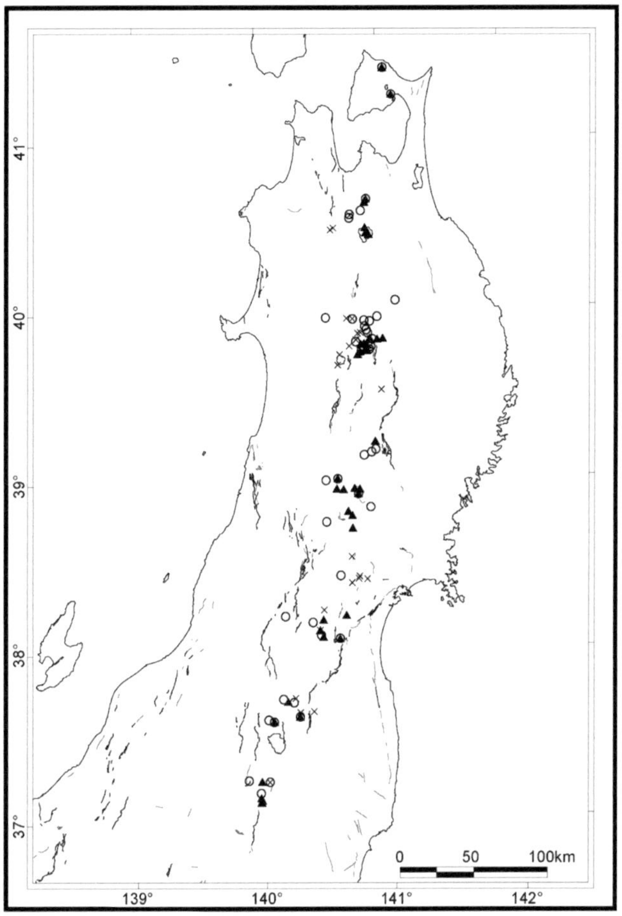

Fig. 6. Distribution of volcanic centers for each stage of Quaternary volcanism (Cross : Stage 1 , Open circle : Stage 2a , Solid triangle : Stage 2b) and active faults.

The distribution of active faults in Northeast Japan (Research Group for Active Faults in Japan, 1991) and volcanic centers formed in the respective stages are shown in Fig. 6. It indicates that the volcanic centers in stage 2b were located in a restricted area between active faults running along the eastern and western margins of the Ou Backbone Range, whereas the volcanic centers in stage 2a are found outside the above area. Fig. 7 shows the uplift and subsidence during the Quaternary (Research Group for Quaternary Tectonic Map, 1968) and the distribution of volcanic centers in the respective stages. Fig. 7 indicates that the volcanic centers formed in stage 2b tend to be distributed more in uplifted areas than those formed in stage 2a. However, there is no active fault near the Hakkoda-Towada V.R. and the Kurikoma-Onikobe V.R., where large-scale felsic pyroclastic flows were erupted in stage 2b and the amount of uplift is less than in other districts. Thus, the distribution of volcanic centers and eruption styles are closely related to the distribution of active faults and amplitudes of uplift suggesting a spatial connection between volcanism and tectonism (Fig. 8).

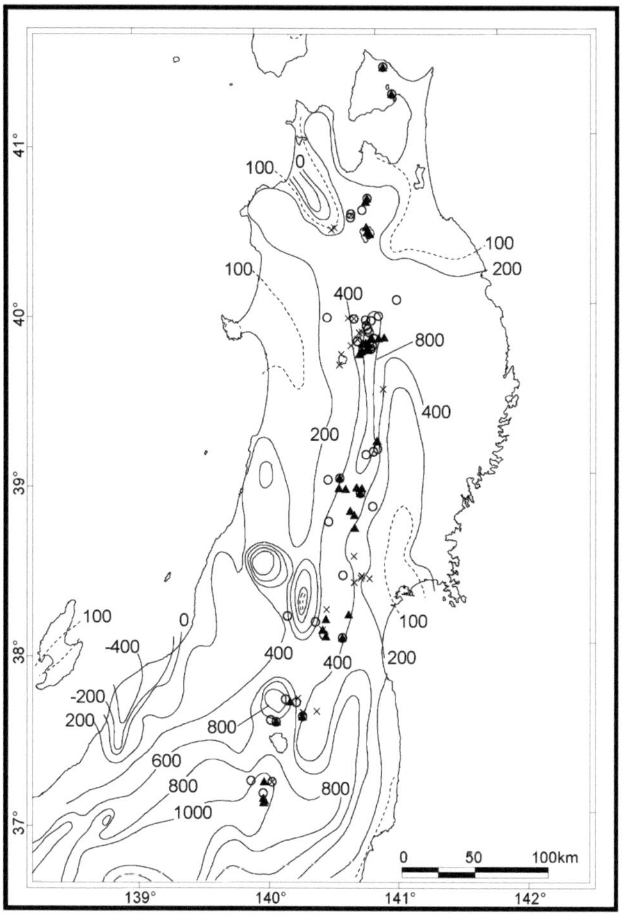

Fig. 7. Distribution of volcanic centers for each stage of Quaternary volcanism (Cross : Stage 1 , Open circle : Stage 2a , Solid triangle : Stage 2b) and amounts of Quaternary uplift and subsidence. Contour interval of uplift/subsidence is 200 m for solid line, and 100 m for broken line.

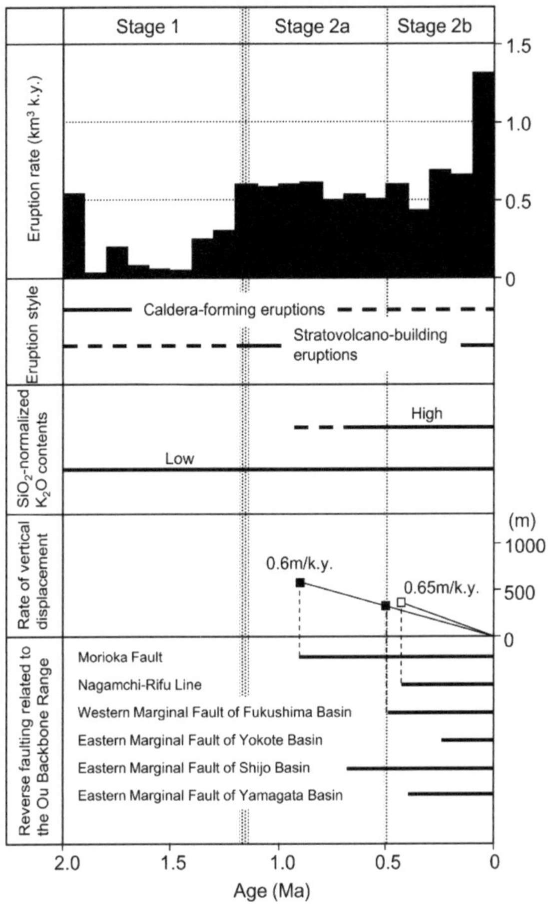

Fig. 8. Summary of volcanism and tectonism since 2.0 Ma along the volcanic front in the Northeast Japan arc.

5. Discussion

In view of the temporal and spatial connections between volcanism and tectonism discussed above, notable changes in eruption style and magma discharge rate occurred around 1.2 Ma. Generally, the crustal stress regime is thought to reflect the eruption style. Caldera formation suggests a tectonic environment facilitating the emplacement of shallow, large-scale felsic magma reservoirs. Therefore, Yoshida et al. (2001) suggested that an intermediate stress field allowing the alternation of weakly compressive and tensile fields is more favourable than a strongly tensile field to develop such a tectonic environment. Similarly, Takahashi (1995) pointed out that the accumulation of a large amount of felsic magma requires a relatively stable tectonic environment with a low crustal strain. In contrast, in stage 2, the crustal stress along the volcanic front is inferred to been changed to a strong compressive stress field with a high crustal strain rate which is favourable for stratovolcano-building volcanism. Although compressive components in the crustal stresses have gradually increased toward the fore-arc side since the Pliocene, the patterns of crustal stress are concordant with the eruption styles in stage 1 and stage 2.

However, it seems to be difficult to interpret a significant increase in erupted magma volume since ca. 1.2 Ma on account of compressional stress regime in fore-arc side. Because, in compressional settings, it is considered that magma cannot ascend so easily, for the reason of magma expanding along horizontal fractures perpendicular to the least principal stress (σ_3) equal to vertical stress (σ_v) (Hubbert and Willis, 1957). For this apparent contradiction, one of the plausible interpretation for this contradiction may be the increase in magma generation in the wedge mantle. Numerical simulations considering fluid migration and melting in the mantle wedge above a subducting plate indicate that melt production rates increase with increasing convergence rate (Cagnioncle et al., 2007). In the Cascade volcanic arc, the convergence rate of the Juan de Fuca plate to the North American plate is thought to control the change in eruption rate (Priest, 1990). Therefore, despite the overall compressive setting along the NE Japan arc, the increase of magma erupted could be interpreted to be due to the product of partial melting in the wedge by significantly faster subducting Pacific slab.

In addition, it is necessary to examine the effect of local crustal stress along the volcanic front on volcanism in stage 2. Local changes in crustal stress are attributable to: 1) the heterogeneity of differential stress caused by thermal structures (Watanabe et al., 1999); 2) a change in crustal stress near the faults caused by faulting (Yoshioka and Suzuki, 1997); and 3) the gravitational instability generated in uplifted mountain blocks (Moriya 1983; Molnar 1986).

5.1 Heterogeneity of differential stress caused by thermal structures

Watanabe et al. (1999) pointed out that the heat spreading from magma reservoirs can produce a horizontal stress heterogeneity which could be lowered locally around the reservoirs, so that the regional crustal stress could be maintained at some distance from the reservoir. In fact, a S-wave reflection horizon correlative to a magma reservoir at a depth of 7 to 12 km below the Kiso-Ontake Volcano has been recognized (Inamori et al., 1992). The focal mechanisms of swarm earthquakes generated near this horizon indicate that σ_{Hmax} is always equal to σ_1, whereas the vertical stress (σ_v) is unstable switching from σ_2 to σ_3 (Hori et al., 1982). A stress field with $\sigma_v = \sigma_2$ could lead to the intrusion of a dike and permit vertical migration of magma. Thus, even though the regional stress field is compressive, magma could still ascend if the adjacent differential stress is lowered by the heat spreading from magma itself. Moreover, it is probable that dikes, conduits for magma are combined with horizontal sheets to form a complicated plexus as indicated by Takahashi (1994). Thus, the local lowering of differential stress by thermal effects is thought to be a factor in magma ascent.

5.2 Change in crustal stress near the faults caused by faulting

It has been noted that the crustal stress around faults changes before and after faulting. According to Yoshioka and Suzuki (1997), when a dislocation is generated by a fault, a reverse fault type of stress field with $\sigma_v = \sigma_3$ develops on upward and downward extensions of the fault plane, whilst a normal fault type of stress field with $\sigma_v = \sigma_1$ occurs immediately above and below it. As mentioned above, active fault systems exist that are believed to reach the lower crust beneath the volcanic front and contribute to the uplifting of the Ou Backbone Range. Dislocations along these faults give rise to a normal fault type of stress field that facilitates the ascent of magma around faults. Therefore, it is probable that the faulting activated all over the Ou Backbone Range since 0.5 Ma resulted in the concentration of volcanic centers in the area between western and eastern marginal active faults during stage 2b.

5.3 Gravitational instability generated in the uplifted mountain block

The uplift of the volcanic front is believed to have been accelerated by the activation of faults all over the Ou Backbone Range since 0.5 Ma. The uplift-related gravitational instability in the mountain blocks in turn is expected to have generated a tensile stress field normal to the elongation of the mountain range (Moriya 1983; Molnar 1986). Based on the fact that the focal mechanism of earthquakes below the volcanoes on the uplifted mountains differs at depth from one another, Takahashi (1994) estimated that the local tensile stress field generated by the gravitational instability may extend down to several km below the mountain top. These facts show that the more severely uplifted areas form a favourable environment for magma ascent due to the local tensile stress field, and provide a reasonable explanation for the increase in eruptive volume and the concentration of volcanic centers in such uplifted areas in stage 2b (Fig. 7). Furthermore, the N-S alignment of volcanic centers in stage 2b might reflect the E-W tensile stress field generated by gravitational instability.

Thus, the faulting and uplifting have presumably generated local lowering of differential stress or the tensile stress field along the volcanic front during stage 2, which in turn affected the amount of magma erupted and the alignment of volcanic centers. In the areas (e.g. the Hakkoda-Towada V.R, the Kurikoma-Onikobe V.R.) where volcanism associated with large-scale felsic pyroclastic flows occurred in stage 2b, a tectonic environment characterized by weak compression and a low crustal strain might have prevailed despite the lack of active faults and uplift.

6. Conclusions

From a compilation and analysis of stratigraphy, radiometric age and eruptive magma volume data for 139 volcanic events along the volcanic front, notable changes in eruption style, magma compositions, variation in eruptive volume, and distribution of volcanic centers can be recognized around 1.2 Ma. Before ca. 1.2, felsic caldera-forming volcanism are thought to occur in regions of neutral stress regime with low crustal strain rate. From ca. 1.2 Ma to the present-day, the crustal stress regime seems to have changed to compression yielding the formation of stratovolcanoes all the volcanic front. It has become apparent that stratovolcanoes lie along major thrust faults associated with uplift of the Ou Backbone Range since the Middle Pleistocene. Although it is widely assumed that magma cannot rise so easily in compressional setting, the increase of erupted magma volume since ca. 1.2 Ma may have been caused by an increase in subduction rate of the Pacific plate between 5 and 2 Ma. In addition, the lowering of differential stress by thermal effects is also thought to facilitate the ascent of magma. On the other hand, the distribution of volcanic centers formed since 0.5 Ma, controlled mostly by the local extensional stress regime in the upper crust, was locally influenced by fault dislocations and gravitational instability.

7. Acknowledgment

The author thanks Drs. R. I. Tilling, S. J. Day and S. Hayashi for many comments that helped us to improve the original manuscript, and Dr. A. J. Martin for editing this manuscript.

8. References

Acocella, V., Yoshida, T., Yamada, R. & Funiciello, F. (2008), Structural control on Late Miocene to Quaternary volcanism in the NE Honshu arc, Japan. *Tectonics*, 27, TC5008. doi:10.1029/2008TC002296.

Aramaki, S. & Ui, T. (1978). *List of geodynamic parameter of Quaternary volcanoes of Japan, Mariana, Kurile and Kamchatka.*, Geodynamics Project 78-2., Japan.

Awata, Y. & Kakimi, T. (1985). Quaternary tectonics and damaging earthquakes in northeast Honshu, Japan. *Earthquake Predict. Res.*, 3, 231-251.

Ban, M., Oba, Y., Ishikawa, K. & Takaoka, N. (1992). K-Ar dating of Mutsu-Hiuchidake, Osoreyama, Nanashigure, and Aoso volcanoes of the Aoso-Osore volcanic zone - The formation of the present volcanic zonation of the Northeast Japan arc -. *J Mineral. Petrol. Econ. Geol.*, 87, 39-49.

Cagnioncle, A.-M., Parmentier, E. M. & Elkins-Tanton, L. T. (2007). Effect of solid flow above a subducting slab on water distribution and melting at convergent plate boundaries. *J. Geophys. Res.*, 112, B09402, doi:10.1029/2007JB004934.

Committee for Catalogue of Quaternary Volcanoes in Japan (1999). *Catalogue of Quaternary Volcanoes in Japan*, Volcanol. Soc. Jpn., Tokyo, Japan.

Cox, A., & Engebretson, D. (1985). Change in motion of Pacific Plate at 5 Myr BP. *Nature*, 313, 472-475.

Hasegawa, A., Nakajima, J., Umino, N. & Miura S. (2005). Deep structure of the northeastern Japan arc and its implications for crustal deformationand shallow seismic activity. *Tectonophys.*, 403, 59-75.

Hori, S., Aoki, H. & Ooida, T. (1982). Focal mechanisms of the earthquake swarm southeast of Mt. Ontake, central Honshu, Japan. *Zisin*, 35, 161-169.

Hubbert, M. K. & Willis, D. G. (1957). Mechanics of hydraulic fracturing, In: *Structural Geology*, M. K. Hubbert, (Ed.), 175-190, Macmillan, New York.

Jolivet, L., Tamaki, K. & Fournier, M. (1994). Japan Sea, opening history and mechanism: A synthesis. *J. Geophys. Res.*, 99 (B11), 22237-22259.

Kimura, J. & Yoshida, T. (2006). Contributions of slab fluid, mantle wedge and crust to the origin of Quaternary lavas in the NE Japan arc. *J. Petrol.*, 47, 2185-2232.

Martin, A. J., Umeda, K., Connor, C. B., Weller, J. N., Zhao, D. & Takahashi, M. (2004). Modeling long-term volcanic hazards through Bayesian inference: An example from the Tohoku volcanic arc, Japan. *J. Geophys. Res.*, 109, B10208, doi: 10.1029/2004JB003201.

Molnar, P. (1986). The structure of mountain ranges. *Sci .Am.*, 255, 64-73.

Nakagawa, M., Shimotori, H. & Yoshida, T. (1988). Across-arc compositional variation of the Quaternary basaltic rocks from the Northeast Japan arc. *J. Miner. Petrol. Econ. Geol.*, 83, 9-25.

Nakamura, K. (1983). Possible nascent trench along the eastern Japan Sea as the convergent boundary between Eurasian and North American plates. *Bull. Earthq. Res. Inst., Univ. Tokyo*, 58, 711-722.

Nakata, T. & Imaizumi, T. (2002). *Digital active fault map of Japan (DVD-ROM)*, Univ. Tokyo Press, ISBN 978-413-0607-40-7, Tokyo, Japan.

Notsu, K. (1983). Strontium isotope composition in volcanic rocks from the Northeast Japan arc. *J. Volcanol. Geotherm. Res.*, 18, 531-548.

Ono, K., Soya, T. & Mimura, K. (1981). *Volcanoes of Japan, 1:2,000,000 map series, no.11, 2nd ed.*, Geol. Surv. Jpn., Tsukuba, Japan.
Otsuki, K. (1990). Neogene tectonic stress fields of northeast Honshu arc and implications for plate boundary conditions. *Tectonophys.*, 181, 151-164.
Otsuki, K., Nakata, T. & Imaizumi, T. (1977). Quaternary crustal movements and block model in the southeastern region of the Northeast Japan. *Earth Sci.*, 31, 1-14.
Pollitz, F. F. (1986). Pliocene change in Pacific Plate motion. *Nature*, 320, 738-741.
Priest, G. (1990). Volcanic and Tectonic Evolution of the Cascade Volcanic Arc, Central Oregon. *J. Geophys. Res.*, 95 (B12), 19583-19599.
Research Group for Quaternary Tectonic Map. (1968). Quaternary Tectonic Map of Japan. *Quaternary Res.*, 7, 182-187.
Sakuyama, M. & Nesbitt, R. W. (1986). Geochemistry of the Quaternary volcanic rocks of the Northeast Japan arc. *J. Volcanol. Geotherm. Res.*, 29, 413-450.
Sato, H. (1994). The relationship between late Cenozoic tectonic events and stress field and basin development in northeast Japan. *J. Geophys. Res.*, 99, (B11), 22261-22274.
Taira, A. (2001). Tectonic evolution of the Japaneses island arc system. *Annu. Rev. Earth Planet. Sci.* , 29, 109-134.
Takahashi, M. (1994). Structure of polygenetic volcano and its relation to crustal stress field 2. P-type • O-type volcano. *Bull Volcanol Soc Jpn.*, 39, 207-218.
Takahashi, M. (1995). Large-volume felsic volcanism and crustal strain rate. *Bull Volcanol Soc Jpn.*, 40, 33-42.
Tamaki, K. & Honza, E. (1985). Incipient subductionand obduction along the eastern margin of Japan Sea. *Tectonophysics*, 119, 381-406.
Tamura, Y., Tatsumi, Y., Zhao, D. P., Kido, Y., & Shukuno, H. (2002). Hot fingers in the mantle wedge: new insights into magma genesis in subduction zones. *Earth. Planet. Sci. Lett.*, 197, 105-116.
Tatsumi, Y. & Eggins, S. (1995). *Subduction zone magmatism*, Blackwell, ISBN 978-086-5423-61-9, Cambridge.
Umeda, K., Hayashi, S., Ban, M., Sasaki, M., Oba, T., & Akaishi, K. (1999). Sequence of volcanism and tectonics during the last 2.0 million years along the volcanic front in Tohoku district, NE Japan. *Bull. Volcanol. Soc. Jpn.*, 44, 233-249.
Watanabe, T., Koyaguchi, T. & Seno, T. (1999). Tectonic stress controls on ascent and emplacement of magmas. *J. Volcanol. Geotherm. Res.*, 91, 65-78.
Yoshioka, S. & Suzuki, H. (1997). Effects of three-dimensional Inhomogeneous viscoelastic structures on quasi-static strain and stress fields associated with dislocation on a rectangular fault. *Zisin*, 50, 277-289.
Yoshida, T. (2001). The evolution of arc magmatism in the NE Honshu arc, Japan. *Tohoku Geophys. J.*, 36, 131-149.

Part 2

Large Igneous Provinces

4

Origin, Distribution and Evolution of Plume Magmatism in East Antarctica

Nadezhda M. Sushchevskaya[1], Boris V. Belyatsky[2] and Anatoly A. Laiba[3]
[1]*Vernadsky Institute of Geochemistry and Analytical Chemistry, Russian Academy of Sciences, Moscow,*
[2]*All-Russian Research Institute of Geology and Mineral Resources of the World Ocean, St.Petersburg,*
[3]*Polar Marine Geological Prospecting Survey, St.Petersburg, Russia*

1. Introduction

According to current models (Dalziel et al., 2000; Lawver et al., 1985; Morgan, 1981), the formation of oceanic crust in the South Atlantic and Indian Ocean was affected by large mantle plumes, such as the Karoo-Maud, Kerguelen, and Parana-Etendeka plumes. The penetration of the Karoo-Maud plume into the upper lithosphere at about 180 Ma affected the southern end of Africa and western part of the East Antarctica and was among the main factors that caused the subsequent breakup of the Gondwana supercontinent (Duncan et al., 1997; Jokat et al., 2003; Storey, 1995; Storey & Kyle, 1997). Later, at about 130 Ma, the Kerguelen plume formed near the spreading zone of the opening Indian Ocean (Coffin et al., 2002; Mahoney et al., 1995; Storey et al., 1989, 1992; Weis et al., 1996) which had a considerable impact on the character of oceanic magmatism and resulted in the formation of numerous volcanic rises (Ninetyeast Ridge, Afanasy Nikitin Rise, Naturaliste Plateau, and, probably, Conrad Rise) (Borisova et al., 1996; Frey et al., 2002; Sushchevskaya et al., 1998). Moreover, it affected the continental margins of India (Rajmahal traps) and Australia (Bunbury basalts) (Curray & Munasinghe, 1991; Frey et al., 1996; Kent et al., 1997, 2002). The plume magmatism of South America and Central Africa was assigned to the activity of another mantle plume, Parana-Etendeka, which caused the formation of a seamount chain, the Walvis Ridge, within the South Atlantic at 130-90 Ma (Renne et al., 1996; Stewart et al., 1996). It is obvious that the interaction of plume and oceanic magmatism has a grate sense for resolving many important problems of marine geology and, primarily, the evolution of the oceanic lithosphere. In addition, plume magmatism provides evidence for deciphering the spatio-temporal spreading of plume materials in the lithosphere (in general sense), determining the timing of plume activity and its evolution under lithospheric conditions, and estimating the influence of plumes on the processes of lithospheric plate disintegrations. In this context, an interesting occurrence of plume activity is the Jurassic magmatism of Antarctica, which has been extensively studied in the past few years (Brewer et al., 1996; Elliot et al., 1999; Elliot & Fleming, 2000; Harris et al., 1990; Hergt et al., 1991; etc.). It is supposed that the Mesozoic plume magmatism of Antarctica propagated along the weakened zones of the Earth's crust at the margins of the East Antarctica, along the

Transantarctic Mountains and Indian Ocean coast (Elliot et al., 1999; Leat et al., 2007; White & McKenzie, 1989). The westernmost occurrences of this magmatism are basalts and dolerites from the western part of the Dronning Maud Land (DML), which have tholeiitic compositions and are geochemically enriched to a varying degree (Harris et al., 1990; Luttinen & Furnes, 2000; Vuori & Luttinen 2003). The magmatic complexes of DML (Vestfjella, Heimefrontfjella, and Kirwanveggen mountains and the Ahlmannryggen Plateau) were formed within a narrow time interval between 183 and 175 Ma, with the maximum magmatic activity at ca. 178 Ma (Belyatsky et al., 2002; Brewer et al., 2003; Riley et al., 2005; Zhang et al., 2003).

In this paper, we consider the results of a comprehensive geochemical study of the Mesozoic dolerites of the Schirmacher Oasis (Fig. 1), which is situated to the east of the previously studied occurrences of flood-basalt magmatism in western DML. The age and compositions of the Schirmacher dolerites indicate their connection to mantle plume activity (Belyatsky et al., 2006), which provides an opportunity to refine the boundaries of the Karoo-Maud plume spreading beneath Antarctica. In contrast to the Karoo-Maud plume, the Kerguelen plume invaded the already open ocean basin, which had to influence its geochemical signature (Doucet et al., 2005; Storey et al., 1989; Weis et al., 1991; Weis & Frey, 1996). In addition, we attempted to compare in this paper the geochemical character of DML plume magmatism and magmatism of the early stages of Kerguelen plume activity and present some geodynamic implications on the basis of this comparison.

2. Geologic setting and composition of the Mesozoic dike complex of the Schirmacher Oasis

The mountainous Schirmacher Oasis is situated in the central part of DML and composes, together with the adjoining nunataks, the northernmost exposed segment of a large mountain chain extending through the whole region (Fig. 1). The oasis is a belt of hilly outcrops, which extends for 20 km in an E-W direction at a maximum width of 4 km. The outcrops of the oasis are completely composed of Precambrian metamorphic rocks, which underwent at least two stages of metamorphic transformations, at about 1000 and 500–450 Ma. The oldest rocks of Late Proterozoic age are alaskite metagranites, metadiorites, metadolerites, metamorphosed gabbronorites, and biotite granites. The Early Paleozoic epoch of granite formation produced three compositional series of pegmatites, diorite dikes, and pegmatoid granites. There is a separate group of Silurian dikes and small bodies of alkaline lamprophyres (Hoch & Tobschall, 1998; Hoch et al., 2001). The complex of Mesozoic dolerites (basic rocks) occurs over the entire Schirmacher Oasis (Fig. 1). The dolerite dikes cut all of the known metamorphic sequences, metagranites and metabasic rocks, veins of pegmatite series, and alkaline lamprophyres occurring within the oasis, which indicates that the dolerites are the youngest igneous rocks of the region. The dikes strike mainly NW-SE and NE-SW and dip 25–90°. The thickness of the dikes is 0.1–1.7 m, occasionally up to 8 m. Their lengths are up to 250–270 m and usually a few tens of meters. With respect to petrographic composition, the dikes are made up of olivine and olivine- free dolerites and gabbro-dolerites affected to a varying degree by secondary alteration.

The dolerites are porphyritic or equigranular rocks with microdoleritic, microgabbroic, or vitrophyric groundmass textures. Sometimes olivine and clinopyroxene phenocrysts account for 10–25% of the rock by volume (e.g. olivine-phyric dolerite). Chilled margins were observed in the thickest dikes. With respect to chemical composition, the majority of basalts and dolerites of the Schirmacher Oasis can be classified as weakly alkaline, but three of them even nepheline normative alkaline, magnesian basalts with 0.6–1.6 wt% K_2O, 0.7–2.0 wt% TiO_2, and 10–17 wt% MgO. Figure 2 shows variations in TiO_2, K_2O, Na_2O, and SiO_2 as functions of MgO content for the dolerite samples those compositions are given in Table 1 (oxide compositions were determined by XRF at Vernadsky Institute, detection limits ranged from 0.001 to 0.02 wt%, RSD from 1.5 to 13 wt%; for detailed description of analytical procedure see (Sushchevskaya et al., 2009)). Three samples from our data set (47240-2, 47235-19, and 47201-4) show elevated potassium contents, and sample 47240-2 is also depleted in silica and enriched in alkalis and titanium, which allows classify them as alkali high potassic basalts. Sample 47201-4 is a strongly altered amygdaloidal dolerite, and the presence of amphibole and biotite in sample 47235-19 indicates plausible lamprophiric input, but in any case theirs compositions can be regarded as primary with some caution. Thus, two of the freshest olivine dolerite samples, 47225-7 and 47139-7, were selected for the investigation of major minerals. The Mesozoic metabasites of the Schirmacher Oasis (Fig. 2) are identical in composition to the dolerites of the dike complex of the Muren massif in western DML (Vuori & Luttinen, 2003) and slightly different from the ancient anorthosite dikes occurring at the Schirmacher Oasis and often showing similar strikes (Belyatsky et al., 2002). In general, the crystallization sequence of the basalts is magnesian olivine-plagioclase–clinopyroxene. This is supported by the analyses of minerals, in particular, more magnesian compositions of olivines compared with clinopyroxenes. The olivine composition varies strongly from $Fo_{91.5}$ to Fo_{55} (~400 grains), which suggests the occurrence of crystal accumulation processes during melt differentiation. The most widespread olivine compositions, within Fo_{88-89}, are similar to the most magnesian olivines from the least

Fig. 1. Simplified geological map of the Schirmacher Oasis region showing sampling sites (asterisks). The distribution of the ice cover is shown by light blue colors, and outcrops of old continental rocks (granulites, gneisses, plagiogranites, schists, etc.) are colored in accordance with insert legend. Tectonic dislocations, ancient dikes of lamprophyres, dolerites, and metagabbroids as well as Jurassic olivine-bearing dolerites and pegmatite veins are shown as lines of corresponded colors. Sample numbers are shown on the map.

component	47133-1	47137-20	47139-7	47201-4	47206-3	47225-6	47225-7	47235-19	47240-2
SiO2	49.70	45.05	44.80	51.39	45.75	42.05	44.70	51.76	36.13
TiO2	1.00	1.29	0.82	1.09	1.30	2.22	1.56	1.00	3.25
Al2O3	15.84	12.13	9.96	10.57	12.07	11.45	12.25	14.87	11.25
Fe2O3	10.18	12.51	11.56	9.31	12.87	16.13	14.15	8.39	14.09
MnO	0.16	0.19	0.36	0.15	0.19	0.21	0.24	0.13	0.23
MgO	9.91	11.34	16.49	11.25	11.89	10.95	13.05	8.01	7.56
CaO	8.96	10.35	9.68	7.21	10.88	10.89	10.06	7.50	10.21
Na2O	2.56	3.18	2.14	1.04	2.21	2.94	2.68	2.72	5.39
K2O	0.77	0.88	0.55	5.33	0.32	1.03	0.76	3.32	3.46
P2O5	0.18	0.30	0.20	0.91	0.18	0.31	0.25	0.51	0.64
LOI	0.88	2.62	2.62	1.16	2.07	1.57	0.28	1.30	7.35
Total	100.14	99.84	99.18	99.41	99.73	99.75	99.98	99.51	99.56
Ba	304	342	283	4216	129	313	258	2765	1266
Th	2.08	1.35	0.69	15.02	1.02	1.34	1.17	21.08	1.49
U	0.43	0.34	0.16	1.58	0.16	0.24	0.21	2.47	0.44
Nb	3.25	9.79	3.66	15.54	3.35	10.89	8.38	11.07	60.91
Ta	0.20	0.61	0.21	0.75	0.20	0.84	0.66	0.70	7.24
La	12.70	10.24	5.34	63.72	7.06	11.92	10.15	76.17	15.58
Ce	30.0	21.6	11.7	123.8	16.4	28.1	23.1	142.9	30.2
Pb	5.00	2.67	2.17	9.43	1.50	4.05	2.93	49.0	5.54
Pr	4.21	2.95	1.65	14.38	2.42	4.14	3.30	16.20	3.82
Sr	154	298	286	831	221	352	321	1064	683
Nd	19.7	14.1	8.22	58.1	12.4	20.6	16.1	63.6	17.1
Sm	5.27	3.70	2.34	10.38	3.61	5.59	4.36	11.06	4.04
Zr	20	83	52	249	83	149	101	177	187
Hf	0.95	2.16	1.36	6.35	2.25	3.86	2.70	4.94	4.97
Eu	1.18	1.23	0.76	2.30	1.19	1.75	1.44	2.65	1.38
Gd	5.94	3.91	2.55	7.31	3.98	5.66	4.63	8.16	4.35
Dy	6.51	3.49	2.49	5.16	3.68	4.78	4.01	5.53	3.99
Y	36.5	17.8	12.8	26.5	18.1	22.6	18.8	26.4	18.9
Er	3.98	1.83	1.30	2.53	1.83	2.28	1.98	2.65	1.94
Yb	3.76	1.60	1.15	2.26	1.56	1.84	1.69	2.34	1.65
Ho	1.39	0.68	0.49	0.97	0.70	0.88	0.77	1.00	0.75
Lu	0.55	0.23	0.17	0.33	0.22	0.27	0.24	0.33	0.23
La/Nb	3.91	1.05	1.46	4.10	2.11	1.09	1.21	6.88	0.26
La/Sm	2.41	2.77	2.28	6.14	1.96	2.13	2.33	6.89	3.85
La/Yb	3.38	6.40	4.67	28.24	4.54	6.49	6.00	32.49	9.46
Sr/Nd	7.79	21.07	34.73	14.29	17.82	17.10	19.91	16.74	40.02
Gd/Yb	1.58	2.44	2.23	3.24	2.56	3.08	2.74	3.48	2.64
La/Ce	0.42	0.47	0.46	0.51	0.43	0.42	0.44	0.53	0.52
^{147}Sm/^{144}Nd	0.16061	0.15862	0.16814	0.10985	0.17377	0.16212	0.16068	0.10522	0.14251
^{143}Nd/^{144}Nd	0.512536	0.512569	0.512649	0.511578	0.512720	0.512656	0.512585	0.511953	0.512695
err (2S)	0.000004	0.000004	0.000004	0.000004	0.000002	0.000003	0.000008	0.000006	0.000005
^{87}Rb/^{86}Sr	0.59815	0.46261	1.13929	0.88193	0.14066	0.24536	0.23932	0.29299	0.77902
^{87}Sr/^{86}Sr	0.711137	0.706852	0.707522	0.714170	0.704928	0.705283	0.705227	0.709301	0.705929
err (2S)	0.000015	0.000006	0.000019	0.000017	0.000015	0.000021	0.000014	0.000008	0.000005
eNd	-1.21	-0.52	0.83	-18.80	2.10	1.10	-0.25	-11.38	2.29
(^{87}Sr/^{86}Sr)t	0.70969	0.70573	0.70477	0.71204	0.70459	0.70469	0.70465	0.70859	0.70405
^{206}Pb/^{204}Pb	18.202	18.936	18.149	18.197	17.969	18.062	17.925	17.360	18.619
^{207}Pb/^{204}Pb	15.559	15.573	15.494	15.535	15.508	15.511	15.494	15.511	15.510
^{208}Pb/^{204}Pb	38.118	38.274	37.934	39.621	38.176	38.208	38.137	37.810	38.192

Table 1. Major (vol%) and trace element (ppm) concentrations, and Sr, Nd, and Pb isotope data for rocks of basalts and dolerites from Schirmacher Oasis. LILE concentration and isotope analysis were determined at Karpinsky Geological Institute (St.Petersburg, Russia) by ICP MS (detailed in Sushchevskaya, et al. 2009). err (2S) - corresponds to isotope ratios error at 95% confidence level. eNd and (^{87}Sr/^{86}Sr)t - initial isotope composition corresponding sample at the time of dike emplacement (170 Ma).

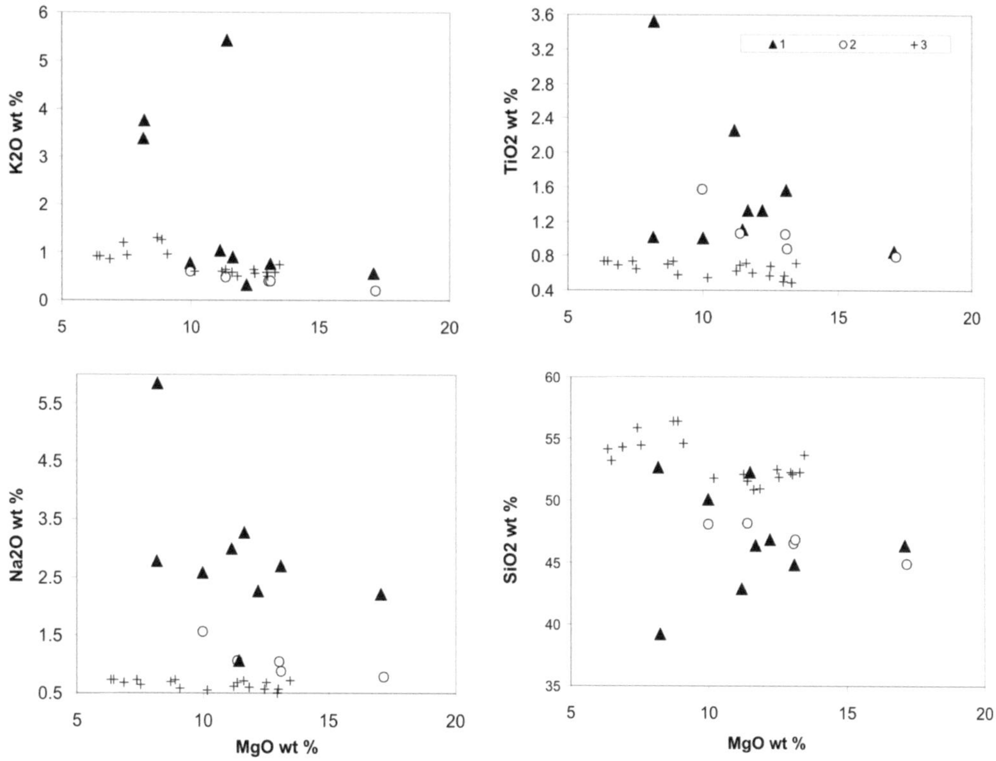

Fig. 2. Variations in major element contents for the Schirmacher dolerites. (*1*) Dolerites of the Schirmacher Oasis (Table 1), (*2*) dikes of the Muren region of DML after (Vuori & Luttinen, 2003), and (*3*) ancient anorthosite dikes from the Schirmacher Oasis region (Belyatsky et al., 2002).

differentiated Jurassic lavas of the western DML (Sushchevskaya et al., 2004). The distinctive feature of the Schirmacher dolerites is the presence of Mg-rich olivines (Fo_{90-91}), which may indicate rapid magma ascent from the generation zones (Fig. 3). Variations in Ca and Ni as a function of forsterite content reflect a decrease in Ni and an increase in Ca contents during primary magma crystallization. In addition, it can be clearly seen in the diagrams that the points of olivine compositions from the two studied samples form two independent trends of NiO and MnO variations with different slopes. The average NiO content in the most magnesian olivines is ~0.35 wt % (N = 340) for sample 47139-7, ~0.4 wt % (N = 46) for sample 47225-7, and ~0.5 wt % for olivines from the Vestfjella Mountains (western DML) (Fig. 3). The differences in Ni content of the liquidus olivines are primarily related to different Ni contents in the initial melts. According to the suggested model of Sobolev with colleagues (Sobolev et al., 2007), an increase in NiO and a decrease in MnO content in magnesian olivines indicate the presence of pyroxenite blobs and veins in the peridotite source. The highest contribution of such a crustal component was inferred for olivines from DML rocks (Sushchevskaya et al., 2004). The model was initially proposed for Hawaiian magmas (Sobolev et al., 2005) and assumed that their plume source contained fragments of crustal eclogites (i.e. lower crustal substance), whose melting began during the ascent of the heterogeneous mantle material at a depth of about 150 km. The produced melts reacted with the peridotite matrix, which resulted in the formation of pyroxenites with Ni-rich pyroxenes. The further melting of pyroxenites at a depth of about 100 km resulted in the formation of melts enriched in Ni relatively the liquids that could be produced by the melting of a peridotite source. The question on the reasons of the heterogeneity of the plume mantle remains unresolved. It could be related to processes accompanying material ascent from the core-mantle boundary, where subducted fragments of the early oceanic crust probably are accumulated (Chase & Patchett, 1988; Christensen & Hofmann, 1994; Hofmann, 1988; Ono et al., 2001; etc.), or to the interaction of the ascending hot peridotitic mantle with the lower parts of continental blocks at depths of 170-220 km (O'Reilly & Griffin, 2010). In either case, eclogite melting could result in the appearance of pyroxenites and their subsequent involvement into the derivation of basaltic magmas (Sobolev et al., 2007). It should be pointed out that, during the initial activity of the plume that was located in the DML region of Antarctica (Leitchenkov & Masolov, 1997; Leitchenkov et al., 2003), this process was more intense, which is reflected in the higher NiO contents of magnesian olivines (Fig. 3).

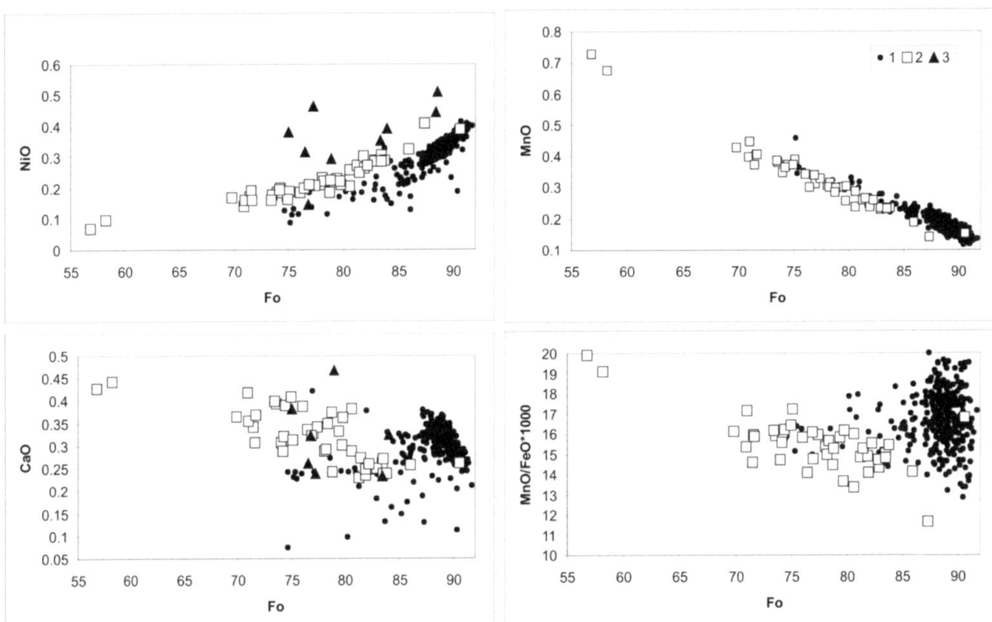

Fig. 3. Contents of NiO, CaO, and MnO in olivines from the Schirmacher Oasis dikes as a function of forsterite mole percentage. Olivines from (*1*) sample 47139-7, (*2*) sample 47225-7, and (*3*) DML basalts (Sushchevskaya et al., 2004).

Clinopyroxene is the third phase crystallizing after olivine and plagioclase. It shows variable Mg value, from 69 to 81 (Fig. 4), and low Cr/Al (0.01–0.14) and Na/Al ratios (0.10–0.15). In all compositional parameters, including the content of lithophile elements (Migdisova et al., 2004), it is similar to clinopyroxenes from the basalts of the Vestfjella Mountains (western DML). The majority of magnesian olivines contain chrome spinel inclusions (Fig. 5a), whose compositions were used to estimate the redox conditions of crystallization. The obtained values suggest that the early crystallization of magmas occurred near the quartz-fayalite-magnetite buffer (QFM), ΔQFM (deviation of $-\lg(f_{O_2})$ from the QFM value) is 1.0–1.2. The conditions of magma fractionation were estimated for the Schirmacher Oasis according the compositions of clinopyroxene by the method of (Nimis & Ulmer, 1998). Using a database of experiments with basanite and picrite - basalt starting materials (more than 100 references and 16 unpublished experiments), empirical pressure dependence was derived for the unit-cell volume and the volume of the M1 site of clinopyroxene. At a given pressure, the unit-cell volume (V_{cell}) is almost linearly correlated with the volume of the M1 site (V_{M1}), and, for a given melt composition, V_{cell} and V_{M1} decrease linearly with increasing pressure. Using these correlations, pressure can be expressed as a linear function of V_{cell} and V_{M1}. For anhydrous and water-saturated magmas, the uncertainty is no higher than 1.70 kbar (the highest discrepancy is 5.4 kbar, N = 157) (Nimis & Ulmer, 1998). Our estimations suggest shallow depths of magma crystallization in a transitional magma chamber at pressures of about 1-2 kbar. The histograms of model depths of melt crystallization (Fig. 5b) calculated for the temperature T=1100°C (average temperature of crystallization in a transitional chamber estimated by the COMAGMAT program (Migdisova et al., 2004)) show that the same range of pressures is also typical for the fractionation of basaltic magmas from the DML region.

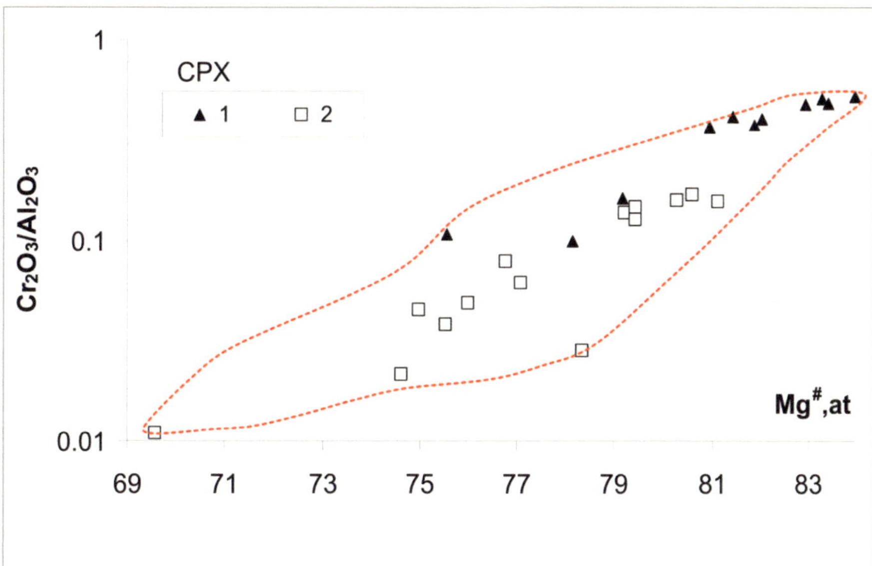

Fig. 4. Composition of clinopyroxene from (*1*) Schirmacher Oasis dolerites and (*2*) DML basalts (Sushchevskaya et al., 2004).

Emplacement 130-105 m.y. ago of dikes and sills of alkaline-ultrabasic composition within Jetty oasis (or Jetty Peninsula, Fig. 6a) is suggested as a later appearance of plume magmatism within the East-Antarctic Shield (Andronikov et al., 1993, 2001; Laiba et al., 1987). This region is located opposite Kerguelen Islands and possibly could be properly connected with activity of the Kerguelen-plume (Foley et al., 2001, 2006). Jurassic-Cretaceous dikes, stocks and sills of alkaline-ultrabasic rocks, relatively close to kimberlite-type, are exposed within Jetty oasis and on the southern shore of the Radock Lake (Mikhalsky et al., 1992). This alkaline-ultrabasic magmatism has appeared to be connected with the main Mesozoic stage of the evolution of the Lambert and Amery glaciers riftogenic structure (Kurinin et al., 1980, 1988). The alkaline-ultrabasic dikes and sills within Jetty oasis

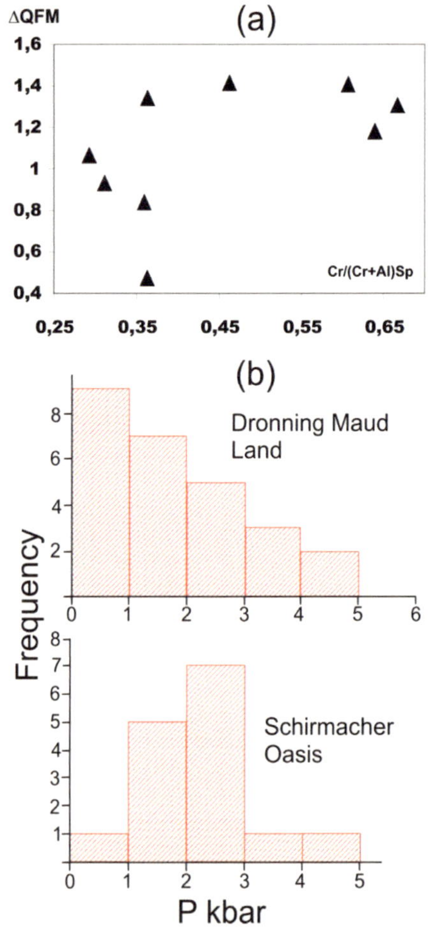

Fig. 5. Estimation of (a) oxygen fugacity, Cr/(Cr + Al) ratio of spinel, and (b) pressure of magma crystallization in the intermediate magma chambers of the Schirmacher Oasis and DML. The data for pyroxenes from the DML rocks are after (Sushchevskaya et al., 2004).

cut the rocks of the Beaver complex, Permo-Triassic terrigeneous successions of the Amery complex, and late Paleozoic low-alkaline basic dikes as well. Dashed chain of 6 stock bodies spread out on 15 km along the eastern shore of the Beaver Lake, marked their allocation with submeridianal zone of the deep cracks, boarded of the eastern side of the Beaver Lake trough. Some of the polzenite and biotite-pyroxene alkaline picrite dikes were discovered on Kamenistaya Platform (northern extremity of Jetty Peninsula, Fig.6a). They have north-north-eastern strike, subvertical dip and are 0.5-1.0 m thick. Oasis Jetty alkaline bodies on the present-day erosion surface have oval, rare isometric forms from 10x25 to 80x120 m; dike bodies reach up to 180 m long and thickness is about 2 m (Fig. 6b). All of these dike rocks are abundant in mantle nodules, mainly peridotites, and numerous xenoliths of the host Permo-Triassic sediments and Precambrian metamorphics as well (Fig. 6c). It has been suggested that the alkaline-ultrabasic bodies were intruded into two phases. In the elder bodies (130-120 Ma, Yuzhnoe and Severnoe stocks) biotite-pyroxene alkaline picrites prevail, at the same time the younger (120-105 Ma, Novoe and Ploskoe stocks) is composed by polzenites and/or melanephelinites (Laiba et al., 1987; Mikhalsky et al., 1998) (Fig. 6).

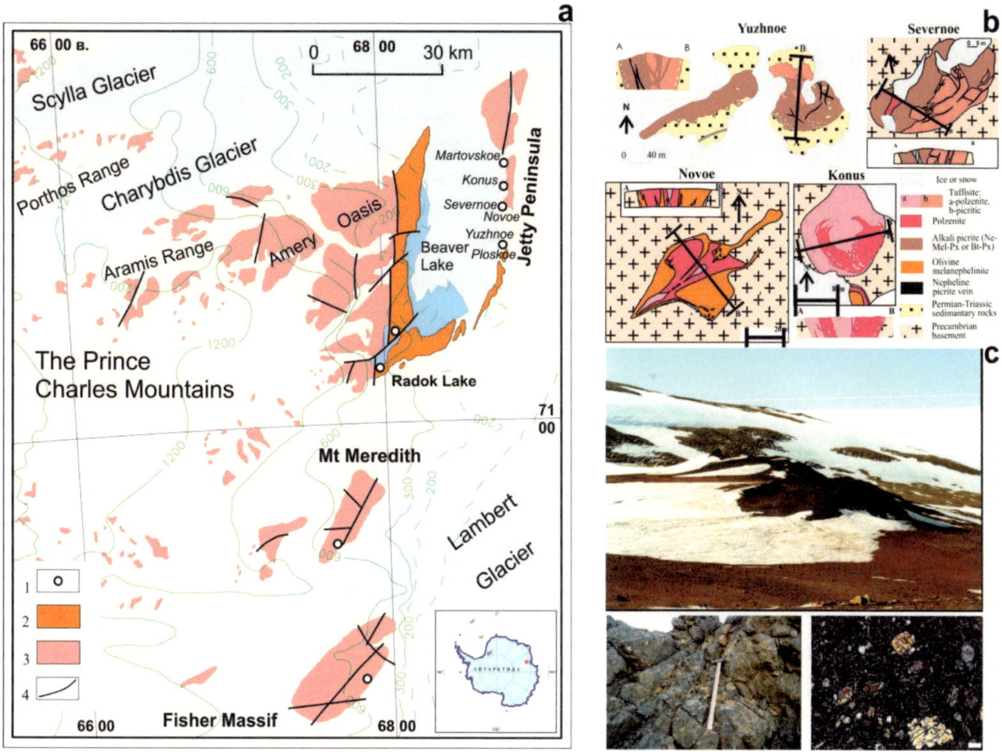

Fig. 6. Geological sketch map showing occurrences of alkaline-ultramafic rocks in the northern and central parts of the Prince Charles Mountains (a): 1- Jurassic-Cretaceous alkaline picrites, 2 - Permian-Triassic coal-dearing deposits, 3 - Precambrian rocks, 4 - faults; (b) schematic structure of the alkaline-ultramafic stock-like bodies from Jetty Oasis; (c) general view of the Severnoe stock, detail of outcrop and optic microfoto of tuffisite in crossed nicols.

The investigated samples are characterized by average composition: SiO_2 = 37.44, MgO = 19.38, TiO_2 = 2.26, CaO = 15.42, Fe_2O_3 = 11.65 %%, and high enough alkalis: Na_2O = 2.33 and K_2O = 2.49 %. The initial Sr and Nd isotope compositions of samples are close to the ranges of the values for dolerites from the Schirmacher Oasis. The alkaline picrites show initial (about 120 Ma) radiogenic to highly radiogenic $^{87}Sr/^{86}Sr$: 0.7048 - 0.7078 and moderately radiogenic to unradiogenic εNd: from +4 to −7 (Table 2).

3. Geochemical characteristics of the Mesozoic basites of the Schirmacher Oasis

The contents of lithophile elements in the samples are given in Table 1 and illustrated in the spider diagram (Fig. 7). The primitive mantle-normalized (Sun & McDonough, 1989) element contents (they are arranged in the diagram in the sequence of decreasing incompatibility) indicate different degrees of lithophile element enrichment in the dolerites. Most of them demonstrate distinct positive Pb, and negative Ta and Nb anomalies (Fig. 7a). Similar distribution patterns of lithophile elements were previously observed by us in the tholeiitic basalts of the ancient volcanic rise of Afanasy Nikitin (90–80 Ma), which is situated in the central part of the Indian Ocean (Borisova et al., 2001; Sushchevskaya et al., 1996) (Fig. 7d). Noteworthy that alkaline picrites of Jetty Oasis (Fig. 7b) have similar enrichment patterns also. Some of them show pronounced negative Nb and Ta anomalies, less apparent positive Pb anomaly, and negative Zr and Hf anomalies. All such samples are from lamprophyre dykes from the outcrop near the Beaver Lake (32R-A101, 32R-A58) and Yuzhnoe body (U-22) (Fig. 6). As these samples do not differ in main composition from the other samples (Table 2) there could be some heterogeneity in the melting source which provides local anomalies for these elements, but in any case the geochemistry points to a high lamprophyric contribution that means very deep melts linked to intraplate plume activity.

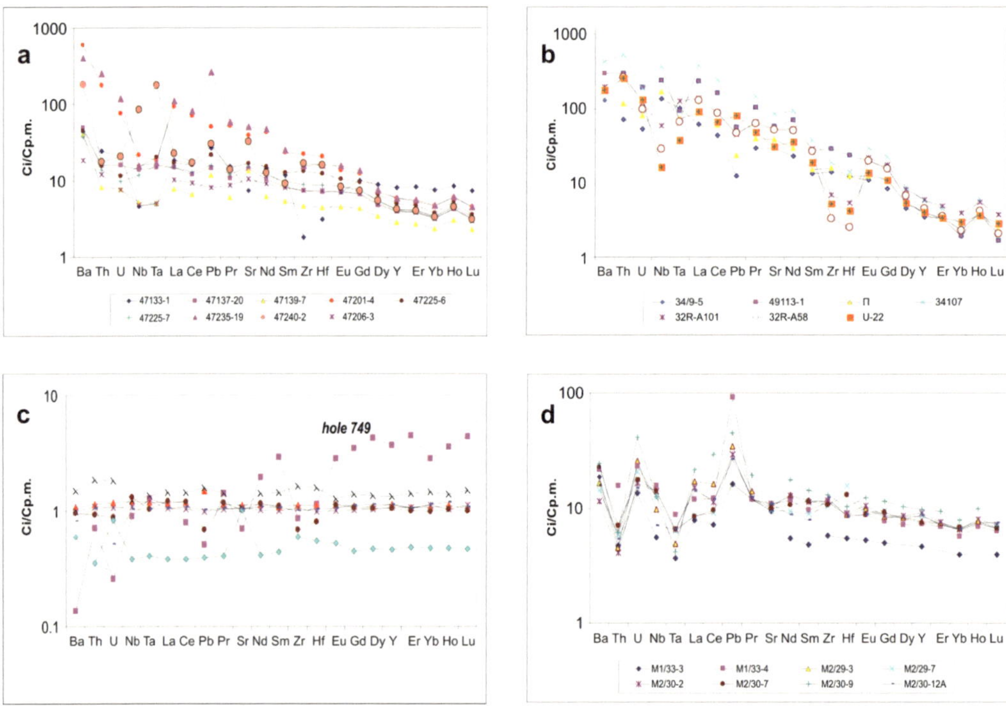

Fig. 7. Primitive mantle-normalized (Sun & McDonough, 1989) distribution of lithophile element patterns in the Schirmacher Oasis dolerites (a), Jetty Oasis alkaline picrites (b), Kerguelen Plateau tholeiites (Frey et al., 2000) (c) and the Afanasy Nikitin Rise tholeiites (Borisova et al., 2001) (d).

Two samples from our data set show a peculiar lithophile element distribution. In particular, sample 47235-19, which was classified as a derivative of alkaline magmas (lamprophyre-like), shows high contents of all elements, but retains all of the specific geochemical anomalies. With respect to these characteristics, it is most similar to altered dolerite sample 47201-4 that means the both of specimens are from alkaline lamprophyre deep-mantle source. The normalized distribution patterns of lithophile elements in specimen 47420-2 show a strong positive Ta–Nb anomaly but at the same time the mineral composition of the sample is characterized by absent of any signs of titanates (Fig. 7a), which emphasizes the different origin of the geochemical enrichment of its source compared with the majority of studied samples.

NN locality	34/9-5 Yuzhnoe stock	49113-1 Meredit massif	П-4 Ploskoe stock	34107 Fisher Massif	32R-A101 Beaver Lake	32R-A58 Beaver Lake	U-22 Yuzhnoe stock
SiO_2	40.6	31.4	39.3	19.3	33.65	32.79	35.76
TiO_2	2.24	2.34	2.25	1.50	2.00	2.20	1.5
Al_2O_3	9.94	4.59	9.21	2.88	10.75	7.58	5.34
Cr_2O_3	0.042	0.069	0.048	0.059	0.06	0.08	0.1
Fe_2O_3	11.5	10.6	10.3	14.2	1.3	4.79	2.31
FeO					7.88	5.3	6.7
MnO	0.161	0.163	0.154	0.199	0.19	0.16	0.17
MgO	14.1	25.0	13.6	15.1	14.66	14.35	23.73
CaO	10.8	10.9	11.4	22.9	15.20	16.02	8.69
Na_2O	2.21	<.05	2.5	<.05	2.29	1.91	1.42
K_2O	2.09	2.63	2.3	0.612	3.53	2.52	1.8
P_2O_5	0.51	0.818	0.657	1.95	0.83	0.73	0.63
LOI	5.65	11.2	8.14	20.8	7.41	11.83	12.18
Total	99.8	99.6	99.8	99.5	99.75	100.26	100.33
CO_2	3.45	6.94	6.31	16.59	3.87	6.18	8.9
Ba	903	2080	1250	2950	1373.6	1061.8	1233
Th	6.13	24.9	9.94	45.5	23.83	22.22	21.8
U	1.14	3.99	1.72	3.96	2.512	2.107	2.808
Nb	96.4	171	121	260	42.23	21.02	11.87
Ta	4.18	3.92	3.07	2.81	5.141	2.782	1.585
La	42.7	159	60.9	261	96.33	89.06	63.65
Ce	79.3	285	108	434	161.3	154.3	120.3
Pb	2.39	10.4	4.47	14.3	8.438	8.884	14.96
Pr	8.37	28.8	11	40.1	18.61	17.56	13.44
Sr	642	1220	828	1830	1141.4	1119.3	667.8
Nd	32.3	95.9	40.6	128	72.98	69.02	50
Sm	6.17	12.3	6.95	17.1	13.11	12.19	8.715
Zr	162	328	182	216	81.25	38.87	61.78
Hf	3.87	7.55	4.02	4.54	1.747	0.8077	1.351
Eu	1.91	3.47	2.15	5.05	3.783	3.462	2.38
Gd	5.2	10.1	6.82	13.9	10.69	9.484	6.733
Dy	3.53	4.65	4.18	6.77	6.341	5.129	4.155
Y	16.4	17.2	18.7	27.5	28.52	21.43	19.32
Er	1.71	1.62	1.67	2.28	2.465	1.788	1.71
Yb	0.99	0.99	1.21	1.39	2.016	1.196	1.535
Ho	0.63	0.72	0.67	1.03	0.9395	0.7165	0.6183
Lu	0.16	0.13	0.18	0.22	0.2862	0.1622	0.2205
Sm, ppm	6.736	15.17	9.586	20.42	12.79	10.09	7.096
Nd, ppm	34.28	102.7	40.45	133.3	74.27	59.49	46.73
$^{147}Sm/^{144}Nd$	0.11876	0.08930	0.14323	0.09257	0.10442	0.10281	0.09209
$^{143}Nd/^{144}Nd$	0.512656	0.512426	0.512630	0.512497	0.512783	0.512721	0.512240
2s, abs	0.000003	0.000003	0.000003	0.000006	0.000018	0.000008	0.000016
Rb, ppm	46.53	96.96	113.3	21.81	124.5	110.8	65.87
Sr, ppm	734.8	1386.7	1204.9	2007.9	1048.7	1778.7	674.1
$^{87}Rb/^{86}Sr$	0.18319	0.20229	0.27200	0.03142	0.34345	0.18037	0.28268
2s, %	1.00	1.27	7.03	1.56	1.00	1.72	0.98
$^{87}Sr/^{86}Sr$	0.705167	0.706082	0.705502	0.704926	0.705148	0.720677	0.707207
2s, abs	0.000009	0.000006	0.000005	0.000005	0.000028	0.000014	0.000023
$^{206}Pb/^{204}Pb$	18.702	18.558	18.621	18.446			
2s, abs	0.004	0.001	0.0004	0.002			
$^{207}Pb/^{204}Pb$	15.563	15.618	15.574	15.623			
2s, abs	0.003	0.001	0.0004	0.002			
$^{208}Pb/^{204}Pb$	39.227	39.068	38.895	39.292			
2s, abs	0.008	0.002	0.001	0.002			

Table 2. Chemical and isotope composition of whole-rock alkaline picrite samples from Jetty Oasis. Major oxides express as wt % and trace elements - as ppm. All analyses were done at Karpinsky Geological Institute (St.Petersburg, Russia) by ICP MS and solid-source HR MS (isotope analysis) (detailed in (Sushchevskaya et al. 2009). 2s, abs and 2s, % - correspond to isotope ratio errors (absolute and relative) at 95% confidence level (2 sigmas).

Tholeiites exposed in drill hole 749 (Southern Kerguelen Plateau) have the age about 114 Ma and belong to the early stages of Kerguelen Plateau formation which maximum age is estimated as 120 Ma (Coffin et al., 2002). Normalized lithophile element patterns of the tholeiites reflect the presence of weakly enriched source and enriched one with pronounced Zr and Hf negative anomaly for a part of samples as well (Fig. 7c) which mantle enrichment processes supposed to be connected with metasomatic acting of lamprophyric melts (Ingle et al., 2002).

The correlation analysis of lithophile element contents in the basites of the Schirmacher Oasis (as an example, element–Th correlation diagrams are shown in Fig. 8) reveals close relations between U and Th, Pb and Th, La and Th, and Yb and Th and poor correlations between other lithophile elements. These relations are less obvious for sample 47133-1, which has relatively low contents of Ta, Nb, Hf, Zr, and Sr. For the sake of comparison, Figure 8 shows the fields of enriched tholeiites from the Kerguelen Plateau recovered at ODP site 749 (Frey et al., 2000). It should be noted that the Schirmacher basites are distinguished by lower contents of Yb and Hf, and, probably, Sr, but the similar ratios of the majority of incompatible elements (especially, Ta/Th, U/Th, and Pb/Th) suggest similar geochemical nature of the enriched sources of both of the magma types. On the other hand, dolerites of the Schirmacher Oasis show higher contents of the majority of lithophile elements compared with tholeiitic basalts of the Kerguelen Plateau. At this time dolerites of the Schirmacher Oasis by many geochemical characteristics are close to alkaline picrites of Jetty Oasis. Origin of the latter is directly connected with the melting of continental lithospheric mantle (Andronikov & Egorov, 1993; Andronikov & Foley, 2001).

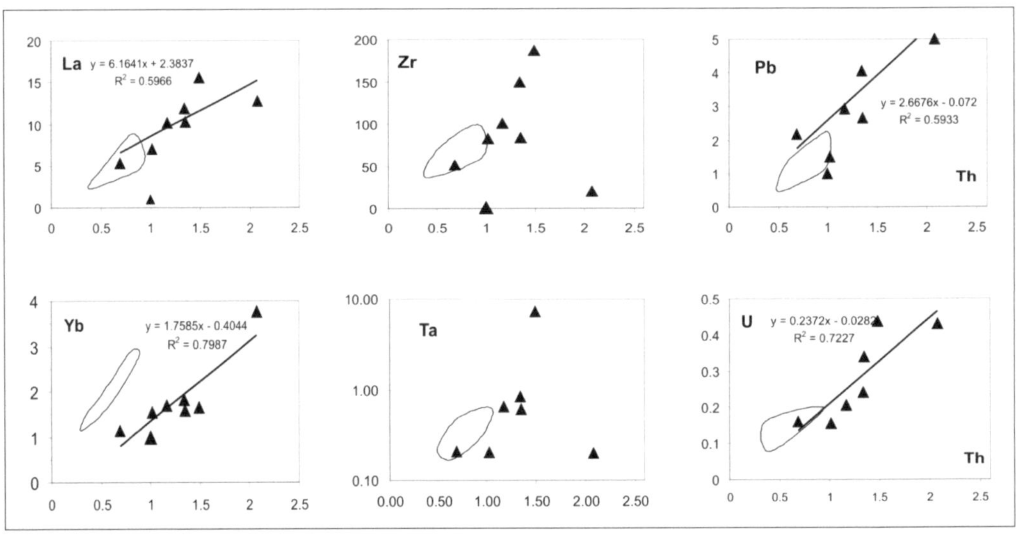

Fig. 8. Correlation between lithophile elements in the dolerites of the Schirmacher Oasis. The fields show lithophile element variations in the basalts of ODP Site 749 according to (Frey et al., 2000).

Figure 9a demonstrates the closeness of characteristic ratios (Th/Nb)n *vs* (La/Nb)n for magmatic rocks in two provinces of eastern Antarctica. High values (Th/Nb)n: 15 -16 and (La/Nb)n: 6 - 7 reflect continental nature of their sources. Basing on the data presented in

Figure 9b we can suggest that such source could be sub-continental lithospheric mantle. Values of Sr/Pb ratio for alkaline picrites from oasis Jetty and dolerites from Schirmacher Oasis are below NMORB and OIB signatures. Presented MgO content in Jetty Oasis alkaline picrites shows that high Sr/Pb and Nb/Nb* are not connected with the process of olivine fractionation or accumulation in the course of which MgO content should regularly decrease but rather reflect the melting of heterogeneous continental mantle.

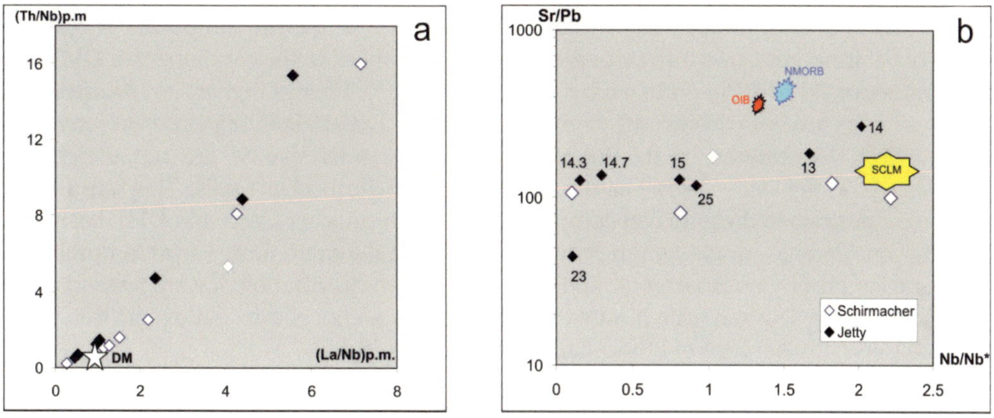

Fig. 9. Diagrams (Th/Nb)n *vs* (La/Nb)n (a) and Sr/Pb *vs* Nb/Nb* (b) for Schirmacher and Jetty oasises. Nb/Nb* was defined by Eisele et al. (2002) as $[Nb_n/\sqrt{(Th_n*La_n)}]$ normalized to primitive mantle. SCLM - sub-continental lithospheric mantle (McDonough, 1990), OIB and NMORB signatures according to (Thompson et al., 2007). MgO values are marked for Jetty oasis rocks only.

The isotopic analysis of dolerites from the Schirmacher Oasis (Table 1, Fig. 10) revealed the following strontium radiogenic composition: $^{87}Sr/^{86}Sr$ of 0.7045-0.7047, and lead: $^{208}Pb/^{204}Pb$ of 37.98-38.2 and $^{207}Pb/^{204}Pb$ of 15.45-15.52. The relatively high $^{206}Pb/^{204}Pb$ (17.9-18.2) and low $^{143}Nd/^{144}Nd$ values (0.51275-0.51255) also indicate the enrichment of the dolerite source compared with the depleted oceanic mantle (Hauri et al., 1994). Two samples from our collection (47133-1 and 47201-4) are strongly enriched in radiogenic strontium: their $^{87}Sr/^{86}Sr$ ratios are 0.7099 and 0.7120, respectively. Dolerite sample 47201-4 is a strongly altered variety and shows a negative Zr-Hf anomaly (Figs. 7 and 8). This sample was not used for the construction of correlation diagrams. It is important that the isotopic compositions of basic rocks from the Schirmacher Oasis are identical to the composition of tholeiites from the Afanasy Nikitin Rise and ODP Site 749. On the other hand, they are significantly different from the tholeiites of the 39°-40°W anomaly of the Southwest Indian Ridge and anomalous olivine tholeiites from the base of the Afanasy Nikitin Rise, which have low values of $^{206}Pb/^{204}Pb$ (17.2-17.6) and $^{143}Nd/^{144}Nd$ (0.5123-0.5124) and high values of $^{87}Sr/^{86}Sr$ (0.705-0.706) and $^{208}Pb/^{204}Pb$ (37.1-38.01) for the given $^{206}Pb/^{204}Pb$ (Fig. 10). This is typical for EM-I enriched mantle sources (Hauri et al., 1994). In general, the isotope characteristics of enriched magmas from DML and the Schirmacher Oasis are similar to those of basalts derived from the Kerguelen plume, but are significantly different from the composition of enriched magmas from the North Atlantic, which are also related to plume activity (Sushchevskaya et al., 2005).

Figure 10 presents isotope data for the rocks connected with Karoo-Maud plume (Dronning Maud Land, Schirmacher) and Kerguelen plume (basalts of Afanasy Nikitin Rise and drill holes 749, 747) and also Jetty Oasis alkaline picrites. On the base of these diagrams we should mark put the following: 1) In spite of the closeness of enrichment character of plume magmas, for the basalts from Schirmacher Oasis there are not noted the presence of depleted types which are found within Dronning Maud Land (Riley et al., 2005). They are especially well traced in Sr-Nd isotope diagram (Fig. 10c). By isotope data the Schirmacher Oasis magmas are more homogeneous and for them the enriched component with low $^{206}Pb/^{204}Pb$ and high $^{87}Sr/^{86}Sr$ is not determined but it is fixed in some basalts of Dronning Maud Land, Afanasy Nikitin Rise and in drill-hole 747 of Kerguelen Plateau. 2) Enriched component revealed in basalts of Schirmacher Oasis differs, perhaps, insignificantly from the component of DML with lower values of $^{208}Pb/^{204}Pb$, $^{207}Pb/^{204}Pb$ at the given $^{206}Pb/^{204}Pb$ and higher $^{87}Sr/^{86}Sr$ (the most altered samples are not considered). It should be marked, that alkaline picrites of Jetty Oasis which reflect the features of the East Antarctic continental mantle are not enriched in radiogenic Sr at the same degree as the basalts of the Schirmacher Oasis. They are close in isotope composition to the enriched component of the Schirmacher Oasis and DML basites but differ by considerable enrichment in radiogenic ^{208}Pb. At the same time, Ferrar lamproites (or ultramafic lamprophyres, Riley et al., 2003) connected with distribution of Karoo-Maud plume to the south along Transantartic mountains are even more enriched in radiogenic lead which reflects the specific character of lithospheric mantle in this region.

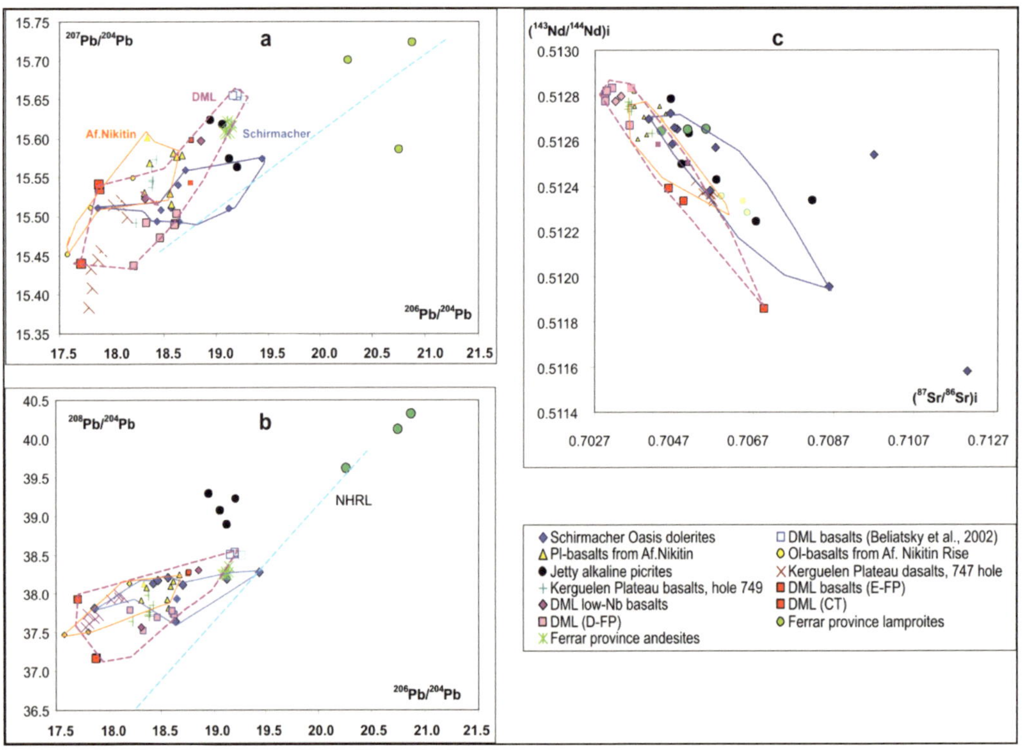

Fig. 10. Comparative isotope (Pb, Sr, Nd) characteristics of Karoo-Maud and Kerguelen plume-related magmatic rocks, Jetty Oasis alkaline picrites and Ferrar province andesites and lamproites (Table 1, 2; Antonini et al., 1999; Belyatsky et al., 2002; Borisova et al., 2001; Frey et al., 2000; Heinonen et al., 2010; Sushchevskaya et al., 1996; Sushchevskaya et al., 2003). Chemical types of DML basalts (low-Nb, D-FP, E-FP, CT) refer to (Heinonen et al., 2010).

4. Discussion

The geochemical data for the basic rocks of the Schirmacher Oasis, which is located to the east of the previously investigated occurrences of plume magmatism, provide new insight into the spatial temporal evolution of this plume. The area of occurrence of basaltic lavas related to the arrival of the Karoo-Maud plume beneath the lithosphere of Central Gondwanaland is up to 2000 km in diameter (Elliot et al., 1997; Elliot & Fleming, 2000; Leitchenkov & Masolov, 1997), which is well consistent with the typical size of superplume heads projected to the surface (White & McKenzie, 1989). Perhaps, the emplacement of the plume and subsequent thermal erosion of the lithosphere under the influence of laterally flowing plume material triggered the breakup of Gondwana: Antarctica separation from the southern Africa was about 165–155 Ma (Martin & Hartnady, 1986). Nevertheless, it should be noted that the Karoo-Maud plume probably was different in some respects from what is considered to be a typical mantle plume (Courtillot et al., 2003): the volume of magmatic material supplied to the surface is minor, no higher than 60 000 km^3, even if the partly overlain volcanics of the Karoo basin with accompanying dikes, sills, and small intrusions are accounted for; the plume-lithosphere interaction occurred in several stages over a long time interval, in contrast to the short-term (1–4 Ma) magmatism that usually accompanies the penetration of mantle plumes into the lithosphere (White & McKenzie, 1989). The investigation of the character of flood basalt magmatism in Antarctica, which was performed within the MAMOG Project (Leat et al., 2007), confirmed that it lasted for at least 20–30 Ma, when a series of dikes and flows were formed around the plume center over an area of about 145 000 km^2 along the Antarctic coast (DML). The eastward spreading of the Karoo-Maud plume can be indirectly supported by the existence of a large basic intrusion, which extends along the DML coast and is marked by a high-amplitude (×100 nT) magnetic anomaly (Golynsky et al., 2002, 2006).

Figure 11 shows the possible eastward direction of plume head spreading in the sublithospheric mantle of Gondwana at 130 Ma. During that time, the activity of the Kerguelen plume began within the already existing Indian Ocean and affected a region from the western margin of Australia to India. Its activity was manifested 40–50 Ma after the maximum magmatic activity related to the Karoo - Maud plume (Coffin et al., 2002; Frey et al., 1996; Mahoney et al., 1995). It is interesting that the activity of the Etendeka-Parana plume, which preceded and accompanied the opening of the South Atlantic, began at the west of the Karoo-Maud plume, in South America and Central Africa at about 130 Ma (Deckart et al., 1998). The lead isotopic ratios of basalts from the Schirmacher Oasis are different from those of the DML basalts, which have even more radiogenic values of $^{206}Pb/^{204}Pb$ ratios, up to 18.7–18.8. Nonetheless, in the lead isotope diagram, all these rocks plot along a common trend, which extends from the relatively depleted dolerites of the Schirmacher Oasis to the more enriched basalts of western DML (Fig. 10) (Belyatsky et al., 2006) and the Transantarctic Mountains, which underwent extensive crustal contamination (Elliot et al., 1999). This diagram also shows the isotopic compositions of ancient Proterozoic anorthosite dikes recalculated to the emplacement age of Mesozoic dolerites. It is clearly seen that they plot in the field of the Mesozoic basalts of Antarctica. As was noted above, the ancient dikes often spatially associate with plume related dikes and have similar strike and dip angles, at least in the Schirmacher Oasis.

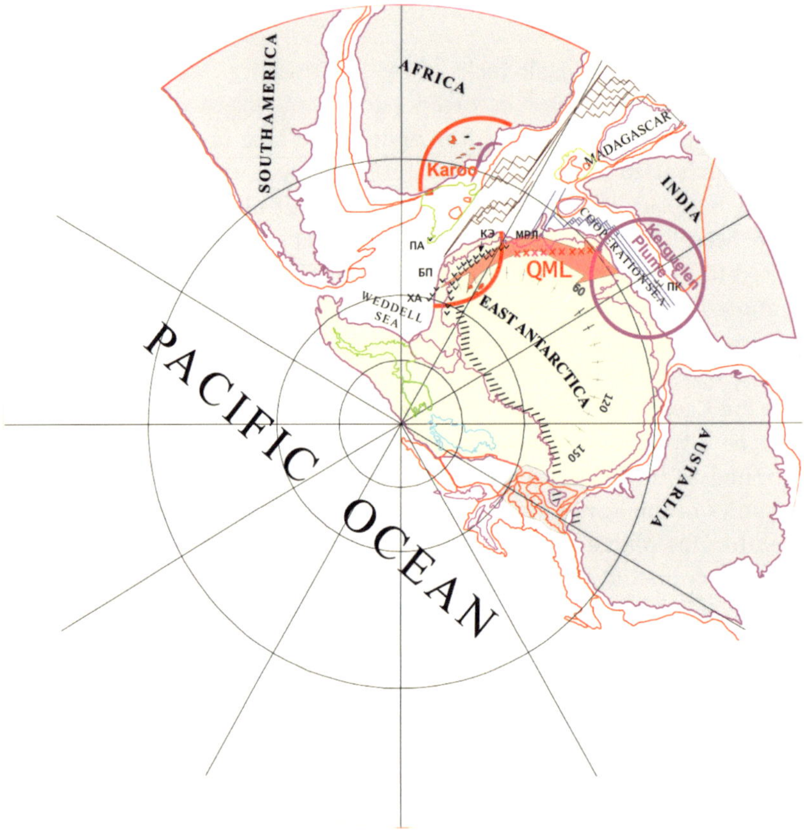

Fig. 11. Eastward propagation of the Karoo-Maud plume according to the reconstruction (Leitchenkov et al., 2003) for 130 Ma. The symbols are: oblique crosses - area of basic intrusion distribution within QML according magnetic data, slashed area - Ferrar dolerites on the geological data, tick marks - volcanic complexes (ridges, plateaus), ПА - Agulhas Plateau, БП - Polarstern Bank, ХА - Andenes Escarpment, КЭ - Explora complex, ПК - Kerguelen Plateau, МРЛ - Riiser-Larsen Sea. The brown and blue lines within ocean display spreading centers, red circles outline plume-related magmatism (suggested).

The spreading of Mesozoic plume magmas evidently occurred mainly along highly permeable weakened zones in the lithosphere (Leitch et al., 1998) and the ancient dike zones could serve as magma conduits. Such a process of plume melt penetration along the zones of ancient dikes was also typical for the Karoo plume distribution in the southeastern part of the African continent at 180–173 Ma (Jourdan et al., 2004, 2006).

The inherited old zircon grains (500–850 Ma) discovered in the dolerites of the Schirmacher Oasis could reflect the contamination of higher temperature Jurassic tholeiitic magmas by the material (zircons) of ancient andesites and/or enclosing metamorphic rocks (Belyatsky et al., 2007). Salient features of the Karoo-Maud plume magmatism in Antarctica are the presence of high-magnesium volcanics and the wide occurrence of dolerite dikes and sills and olivine-rich gabbroid intrusions. Many of the dolerites are chemically similar to slightly enriched tholeiites (Luttinen & Furnes, 2000), which suggests that the plume magma source was similar to the depleted oceanic mantle (Fig. 12). The plagioclase K-Ar age of basaltic lavas from the Vestfjella Mountains is 180 Ma, which is very close to the ages of the basalts of the Kirwanveggen (Luttinen & Furnes, 2000) and mafic dikes and flows of southeastern Africa (183 ± 1 Ma) (Duncan et al., 1997; Jourdan et al., 2006). The Sm-Nd mineral (pyroxene and plagioclase) and bulk rock isochron age of the dolerites of the Schirmacher Oasis is 171±24 Ma, which coincides within the uncertainty with the age of the Jurassic magmatism of DML and southeastern Africa (Belyatsky et al., 2006).

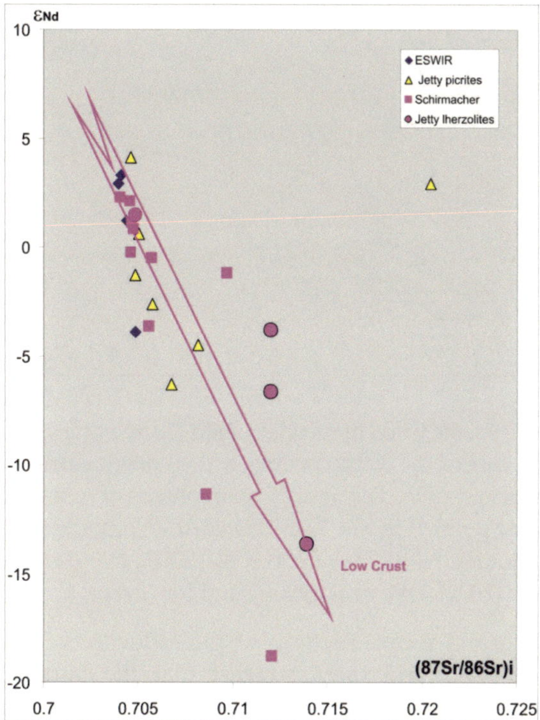

Fig. 12. Isotope characteristics of Schirmacher Oasis dolerites, Jetty Oasis alkaline picrites and mantle lherzolite inclusions and SWIR tholeiites as well reflect mainly the effect of lower crust contamination (enriched crustal component) of Antarctic plume primary melts. The most isotope enriched sample 47201-4 (Schirmacher Oasis) is a highly carbonatized and potassium enriched altered dolerite.

Thus, based on the geochemical, isotopic, and petrological similarity and simultaneous formation, the Mesozoic basic magmatism of the Schirmacher Oasis can be interpreted with a high degree of certainty as a derivative of the activity of the Karoo-Maud mantle plume (by analogy with the western DML of Antarctica and the Karoo province of the southeastern Africa). The youngest Mesozoic-Cenozoic occurrences of intraplate magmatism (alkaline ultrabasic intrusions, kimberlite sills and dikes, and leucite basalts of the Manning massif, Fig. 6) in East Antarctica are localized at the region of the Prince Charles Mountains (McRobertson Land) and Princess Elizabeth Coast (alkaline volcanics of Gaussberg), 3 000–4 000 km east of DML. Their formation is supposedly related to the activation of the sublithospheric mantle of Antarctica under the influence of the Kerguelen plume, whereas the emplacement of the alkaline ultrabasic dikes and sills of the Jetty Oasis is directly correlated with the initial stages of Kerguelen plume interaction with the lithosphere of Antarctica at 130–105 Ma (Laiba et al., 1987).

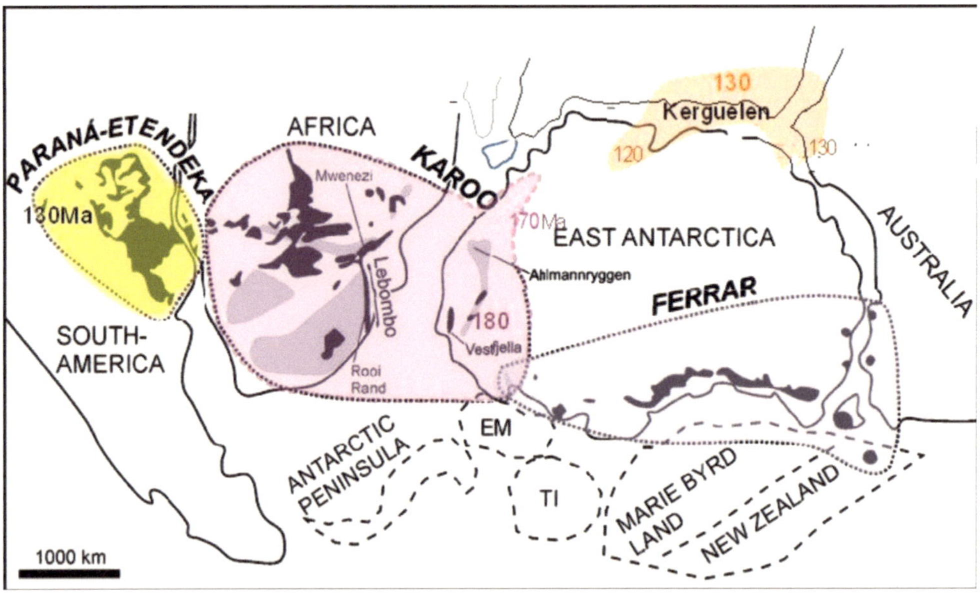

Fig. 13. Distribution of Mesozoic CFBs in reconstructed Gondwana supercontinent at the age about 170 Ma ago. In the case of the Karoo province, the known extent of intrusive equivalents (found outside CFBs) is also shown. The ages of main phases of magmatic activities of Karoo and Kerguelen plumes are pointed in Ma. EM=Ellsworth-Whitmore Mountains, TI=Thurston Island. Reconstruction modified after Heinonen et al. (2010), Hergt et al. (1991), Kent (1991), Storey et al. (1992), Segev (2002), Leat et al. (2006), and Jourdan et al. (2004).

Numerous investigations and reconstructions of the influence of the Kerguelen plume on the lithosphere of Antarctica and India suggest that the plume head spreads in the northeastern direction (Coffin et al., 2002; Frey et al., 1996, 2000; Kent, 1991; Mahoney et al., 1995). The possible early occurrence of the plume at 130 Ma is reconstructed from the magmatism of continental margins (Fig. 13). This concerns primarily the Bunbury Province of the southwestern coast of Australia and the magmatism of the eastern coast of India (Rajmahal traps). On the other hand, considerable areas of igneous rocks related to the Kerguelen plume are currently located below the sea level and detected only on the basis of seismic data for the northern Bay of Bengal and the Naturaliste Plateau (Kent, 1991). It is supposed that the Kerguelen plume also extended far to the northeast (i.e., it differed from the classical concentrically isometric shape during emplacement in the lithosphere).

Figure 13 shows the area of Kerguelen plume development during the early stages of ocean opening, which were accompanied by the extensive occurrence of enriched tholeiites. The distribution of enriched basalts with high $(La/Sm)_n$ (Sushchevskaya et al., 1998) reflects the possible consequences of plume activity (formation of enriched tholeiitic magmas and formation of large rises at a fossil spreading zone). The depleted compositions of igneous rocks exposed at the flanks indirectly support the emplacement of the Kerguelen plume in the already open ocean.

The Karoo-Maud superplume is probably an example of deep plumes ascending from the core-mantle boundary. Its penetration into the upper parts of the lithosphere resulted in the breakup of Gondwana and the separation of Africa from Antarctica. Then, for a considerable time (about 40 Ma) it could flow laterally along preexisting boundaries in the upper parts of the lithosphere, which caused the formation of the younger Parana–Etendeka and Kerguelen mantle plumes. The formation of two large igneous provinces, Parana–Etendeka in the Atlantic and Kerguelen in the Indian Ocean, significantly influenced oceanic magmatism and resulted primarily in the appearance of geochemical anomalies within mid-ocean ridges. The spreading of plume material along the weakened zones of Gondwana to the south (Ferrar dolerites in the Transantarctic Mountains) and east (from western DML to the Prince Charles Mountains) of Antarctica reflects possible deep plume motions, which probably occurred at different levels of the lithosphere and sublithospheric mantle (Fig. 11). Geological and geophysical data for the basement of the basins of the Cosmonaut, Cooperation, and Davis seas suggest moderate spreading rates at the early stages of ocean opening (Hinz & Krause, 1982; Leitchenkov et al., 2006). The early stages of this process produced volcanic rises in the western part of the Riiser–Larsen Sea (Astrid Ridge), in the Lazarev Sea, in the southeastern Weddell Sea (Leitchenkov & Guseva, 2006), and in the southeastern Davis Sea (Bruce Bank, in the east of the Kerguelen Plateau). These structural and tectonic features could be related to the influence of the Karoo–Maud superplume on the formation of the oceanic crust in this region, if one supposes its eastward spreading up to the old deep rift zone (region of the Lambert-Amery Glacier and Prydz Bay), which inherited to some extent the structures of Paleozoic grabens, was filled with sediments up to 10 km thick (Fedorov et al., 1982; Leitchenkov et al., 1999) and separated different blocks of the continental lithosphere of Antarctica (probably of different thickness) to the west and east of it. The further lateral spreading of the plume occurred towards the already existing ocean basin, which resulted in the formation along the proto-Southeast Indian Ridge of the Conrad, Kerguelen, and Afanasy Nikitin volcanic rises and the Southeast Indian Ridge (Sushchevskaya et al., 1998, 2003). Blocks of metamorphosed old subcontinental mantle could be retained within and near the spreading zone and subsequently involved into melting processes (Fig. 12).

Geophysical data obtained by Russian expeditions in 2003–2004 demonstrated that, before the beginning of the Australia–Antarctica breakup, a continental block existed at the margin of the Australia–Antarctica continent and was subsequently divided by the spreading zone into the Bruce Bank (remaining near Antarctica) and the Naturaliste Plateau (Leitchenkov & Guseva, 2006). The layered basement of the Bruce Bank is a pile of volcanic rocks, up to 3.5 km thick, underlain by the thinned continental crust (Leitchenkov & Guseva, 2006). The results of deep-sea drilling in the Elan Bank in the western spur of the Kerguelen Plateau also suggest its continental origin (Ingle et al., 2002). Thus, the complex formation of an ancient spreading zone during the initial stages of ocean opening resulted in the detachment of continental lithospheric blocks of different size, which could significantly affect the geochemical character of tholeiitic magmatism within rift zones. This effect is observed in volcanic rises, such as the Kerguelen Plateau, Afanasy Nikitin Plateau, and Ninetyeast Ridge. The eastward propagation of plume magmatism in Antarctica along the ancient collision zone reflects the processes that occurred in the apical part of the plume at about 180–110 Ma. The established geochemical similarity between the basalts of the continental margin of eastern India (Rajmahal traps) and Southwestern Australia (Bunbury basalts) and the tholeiites of the Kerguelen Plateau implies the assimilation of the ancient continental crust of Gondwanaland by the basaltic melts (Ingle et al., 2002). The isotopic systematics of the dolerites of the Schirmacher Oasis suggests the following average initial isotopic ratios for their primary magmas: $^{207}Pb/^{204}Pb$ = 15.502, $^{208}Pb/^{204}Pb$ = 38.114, $^{206}Pb/^{204}Pb$ = 18.026, $^{87}Sr/^{86}Sr$ = 0.70568, $^{143}Nd/^{144}Nd$ = 0.512629 (Fig. 12). Similar isotopic ratios are characteristic of the tholeiites of the Afanasy Nikitin Rise (Borisova et al., 2001; Sushchevskaya et al.,

1996), which were formed 90 Ma near the proto-Southeast Indian Ridge, and of the 115 Ma tholeiites of the central part of the Kerguelen Island (Frey et al., 2002).
The problem of the deep origin of hotspots (and low-velocity zones) in the Earth's shells has acquired special significance, because it has a direct influence on the possible spatial movement (both vertical and horizontal) of convective mantle flows (Burov et al., 2007). For instance, Pushcharovskii (Pushcharovskii & Pushcharovskii, 1999) emphasized that the Earth's shells, including the upper, middle, and lower mantle, with boundaries at depths of 670–900 km, 1700–2000 km, and 2900 km, have heterogeneities of various scales, which reflect possible lateral movement of materials. Ruzhentsev with colleagues (Ruzhentsev et al., 1999) investigated the deep structure of the India–Atlantic segment of the Earth using the deep seismic tomography data and concluded that heated zones exist in the mantle at different depths. These zones are not traced into deeper levels which suggests that some of them are detached from their roots and have a lateral distribution (Ruzhentsev et al., 1999). According to recent data (Class & le Roex, 2008), the possible horizontal movement of sublithospheric flows from beneath Africa over considerable distances can be exemplified by the formation of peculiar enriched magmas in the central part of the Walvis Ridge.

5. Conclusions

The investigation of the Mesozoic (about 170 Ma) basaltic magmatism of the Schirmacher Oasis showed that the basalts and dolerites are petrologically identical to the previously studied rocks of western Dronning Maud Land (Vestfjella Mountains region), which are interpreted as the manifestation of the Karoo–Maud plume in Antarctica. The spatial distribution of the dikes indicates the eastward spreading of the plume material from DML to the Schirmacher Oasis within at least 10 Ma (up to ~35 Ma, taking into account the uncertainty of age determination). On the other hand, the considerable duration and multistage character of plume magmatism related to the activity of the Karoo–Maud plume in Antarctica and Africa (Leat et al., 2007; Luttinen et al., 2002) may indicate that the Mesozoic dikes of the oasis correspond to a single stage of plume magmatism. In such a case, the rate of eastward plume propagation can be considered only as a rough estimate, and the time of 10 Ma, as the upper limit.
The geochemical characteristics (relatively radiogenic Sr and unradiogenic Nd isotope composition, high Th/Nb and Ta/Nb ratios) of Schirmacher Oasis magmas indicate crustal contamination, which occurred during plume ascent and lateral spreading. The magmas of the initial stage of plume activity (western DML region) appeared to be the most contaminated. A peculiar feature of the opening of the Indian Ocean, which was triggered by the influence of the Karoo–Maud plume, is its occurrence in the presence of nonspreading blocks of varying thickness, such as the Elan Bank in the central part of the Kerguelen Plateau, and it was accompanied by the formation of intraplate volcanic rises, which were detected in the seafloor relief around Antarctica. The geochemical characteristics of some of these highs (Afanasy Nikitin, Kerguelen, Naturaliste, and Ninetyeast Ridge) as demonstrated by (Borisova et al., 2001) were mainly affected by crustal assimilation processes. The identical geochemical characteristics of the Mesozoic magmas of the Schirmacher Oasis, the lavas of the Afanasy Nikitin Rise, and the rocks of the central Kerguelen Plateau (ODP Site 749) suggest that the enrichment of all these magmas was related to the old continental rocks of the Gondwanaland. The magmatism that occurred 40

Ma after the main phase of the Karoo volcanism at the margins of the adjoining continents of Australia (Bunbury basalts) and India (Rajmahal traps) could be initiated by the Karoo-Maud plume, which propagated along the developing spreading zone and subsequently moved toward the Kerguelen Plateau, where it occurs currently as an active hotspot.

6. Acknowledgment

This work was financially supported by the Russian Foundation for Basic Research, project no. 09-05-00256. Editor Dr. Francesco Stoppa is sincerely thanked for reviewing and constructive comments that helped to improve the quality of the paper.

7. References

Andronikov, A. V. & Egorov, L.S. (1993). Mesozoic alkaline-ultrabasic magmatism of Jetty Peninsula. In: *Gondwana Eight: Assembly, Evolution and Dispersal.* Findlay, R.H., Unrug, R., Banks, M.R. & Veevers, J.J. (eds). Balkema. Rotterdam. pp. 547–557.

Andronikov, A.V. & Foley, S.F. (2001). Trace element and Nd-Sr isotopic composition of ultramafic lamprophyres from the East Antarctic Beaver Lake area. *Chemical Geology,* Vol.175, pp. 291-305.

Antonini, P., Piccirillo, E.M., Petrini, R. & et al. (1999). Enriched mantle - Dupal signature in the genesis of the Jurassic tholeiites from Prince Albert Mountains (Victoria Land-Antarctica). *Contributions to Mineralogy and Petrology,* Vol.136, pp. 1-19.

Belyatsky, B.V., Prasolov, E.M., Sushchevskaya, N.M. & et al. (2002). Specific Features of the Isotopic Composition of Jurassic Magmas in the Dronning Maud Land, Antarctica. *Doklady Earth Sciences,* Vol.386, pp. 855–858.

Belyatsky, B.V., Sushchevskaya, N.M., Leichenkov, G.L. & et al., (2006). Magmatism of the Karoo–Maud Superplume in the Schirmacher Oasis, East Antarctica. *Doklady Earth Sciences,* Vol.406, pp. 128– 131.

Belyatsky, B., Rodionov, N., Savva, E. & Leitchenkov, G. (2007). Zircons from mafic dykes as a tool for understanding of composition and structure of continental crust: on the example of Mesozoic olivine dolerite dykes. Schirmacher oasis. Antarctica. *Geophysical Research Abstracts,* Vol.9, 10509. SRef-ID: 1607-7962/gra/EGU2007-A-10509.

Borisova, A.Yu., Nikulin, V.V., Belyatskii, B.V. & et al. (1996). Late Alkaline Lavas of the Ob and Lena Seamounts (Conrad Rise, Indian Ocean): Geochemistry and Characteristics of Mantle Sources. *Geochemistry International,* Vol.34, pp. 503–517.

Borisova, A.Yu., Belyatsky, B.V., Portnyagin, M.V. & Suschevskaya, N.M. (2001). Petrogenesis of an olivine-phyric basalts from the Aphanasey Nikitin Rise: evidence for contamination by cratonic lower continental crust. *Journal of Petrology,* Vol.42, pp. 277-319.

Brewer, T.S., Rex, D., Guise, P.G. & Hawkesworth, C.J. (1996). Geochronology of Mesozoic tholeiitic magmatism in Antarctica: implications for the development of the failed Weddell Sea rift system. In: *Weddell Sea: Tectonics and Gondwana break-up.* Storey, B., King, E. & Livermore, R. (eds). 1996. Geol. Soc. Spec. Publ., Vol.108, pp. 45-62.

Burov, E., Guillou-Frottier, L.,. d'Acremont, E. & et al. (2007). Plume head - lithosphere interactions near intra-continental plate bounadries. *Tectonophysics,* Vol.434, No.1-4, pp. 15-38.

Chase, C.G. & Patchett, P.J. (1988). Stored mafic/ultramafic crust and early Archean mantle depletion. *Earth and Planetary Science Letters,* Vol.91, pp. 66-72.

Christensen, U.R. & Hofmann, A.W. (1994). Segregation of subducted oceanic crust in the convecting mantle. *Journal of Geophysical Research,* Vol.99 (B10), pp. 19867-19884.

Class, C. & le Roex, A.P. (2008). Continental material in the shallow oceanic mantle–How does it get there? *Geology,* Vol.34, No.3, pp. 129-132.

Coffin, M.F., Pringle, M.S., Duncan, R.A. & et al. (2002). Kerguelen hotspot magma output since 130 Ma. *Journal of Petrology,* Vol.43, pp. 1121-1139.

Courtillot, V., Davaille, A., Besse, J. & Stock, J. (2003). Three distinct types of hotspots in the Earth's mantle. *Earth and Planetary Science Letters,* Vol.205, pp. 295-308.

Curray, J.R. & Munasinghe, T. (1991). Origin of the Rajmahal traps and the 85°E Ridge: preliminary reconstructions of the trace of the Crozet hotspot. *Geology,* Vol.19, pp. 1237-1240.

Dalziel, I.W.D., Lawver, L.A. & Murphy, J.B. (2000). Plumes, orogenesis, and supercontinental fragmentation. *Earth and Planetary Science Letters,* Vol.178, pp. 1-11.

Deckart, K., Feraund, G., Marques, L.S. & Bertrand, H. (1998). New time constraints on dyke swarms related to the Parana-Etendeka magmatic province and subsequent South Atlantic opening, Southeastern Brazil. *Journal of Volcanological and Geothermal Researches,* Vol.80, No.1-2, pp. 67-83.

Doucet, S., Scoates, J.S., Weis, D. & Giret, A. (2005). Constraining the components of the Kerguelen mantle plume: a Hf-Pb-Sr-Nd isotopic study of picrites and high-MgO basalts from the Kerguelen Archipelago. *Geochem. Geophys. Geosys.,* Vol. 6, No.4, Q04007. doi: 10.1029/2004GC000806.

Duncan, R.A., Hooper, P.R., Rehacek, J. et al. (1997). The timing and duration of the Karoo igneous event. southern Gondwana. *Journal of Geophysical Research,* V.102 (B8), pp. 18127-18138.

Elliot, D.H., Fleming, T.H., Kyle, P.R. & Foland, K.A. (1999). Long-distance transport of magmas in the Jurassic Ferrar Large Igneous Province, Antarctica. *Earth and Planetary Science Letters,* Vol.167, pp. 89-104.

Elliot, D.H. & Fleming, T.H. (2000). Weddell triple junction: the principal focus of Ferrar and Karoo magmatism during initial break-up of Gondwana. *Geology,* Vol.28, pp. 539-542.

Fedorov, L.V., Ravich, M.G. & Hofmann, J. (1982). Geologic comparison of southeastern Peninsular India and Sri Lanka with a part of East Antarctica (Enderby Land. MacRobertson Land. and Princess Elizabeth Land). in: *Antarctic Geoscience.* Craddock, C. (ed). The University of Wisconsin Press. Madison. Wisconsin. pp. 73-78.

Foley, S.F., Andronikov, A.V. & Melzer, S. (2001). Petrology of ultramafic lamprophyres from the Beaver Lake area of Eastern Antarctica and their relation to the breakup of Gondwanaland. *Mineralogy and Petrology.* V.74. pp. 361-384.

Foley, S.F., Andronikov, A.V., Jacob, D.E. & Melzer, S. (2006). Evidence from Antarctic mantle peridotite xenoliths for changes in mineralogy, geochemistry and geothermal gradients beneath a developing rift. *Geochemical et Cosmochemical Acta,* Vol.70, pp. 3096-3120.

Frey, F.A., McNaughton, N.J., Nelson, D.R. & et al. (1996). Petrogenesis of the Bunbary basalt. western Australia: ineraction between the Kerguelen plume and Gondwana lithosphere? *Earth Planetary Science Letters,* Vol.144, pp. 163-183.

Frey, F.A., Coffin, M.F. & 18 others. (2000). Origin and evolution of a submarine large igneous province: the Kerguelen Plateau and Broken Ridge, southern Indian Ocean. *Earth and Planetary Science Letters,* Vol.176, pp. 73-89.

Frey, F.A., Nicolaysen, K., Kubit, B.K., & et al. (2002). Flood basalt from Mount Tourmente in the Central Kerguelen Archipelago: the change from transitional to alkaline basalt at ~25 Ma. *Journal of Petrology,* Vol.43, No.7, pp. 1367-1387.

Golynsky, A.V., Alyavdin, S.V., Masolov, V.N. & et al. (2002). The composite magnetic anomaly map of the East Antarctic. *Tectonophysics,* Vol.347, pp. 109-120.

Golynsky, A.V., Chiappini, M., Damaske, D. & et al. (2006). ADMAP - a digital magnetic anomaly map of the Antarctic. in: *Antarctica - contributions to global Earth Sciences.* Futerer, D.K., Damaske, D., Kleinschmidt, G., Miller, H. & Tessensohn, F. (eds). Springer-Verlag. Berlin. pp. 109-116.

Harris, C., Marsh, J.S., Duncan, A.R. & Erlank, A.J. (1990). Petrogenesis of the Kirwan basalts of Dronning Maud Land. Antarctica. *Journal of Petrology,* Vol.31, pp. 341-369.

Hauri, E.H., Whitehead, J.A. & Hart, S.R. (1994). Fluid dynamic and geochemical aspects of entrainment in mantle plumes. *Journal of Geophysical Research,* Vol.99, pp. 24275-24300.

Heinonen, J.S., Carlson, R.W. & Luttinen, A.V. (2010). Isotopic (Sr, Nd, Pb and Os) composition of highly magnesian dikes of Vestfjella, western Dronning Maud Land, Antarctica: a key to the origins of the Jurassic Karoo large igneous province? *Chemical Geology,* Vol.277, pp. 227-244.

Hergt, J.M., Peate, D.W. & Hawkesworth, C.J. (1991). The petrogenesis of Mesozoic Gondwana low-Ti flood basalts. *Earth and Planetary Science Letters,* Vol.105, pp. 134-148.

Hinz, K. & Krause, W. (1982). The continental margin of Queen Maud Land. In: Antarctica seismic sequences. structural elements and geological developments. *Geologisches Jahrbuch,* Vol.E23, pp. 17-41.

Hoch, M. & Tobschall, H.J. (1998). Minettes from Schirmacher Oasis. East Antarctica - indicators of an enriched mantle source. *Antarctic Science,* Vol.10, No.4, pp. 476-486.

Hoch, M., Rehkamper, M. & Tobschall, H.J. (2001). Sr, Nd, Pb and O isotopes of minettes from Schirmacher Oasis. East Antarctica: a case of mantle metasomatism involving subducted continental material. *Journal of Petrology,* Vol.42, No.7, pp. 1387-1400.

Hofmann, A.W. (1988). Chemical differentiation of the Earth: The relationship between mantle, continental crust, and oceanic crust. *Earth and Planetary Science Letters,* Vol.90, pp. 297-314.

Ingle, S., Weis, D., Scoates, J.S. & Frey F.A. (2002). Relationship between the early Kerguelen plume and continental flood basalts of the paleo-Eastern Gondwanan margins. *Earth and Planetary Science Letters,* Vol.197, pp. 35-50.

Jokat, W., Boebel, T.M., Konig, M. & Meyer, U. (2003). Timing and geometry of early Gondwana breakup. *Journal of Geophysical Research,* Vol.108 (B9), No.2428, doi:10.1029/2002JB001802.

Jourdan, F., Feraud, G., Bertrand, H. & et al. (2004). The Karoo triple junction questioned: evidence from Jurassic and Proterozoic $^{40}Ar/^{39}Ar$ ages and geochemistry of the giant Okavango dyke swarm (Botswana). *Earth and Planetary Science Letters,* Vol.222, pp. 989-1006.

Jourdan, F., Feraud, G., Bertrand, H. & et al. (2005). Karoo large igneous province: brevity. origin. and relation to mass extinction questioned by new $^{40}Ar/^{39}Ar$ age data. *Geology,* Vol.33, No.9, pp. 745-748. doi:10.1130/G21632.1.

Jourdan, F., Feraud, G., Bertrand, H. & et al. (2006). Basement control on dyke distribution in Large Igneous Provinces: case study of the Karoo triple junction. *Earth and Planetary Science Letters,* Vol.241, pp. 307-322.

Kent, R. (1991). Lithospheric uplift in eastern Gondwana: Evidence for a long-lived mantle plume system? *Geology,* Vol.19. pp. 19-23.

Kent, R.W., Saunders, A.D., Kempton, P.D. & Ghose, N.C. (1997). Rajmahal basalts. eastern India: mantle sources and melt distribution at a volcanic rifted margin. In: *Large igneous provinces: continental. oceanic and planetary flood volcanism.* Mahoney, J.J. & Coffin, M.F. (eds). *Geophysical Monograph Series,* Vol.100, AGU. Washington DC. pp. 145-182.

Kent, R.W., Pringle, M.S., Mueller, R.D. & et al. (2002). $^{40}Ar/^{39}Ar$ Geochronology of the Rajmahal basalts. India. and their relationship to the Kerguelen plateau. *Journal of Petrology,* Vol.43, pp. 1141-1153.

Kurinin, R.G., Grinson, A.S. & Dhun Zhun Yu. (1988). Rift-zone of Lambert Glacier - as possible alkaline-ultramafic province in the East Antarctica. *Reports of SU Academy of Sciences,* Vol.299, pp. 944-947.

Kurinin, R.G. & Grikurov, G.E. (1980). Structure of the rift zone of the Lambert glacier. *Reports of SAE.* Hydrometeoizdat. Leningrad, Vol.70, pp. 75-86.

Laiba, A.A., Vorobiev, D.M., Gonghurov N.A. & Tolunas, Yu.V. (2002). Preliminary reports of geological studies within Schirmacher oasis during 47 RAE. In: *Research and environment protection in Antarctica.* Abstracts Volume. 2002. St.Petersburg. AARI. pp. 64-66 (in Russian).

Laiba, A.A., Andronikov, A.V., Egorov, L.S. & Fedorov, L.V. (1987). Stock and dyke bodies of alkaline-ultrabasic composition at Jetty Peninsula (Prince Charles Mountains, East Antarctica). In: Geological-geophysical investigations in Antarctica. Ivanov, V.L. & Grikurov, G.E. (eds). Sevmorgeologia. Leningrad. pp. 35–47 (in Russian).

Lawver, L.A., Sclater, J.G. & Meinke, L. (1985). Mesozoic and Cenozoic reconstruction of the South Atlantic. *Tectonophysics,* Vol.114, pp. 233-254.

Leat, P.T., Curtis, M.L., Riley, T.R. & Ferraccioli, F. (2007). Jurassic magmatism in Dronning Maud Land: synthesis of results of the MAMOG project. U.S. Geological Survey and the National Academies. USGS OF-2007- 1047. Short Research Paper 033; doi: 10.3133/of2007-1047 srp033.

Leitch, A.M., Davies, G.F. & Wells, M. (1998). A plume head melting under a rifting margin. *Earth and Planetary Science Letters,* Vol.161, pp. 161-177.

Leitchenkov, G.L. & Masolov, V.N. (1997). Tectonic and magmatic history of the Eastern Weddell Sea Region. In: *The Antarctic Region: Geological Evolution and Processes.* pp. 461-466.

Leitchenkov, G.L., O'Brien, P.E., Ishihara, T. & Gandyukhin, V.V. (1999). The rift structure of Prydz Bay - Cooperation Sea and history of pre-breakup crustal extension between India and Antarctica. In: *Abstracts of 8th International Symp. on Antarctic Earth Sciences,* New Zealand. pp. 188-190.

Leitchenkov, G.L., Sushchevskaya, N.M. & Belyatsky, B.V. (2003). Geodynamics of the Atlantic and Indian Sectors of the South Ocean. *Doklady Earth Sciences,* Vol.391, pp. 675–678.

Leitchenkov, G.L. & Guseva, Yu.B. (2006). Structure and Evolution of the Earth's Crust of the Sedimentary Basin of the Davis Sea, East Antarctica. In: *Scientific Results of Geological and Geophysical Studies in Antarctica.* Leitchenkov, G.L. & Laiba, A.A. (eds). VNIIOkeanologiya, St. Petersburg, 2006. Vol.1, pp. 101–115 (in Russian).

Luttinen, A.V. & Furnes, H. (2000). Flood basalts of Vestfjella: Jurassic magmatism across an Archaen-Proterozoic lithospheric boundary in Dronning Maud Land. Antarctica. *Journal of Petrology,* Vol.41, pp.1271-1305.

Luttinen, A.V., Zhang, X. & Foland, K.A. (2002). 159 Ma Kjakebeinet lamproites (Dronning Maud Land. Antarctica) and their implications for Gondwana breakup process. *Geological Magazine,* Vol.139, No.5, pp. 525-539.

Mahoney, J.J., Jones, W.B., Frey, F.A. & et al. (1995). Geochemical characteristics of lavas from Broken Ridge. the Naturaliste Plateau and southernmost Kerguelen Plateau: Cretaceous plateau volcanism in the southeast Indian Ocean. *Chemical Geology,* Vol.120, pp. 315-345.

Martin, P.K. & Hartnady, C. (1986). Plate tectonic development of the southwest Indian Ocean: a revised reconstruction of East Antarctica and Africa. *Journal of Geophysical Research,* Vol.91, pp. 4767-4785.

Migdisova, N.A., Sushchevskaya, N.M., Lattenen, A.V., & Mikhalskii, E.M. (2004). Variations in the Composition of Clinopyroxene from the Basalts of Various Geodynamic Settings of the Antarctic Region. *Petrology,* Vol.12, pp. 176–194.

Mikhalsky, E.V., Andronikov, A.V. & Beliatsky, B.V. (1992). Mafic igneous suites in the Lambert rift zone. In: *Recent Progress in Antarctic Earth Science*. Yoshida, Y., Kaminuma K. & Shiraishi, K. (eds). Terrapub, Tokyo, pp. 173-178.

Mikhalsky, E.V., Laiba, A.A. & Surina, N.P. (1998). The Lambert Province of alkaline-basic and alkaline-ultrabasic rocks in East Antarctica: geochemical and genetic characteristics of igneous complexes. *Petrology,* Vol.6, pp. 466–479.

Morgan, W.J. (1981). Hotspot tracks and opening of the Atlantic and Indian Oceans. In: *The Sea*. Emiliani, C. (ed.). 1981. Vol.7, New York. Wiley Interscience. pp. 443-487.

Nimis, P. & Ulmer, P. (1998). Clinopyroxene geobarometry of magmatic rocks Part 1: An expanded structural geobarometer for anhydrous and hydrous. basic and ultrabasic systems. *Contributions to Mineralogy and Petrology,* Vol.133, pp. 122-135.

Ono, S., Ito, E. & Katsura, T. (2001). Mineralogy of subducted basaltic crust (MORB) from 25 to 37 GPa, and chemical heterogeneity of the lower mantle. *Earth and Planetary Science Letters,* Vol.190, pp. 57-63.

O'Reilly, S.Y. & Griffin, W.L. (2010). The continental lithosphere-asthenosphere boundary: Can we sample it? *Lithos,* Vol.120, pp. 1-13.

Pushcharovskii, Yu.M. & Pushcharovskii, D.Yu. (1999). Geosphere of the Earth mantle. *Geotectonica,* No.1, pp. 3-14 (in Russian).

Renne, P.R., Glen, J.M., Milner, S.C. & Duncan, R.A. (1996). Age of Etendeka-flood volcanism and associated intrusions in southwestern Africa. *Geology,* Vol.24, No.7, pp. 659-662.

Riley, T.R., Leat, P.T., Storey, B.C. & et al. (2003). Ultramafic lamprophyres of the Ferrar large igneous province: evidence for a HIMU mantle component. *Lithos,* Vol.66, pp. 63-76.

Riley, T.R., Leat, P.T., Curtis, M.L. & et al. (2005). Early-Middle Jurassic dolerite dykes from Western Dronning Maud Land (Antarctica): identifying mantle source in Karoo large igneous province. *Journal of Petrology,* Vol.46, No.7, pp. 1489-1524.

Ruzhentsev, S.V., Melankholina, E.N. & Mossakovsky, A.A. (1999). Phanerozoic geodynamics of the Pacific Ocean and Indian-Atlantic segments of the Earth and mantle structure. in: Some problems of lithospheric geodynamics. 1999. N. 511. Lukianov, A.V. (ed.) Moscow. Nauka. pp. 27-43 (in Russian).

Sobolev, A.V., Hofmann, A.W., Sobolev, A.V. & Nikogosian, I.K. (2005). An olivine-free mantle source of Hawaiian shield basalts. *Nature,* Vol.434, N7033, pp. 590-597.

Sobolev, A.V., Hofmann, A.W., Kuzmin, D.V. & et al. (2007). The amount of recycled crust in sources of mantle-derived melts. *Science,* Vol.316, pp. 412-417.

Stewart, K., Turner, S., Kelly, S. & et al. (1996). $^{40}Ar/^{39}Ar$ geochronology in the Parana continental flood basalt province. *Earth and Planetary Science Letters*, Vol.143, No.1-2, pp. 95-109.

Storey, B.C. (1995). The role of mantle plumes in continental breakup: case histories from Gondwanaland. *Nature*, Vol.377, pp. 301-308.

Storey, B.C. & Kyle, P.R. (1997). An active mantle mechanism for Gondwana break-up. *South African Journal of Geology*, Vol.100, pp. 283-290.

Storey, M., Saunders, A.D., Tarney, J., et al. (1989). Contamination of Indian Ocean asthenosphere by the Kerguelen-Heard mantle plume. *Nature*, Vol.338, pp. 574-576.

Storey, M., Kent, R.W., Saunders, A.D. & et al. (1992). Lower Cretaceous volcanic rocks on continental margins and their relationships to the Kerguelen Plateau. In: *Proceedings of ODP Scientific Results*. Wise, S.W.J. & Schlich, R. (eds). 1992. Vol.120. pp. 33-53.

Sun, S.-S. & McDonough, W.F. (1989). Chemical and isotopic systematics of oceanic basalts: implications for mantle composition and processes. In: *Magmatism in the ocean basins*. Saunders, A.D. & Norry, M.J. (eds). 1989. Geol. Soc. Spec. Publ. Vol.42, pp.313-345.

Sushchevskaya, N.M., Ovchinnikova, G.V., Borisova, A.Yu. & et al. (1996). Geochemical Heterogeneity of the Magmatism of Afanasij Nikitin Rise, Northeastern Indian Ocean. *Petrology*, Vol.4, pp.119–136.

Sushchevskaya, N.M., Belyatskii, B.V., Tsekhonya, T.I. & et al. (1998). Petrology and Geochemistry of Basalts from the Eastern Indian Ocean: Implications for Its Early Evolution. *Petrology*, Vol.6, pp. 528–555.

Sushchevskaya, N.M., Belyatskii, B.V, Dubinin, E.P. & et al. (2003). Geochemical Heterogeneity of Tholeiitic Magmatism in Circum-Antarctic Rift Zones. *Geochemistry International*, Vol.41, pp. 727–740.

Sushchevskaya, N.M., Mikhalskii, E.M. & Belyatskii, B.V. (2004). Magmatic evolution of the South sector of the Earth. In: *Petrology of magmatic and metamorphic complexes*, No.4, 2004. Tomsk. pp. 219-223.

Sushchevskaya, N.M., Cherkashov, G.A., Baranov, B.V. et al. (2005). Tholeiitic Magmatism of an Ultraslow Spreading Environment: An Example from the Knipovich Ridge, North Atlantic. *Geochemistry International*, Vol.43, pp. 222–241.

Sushchevskaya, N.M., Korago, E.A., Belyatsky, B.V. & Sirotkin, A.N. (2009). Geochemistry of Neogene Magmatism at Spitsbergen Island. *Geochemistry International*, Vol.47, pp. 966-978.

Vuori, S.K. & Luttinen, A.V. (2003). The Jurassic gabbroic intrusions of Utpostane and Muren: insights into Karoo-related plutonism in Dronning Maud Land. Antarctica. *Antarctic Science*, Vol.15, pp. 283-301.

Weis, D., Frey, F.A., Saunders, A. & Leg 121 team. (1991). Ninetyeast Ridge (Indian Ocean): a 500 km record of a Dupal mantle plume. *Geology*, Vol.19, pp. 99-102.

Weis, D. & Frey, F.A. (1996). Role of the Kerguelen Plume in generating the eastern Indian Ocean seafloor. *Journal of Geophysical Research*, Vol.101, No.B6, pp. 13831-13849.

White, R. & McKenzie, D. (1989). Magmatism at rift zones: the generating of volcanic continental margins and flood basalts. *Journal of Geophysical Research*, Vol.94, pp. 7685-7729.

Zhang, X., Luttinen, A.V., Elliot, D.H. & et al. (2003). Early stages of Gondwana breakup: the $^{40}Ar/^{39}Ar$ geochronology of Jurassic basaltic rocks from western Dronning Maud Land, Antarctica, and implications for the timing of magmatic and hydrothermal events. *Journal of Geophysical Research*, Vol.108, No.B9, pp. 2449-2466. doi: 10.1029/2001JB001070.

5

Bimodal Volcano-Plutonic Complexes in the Frame of Eastern Member of Mongol-Okhotsk Orogenic Belt, as a Proof of the Time of Final Closure of Mongol-Okhotsk Basin

I. M. Derbeko
Institute of Geology and Nature Management FEB RAS, Blagoveschensk
Russia

1. Introduction

We can ascertain at the example of the modern functioning volcanoes, that their activity accompanies the tectonical rebuilding, that occurred in the Earth's crust. Material and isotope characteristics of volcanological (volcano-plutonic) formations reflect also the processes that occur at such a depth of Earth, which are not reachable for research for today. The decoding of the interdependence assists to better understanding of the evolution of our planet in different temporal intervals. It also helps to imagine the processes of formation or destruction of the continents. One of them is Euro-Asian continent. It's name indicates the binominal structure of Europe and Asia. Once two separated continents united in a single whole, forming by that a superstructure - Central Asian folded belt. And the axis of the structure is the Mongol-Okhotsk orogenic belt (Parfenov et al., 1999). During almost all of the Phanerocoic the formation of the belt was accompanied with the formation of a whole series of the volcanic and volcano-plutonic complexes and by the closure of Mongol-Okhotsk basin. The final stage of the closure of Mongol-Okhotsk basin that occurred at the end of early Cretaceous and was conducted by the formation of the bimodal volcano-plutonic complexes is examined in the article.

The bimodal volcano-plutonic complexes of the basalt-trachybasalt - rhyolitic content of the end of the early Cretaceous (119 - 97 Ma) are separated in the Southern and Northern frame of Eastern member of Mongol-Okhotsk orogenik belt. The rocks of the complexes disappear in the Western direction in the region of wedging of the Eastern member of the belt in the same direction. In eastern direction in the South, their development is limited by the structures of Bureya-Jiamusi terrane, and in the North it is limited by Okhotsk-Koriaksky terrane. The features of the geological structure of bimodal complexes are examined in the very article. The results of geochemical, isotope -geochemical and geo-chronological precision research are also shown in the article (by authors and the literature data). A proposal has been made that the rocks of the examined bimodal complexes correlate with some of the Late Mesozoic - Early Cretaceous intra plate bimodal associations of North-Asian continent by the row of geochemical and isotope geochemical characteristics. And their formation was caused by collision processes that stipulated the closure of the Eastern flank of Mongol-Okhotsk basin by the possible contribution of plume source.

The magmatic formations accompanying the formation of the rift's zone in the frames of different modern continents, reveal considerable variety of geochemical and isotope characteristics. The Euro-Asian continent is not an exception (Types of magma, 2006). Today the models of geodynamic environments for the row of its regions are worked out. They combine the action of the depth plume and the rift's formation (Yarmolyuk & Kovalenko, 2000; Yarmolyuk et al., 2000; Yarmolyuk et al., 2002; Kovalenko et al., 2003; Kozlovsky et al., 2005; Vorontsov et al., 2007). The basic role in formation of magma complexes belongs not only to the source of different nature but also to characteristics of contaminated continental crust that stipulates the formation of contrast volcanism, widely developed in the Central-Asian zone of fold belt (Types of magma, 2006). What is the role of characteristics of lithosphere in the formation of bimodal complexes; does the productivity of the plume remain in the spatiotemporal understanding and can it be the same for the differently aged magmatism? It is possible to come up to the research of these problems if there is the presence of the correct analytic data for the concrete complexes which formation has taken place not only in similar geodynamical environments but also in sequentially changed age intervals. Bimodal volcano-plutonic complexes spatially located in the frames of Eastern link of Mongol-Okhotsk belt are shown in the article (Fig. 1).

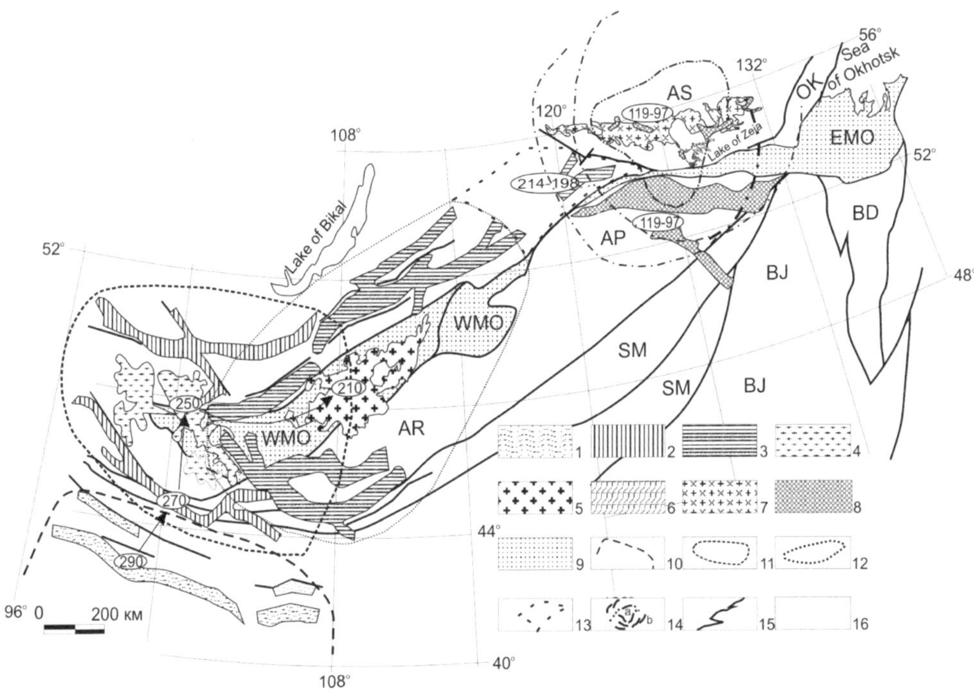

Fig. 1. Structure-tecktonical scheme of the dislocation of volcano-plutonic and plutonic complexes of the Late Paleozoic - Mesozoic in the framed of Mongol-Okhotsk orogenic belt. The scheme of the structure of the Western link of Mongol-Okhotsky orogenic belt (WMO), on the East of meridian 120 is made by (Yarmolyuk et al., 2002): 1-3 - the rocks of the bimodal associations and alkali granites: 1 - Late carboniferous - Erly Permian, 2 - Permian; 3 - Early Mesozoic; 4 - 5 - granitoid complexes; 4 - Permian; 5 - Early Mesozoic; 6 - collisional alkali - moderata alkali volcano-plutonic complex (Kozak et al., 2004; Strikha, 2006) on the duration of Eastren-Transbaikalian riftous zone (Derbeko, 1998). The scheme of

the structure of the Eastern link of Mongol-Okhotsk orogenic belt (EMO) is made by using the data of (Geological map…, 1999; Parfionov et al., 2003): 7 - Late Jurassic - Early Cretaceous plutonic complex; 8 – the field of the separation of the bimodal volcano-plutonic complexes of the end of Early Cretaceous. 9 – Mongol-Okhotsk orogenic belt. 10 – 12 – the borders of the projection of Mongolian plume by (Types of magmas, 2006) during time periods: 10 - Late Carboniferous – Early Permian, 11 – Permian, 12 – Early Mesozoic, 13 - an assumed continuation of the territory of the affection of the plume of Early Mesozoic by author. 14 – Aldan-Zeisky plume by (Petrischevsky & Khanchuk, 2006). 15 – Tectonic borders. 16 – water basins. Superterrains and terrains: AR – Argunsky, SM – South Mongolian, BJ – Bureinsko-Jiamusinsky, BD – Badgialsky (Parfionov and others, 2003); Aldano-Stanovoy (Gusev & Khain, 1995); Okhotsk-Koriaksky (Geodynamics, magmatism…, 2006).

They are met as truncated volcanic fields along the Southern and Northern borders of Mongol-Okhotsk orogenic belt about 600 km, from 50 to 400 km wide. The formation of bimodal late Paleozoic - early Mesozoic formations in the frames of Western link of Mongol-Okhotsk belt is connected with collision of North-Asian and Sino-Corean continents (Types of magma, 2006). Collision processes in that region, combined with the dimensioned intra plated processes, by the beginning of the Early Cretaceous (190 Ma) completed by the closure of the Paleozoic ocean and Western part of Mongol-Okhotsk basin (Yarmolyuk & Kovalenko, 1991; Yarmolyuk & Kovalenko, 2000; Yarmolyuk et al., 2000; Kovalenko et al., 2003; Kozlovsky et al., 2005; Types of magma, 2006; Vorontsov et al., 2007).
During the period terrigenic sea sediments basically were accumulated in the East of Mongol-Okhotsk basin. And only about 150 million years ago the continental volcanic and volcano-plutonic complexes began to form, both in the frames of Eastern belt and it's rim.
Along the borders of Mongol-Okhotsk belt first the formation of calc-alkali andesite complexes goes, and then there the formation of the bimodal basalt-trachybasalt-rthyolitic complexes goes (Martynyuk et al., 1990; Kozyrev, 2000a; Kozak et al., 2004; Sorokin et al., 2004; Derbeko, et al., 2008a; Derbeko et al., 2008b). The substantial characteristics of the rocks of bimodal complexes of Early Cretaceous, geo-chronological definitions of their age, isotope-geochemical data for some types of volcanites, the comparison of their characteristics with the rocks of the rifting system of Central Asia are shown in the article. The data allow us to make an assumption about the connection between the formation of the bimodal complexes and the collision processes, which took place in the region during the Early Cretaceous and about an active manifestation of the depth plume during the period.

2. Geological position of bimodal complexes

2.1 The contrast volcano-plutonic complexes in the rim of Mongol-Okhotsk orogenic belt were recognized in late 70th, by making the field surveying (Martinyuk et al., 1999). They are composed by volcanogenic-sedimentary thickness, subvolcanic and plutonic bodies. The fields of the development of the complexes are often spatially combined with the fields of volcanitcs of calc-alkali series and comagmatic, plutonic formations of parti-coloured (diorite-monocite-granodiorite-granite) composition of the Early Cretaceous (Martinyuk et al., 1990; Kozyrev, 2000a; Kozyrev, 2000b; Kozak et al., 2004; Derbeko, 2004; Antonov, 2007).
Last years a row of geo-chronological dating of the age of volcanic and plutonic rocks of above listed formations was gained. According to the data, the formation of the volcanites of

bimodal complexes occurred in the time interval from 119 to 97 Ma. The building of the formation of calc-alkali series occurred during the time interval about 140 - 120 Ma (Neimark et al., 1996; Kozyrev, 2002a; Sorokin et al., 2003; Sorokin et al., 2006; Strikha, 2006; Antonov, 2007).

The fields of the volcanic and intrusive formations of bimodal series are developed along Southern and Northern borders of Eastern link of Mongol-Okhotsk belt till it's wedging on the West - Aginsky "threshold" (Geodymanic, magmatism..., 2006). They disappear here, but later they show up in the frames of western link of Mongol-Okhotsk belt with the age of the Early Mesozoic and earlier (Yarmoluk & Kovalenko, 1991; Gordienko, 1987; Yarmoluk & Kovalenko, 2000; Yarmoluk et al., 2000; Yarmoluk et al., 2002). By the data of Zhang Hun and others (Zhang Hun et al., 2000), bimodal formations are not registered on the neighboring territory of China, which means that, in Southern direction of their development is limited by the river Amur. In the Northern frames, the rocks are met in the limits of Aldan-Stanovoy terrane (Gusev & Khain, 1995) as rare uncoordinated strongly truncated volcanic fields.

Paleovolcanic structures that are formed with the formations of bimodal complexes, by the character of volcanic activity represent paleovolcanoes of central-claftlike type. They form volcanic fields from 2 to first hundreds km^2. The differences thickness, the percentage ratio of covering, vent and subvolcanic components are stated in their structure, but the consistency of the geologic profiles of the covering facies mostly held out. It consists in being timed to the base of the profile's cover of trachyandesitic basaltes, andesitic basaltes, andesites, trachyandesites, trachybasaltes, interstratified with tuffo-sandstones, tuffogravelites and tuffoconglamerates. The middle part of the profile is composed of rhyolites, rhyolitic dacites, trachydacites, dacites, perlites, tuffs and ignimbrites. In some of the volcanic fields, the profile finishes with the cover of trachyandesites, andesites, andesibasaltes. The percentage ratio of the rocks of the complex out of its general mass is 35-60%, middle - 10-40%, acid - 20-30%, tuffogenous-sedimentary - 10-15%.

In the frames of volcanic fields and close to them, the bodies of vent facies with the square of exit equal 2 km2 are stated. They are represented by agglomerates, lapillies and psammitic tuffs, clastolavas of acid rarely medium content. Subvolcanic bodies that are comagmatic to the covers form not significant shafts and dykes.

2.2 Age

The age of bimodal rocks in Southern frames (Galkinsky volcano-plutonic complex) is determineted by the complex of bivalvia mollusk (pelecipodes) and phyllopodous crustaceas (konchostrak) found in ashes tuffs and tuffalevrolites. They characterize the lowest layers of upper Cretaceous (Kozyrev, 2000b). The data of isotope dating by U-Pb method on the zircons from of rhyolytes correlated to 117.1; 117.6 Ma, and on trachydacites - 117±1 Ma (definitions of the Institute of Geology and Geochronology of Precambrian, Sant-Petersburg). Comparable age was obtained with ^{40}Ar–^{39}Ar dating of analogous rocks, and for the rocks of the basic - medium content lower data was obtained 105 - 97 Ma (Kozyrev, 2000a; Kozyrev, 2000ba; Sorokin et al., 2004; Sorokin et al., 2006). According to the data of the author the age of trachyandesites from the fundamental outcrop in the basin of the Mokhovoy stream (sample d2063-7), identified by ^{40}Ar–^{39}Ar method (the definitions of V.A. Ponomarchuk, Institute of Geology and Mineralogy SD RAS), was 114.7±1.6 Ma (Derbeko, 2009).

Summarizing what was said, it can be stated, that formation of Galkinsky complex occurred in the following order: trachyandesitic basaltes of the lower part of the open-cast - 118.7±0.9 Ma, rhyolites, trachyrhyolites - 118.7±0.4; 117±1; 117.1; 117.6; 115.3±1.5 Ma and trachyandesites 115 Ma (middle part of the section); trachyandesites, andesites of the normal row that are close to trachyandesite types by their content - 105.9; 100; 97 Ma and rhyolites - 97±5 Ma. Thus, it can be assumed almost an uninterrupted stage of magmatic activity, which lasted during 119-97 Ma ago, which means it subsided in the very beginning of the Late Cretaceous. By the dating of plutogenic rocks of the complex by Rb-Sr method (by the data of geological survey 1978 th) for the quartz monocyte an age of 117.2 Ma was obtained. Later, by the Ar-Ar method the age of subalcalic granites was stated like 118 Ma (Sorokin & Ponomarchuk, 2002).

The age of the rocks of bimodal volcano-plutonic complexes in the Northern frames of Mongol-Okhotsk belt (Bomnaksky volcano-plutonic complex) according to the data of geological mapping of 80th - 90th is 112±8 - 99±7 Ma (definition by K/Ar method by trachyrhyolites) and 98±6 by andesites; 101±2 and 101±4 Ma (Neimark et al., 1996; U/Pb method, definitions by Up/Pb method by granocienite and dacites); 117±0.8 and 109 Ma (Antonov et al., 2001, definitions by Rb/Sr method by quartz diorite and quartz monocite); 108.6±1.3 and 110.3±2.9 Ma (Strikha & Rodionov, 2006, definitions by U/Pb method by granites, leikogranites).

Summing the available geochronoligical datings it can be said that, the formation of the rocks of the bimodal complexes in the frames of the Eastern link of Mongol-Okhotsk belt occurred simultaneously during 119 - 97 Ma.

3. Analytical methods of the research

The research of the gross geochemical composition of the rocks took place using the RFA methods (petrogenic components Sr, Zr, Nb) in the Institute of Geology and Nature Management Far Eastern Branch Russian Academy of Science and in the Institute of Geochemistry Siberian Branch Russian Academy of Science (Irkutsk, Russia), ICP-MS - in the Institute of Geochemistry Siberian Branch Russian Academy of Science (Irkutsk) and Institute of Tectonics and Geophysics Far Eastern Branch Russian Academy of Science (Khabarovsk, Russia). For the analysis of micro elements by the ICP-MS technology, a worn-out sample received an acid decomposition in HF and HNO_3 in fluoroplastic containers.

$^{40}Ar/^{39}Ar$ isotope chronological research of the sample d2063-7 was made in the Institute of Geology and Mineralogy SB RAS (Novosibirsk, Russia) with the usage of mass-spectrometer MI-1201V. Correction of the obtained data was made by a standard method taking into account the atmosphere contamination and interfering peak of side neutron-induced reactions. Isochronous constructions in $^{39}Ar/^{40}Ar$ - $^{36}Ar/^{40}Ar$ coordinates were produced by polynominal method of minimal square.

Rb/Sr and Sm/Nd isotope research is made in IGGD RAS. Isotope composition Rb, Sr, Sm and Nd are measured by numerous collectoral mass-spectrometer Finnigan MAT-261 and TRITON TI in static rate. The accuracy of the determination of the concentrations of Rb, Sr, Sm, Nd was ±0.05 % (2δ).

The author used the original samples from authors' collection and data from the published works to analyze the rocks (Kozyrev, 2000a; Sorokin et al., 2003; Sorokin et al., 2006; Strikha, 2006; Antonov, 2007).

4. The composition of the bimodal volcano-plutonic complexes

4.1 The description of the rocks of bimodal complex in Southern frames

Volcano-plutonic complex in the frames of Southern belt (Galkinsky) is formed by the rocks of basic-middle and acid contents. Lavas of basic-medium content are shown by porphyritic differences with almond-rocks and massive structures. By the association of the porphyritic discharges trachybasaltes, trachyandesitic basaltes are separated on plagioclase, pyroxene, pyroxene-amphibole, olivine-pyroxene, olivine-amphibole; trachyandesites and andesites – olivine-pyroxene, plagioclase, pyroxene, amphibole-pyroxene, amphibole. Phenocrysts (to 30%): clinopyroxenes (augite and diopside), orthopyroxene (hypersthene and enstatite), sometimes opacitizated, plagioclase (An_{10-36} labradorite - andesite), usual and basaltic amphibole, biotite, sanidine, olivine. Heightened alkalinity of lavas of basic-medium content is conditioned by the presence of the feldspar or red-brown biotite in the matrix of the rocks, rarely – sanidine in disseminations. The basic mass (90-70%) in plagioclase differences has an intersertal structure, in others – doleritic, phylotacsitic, trachytoid, microophite. It consists of the volcano glass with laths of plagioclase, granules of ortho-and clinopyroxenes, amphiboles, ripples of biotite, ore minerals. Accessory minerals are grothite, apatite, titanmagnetite, magnetite.

Rhyolites, trachytoid rhyolites, rhyolitic dacites with fluidal, massive or amygdaloidal texture form pink and light violet – light gray flows with the interlayers of perlites, tuffs and ignimbrites. Phenocrysts in lavas with porphyritic structure are represented by zonal plagioclase (An_{10-36}), potassic feldspar quartz, biotite, sometimes basaltic amphibole (to 3%). Phenocrysts of sanidine are present in some types of trahyrhyolites. By the data in sections of most high alumina rhyolites, singular granules of cordierite, spherulite of tridymite are found in the basic mass. The basic mass (to 80%) is of quartz- feldspar content with the scales of biotite, sometimes with small granules of amphibole, having perlitic, micropoikilitic, microspherolitic (spherolites of sanidine with albitic center) or microfelsitic structure.

Among tuffs the ashy aleurolites and psammitic varieties of acid, rarely basic-medium content, sometimes an admixture of lappilies up to 60% are registered.

Granitoids, comagmatic to Galkinsky complex, correlate with subalkalic granites, subalkalic leucogranites, granitoids, quartz diorites, quartz monocites by their content. Rock forming minerals of plutonic varieties are identical to volcanites,and among each other, they differ only by percentage of minerals.

The bimodal content of the rocks of the complex is determined by two ranges of the content of SiO_2: 47-64 и 72-78 wt.% with almost absolute absence of intermediate varieties (Fig. 2).

Volcanites, with the content of SiO_2 47-64 wt.% (Fig. 2a), are characterized by mostly moderate alkalinity. The amount of K_2O, same as the content of Fe (Fig. 3), with changing weakly of Na2O content, is unevenly growing proportionally to the content of SiO_2. These are high alumina (Al_2O_3 = 15.22-17.30 wt.%), moderate – low magnesia, low titaniferous (<2 wt.%) formations. They belong to high potassic calcareous-alkaline series (in singular cases - shoshonite) (Fig. 2b).

Volcanites with the content of SiO_2 72-78 wt.% are characterized by normal, rarely moderate alkalinity, with the growth of the content of SiO_2 generate alkalinity is getting lower (from 9.13 to 7.13 wt.%). The content of Na_2O reduces by slowly changing concentration of K_2O. These are low alumina (Al_2O_3 = 11.15-13.96 wt.%), low magnesia, mostly ferruginous (Fig. 3), low titaniferous formations. They belong to high potassic calcareous-alkaline series (Fig. 2b).

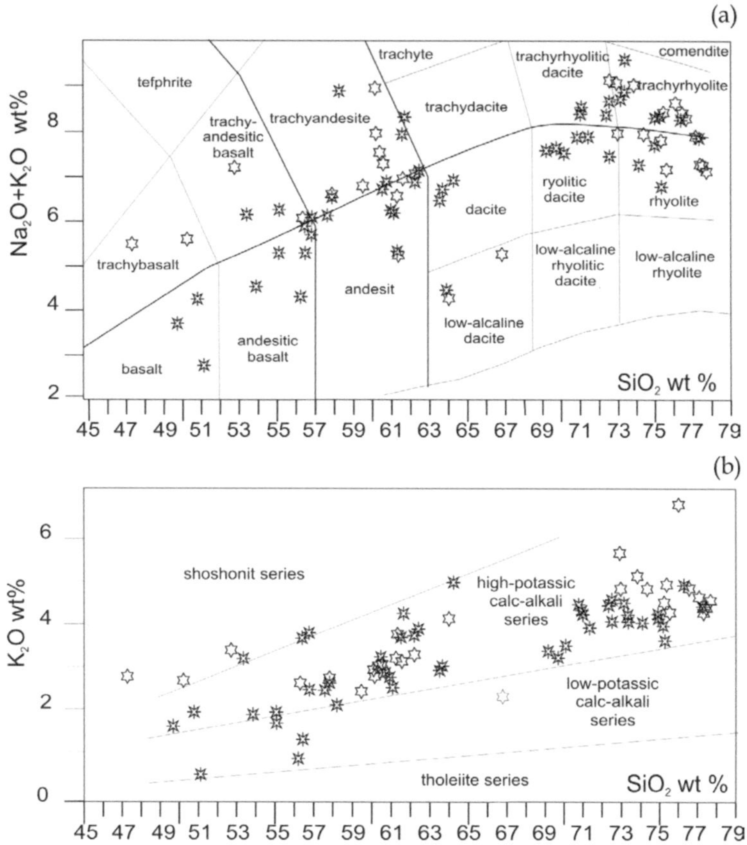

Fig. 2. Classification diagrams for the rocks of bimodal volcano-plutonic complexes of the framing of Mongol-Okhotsk orogenic belt. Northern framing - Bomnaksky complex (1) and on the South - Galkinsky (2): a) (Na_2O+K_2O) - SiO_2 (Magmatic rocks, 1983); b) K_2O+SiO_2 (Le Bas et al., 1986).

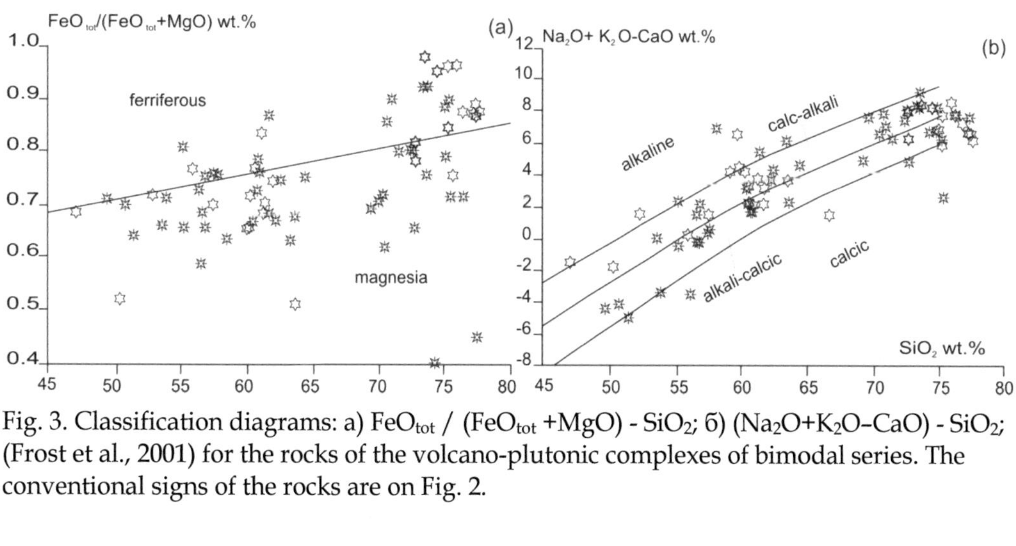

Fig. 3. Classification diagrams: a) FeO_{tot} / (FeO_{tot} +MgO) - SiO_2; б) (Na_2O+K_2O–CaO) - SiO_2; (Frost et al., 2001) for the rocks of the volcano-plutonic complexes of bimodal series. The conventional signs of the rocks are on Fig. 2.

4.2 The description of the rocks of bimodal complex in northern frames

The volcano-plutonic complex in Northern frames of the belt (Bomnakskiy) is formed by the rocks of basic-medium and acid contents. Lavas of basic- medium contents are represented by porphyritic differences with massive, rarely amygdaloidal textures. By the association of porphyritic discharge, the rocks of basic-middle content are represented by basalts, trachyte andesite basalts and andesite basalts, plagioclase, plagioclase-pyroxene, pyroxene-amphibolic, rarely olivine-pyroxene. Among trachyandesites and andesites following types are separated: plagioclase, pyroxene, amphibolic- pyroxene, amphibolic. Particularly, porphiric phenocrysts (to 45-50%) are represented by clinopyroxene (diopside), orthopyroxene (hypersthene, ferrohypersthene), plagioclase ($An_{61\text{-}35}$), usual amphibolic, biotite, rarely - olivine. Often, glomeroporphyritic joints of plagioclase or pyroxenes with titanmagnetites are found. The basic mass (up to 55 %) has got a pilotaxitic, doleritic, trachytoid or mikrophytal structures. It consists of volcanic glass with laths of plagioclase ($An_{50\text{-}35}$), granules of ortho- and clinopyroxenes, amphibolics, ripples of biotite, ore mineral. Accessory minerals are sphene, apatite, titanomagnetite, magnetite.

Rhyolitic dacites, rhyolites, trachyrhyolitic dacites, trachyrhyolites with fluidal or massive texture are forming flows with intercalation of perlites, tuffs and ignimbrites. Phenocrysts in lavas with porphiric structure are represented by zonal plagioclase ($An_{10\text{-}28}$), potassic feldspar, quartz, biotite, rarely usual amphibolic (less than 5%), singular and very small granules of pyroxene. Phenocrysts of sanidine are present in some types of trachyrhyplites. The basic mass (up to 90%) of quartz - feldspar content with the ripples of biotite, rarely with small granules of feldspar, has got perlitic, micropoicilitic, microfelsitic or microferrolitic (spherulite of potassic feldspar with albite or quartz center) structures.

The bimodal content of the rocks of Bomnaksky complex is determined by two diapasons of the content of SiO_2: 49-64 and 69-78 wt.% (Fig. 2).

Volcanites with the content of SiO_2 49-64 wt.% (Fig. 2a) are characterized by moderate and normal alkalinity, with straight correlation between general alkalinity and the content of SiO_2. The content of Na_2O is unstable, and the amount of K_2O irregularly changes from 1.7 to 4.2 wt.%. This are highalumina (Al_2O_3 = 15.65-18.86 wt.%), mostly magnesial (moderate-to low magnesial, Fig. 3a), low titanoferous (TiO_2 < 2 wt.%) formations. They belong to low-high potassic calcareous-alkaline series; singular trials get into the field of shoshonite series (Fig. 2b).

Volcanites with the content of SiO_2 69-78 wt.% are characterized by normal - moderate alkalinity (Fig. 2a). With the content of SiO_2 = 73 wt.%, general alkalinity is decreased. These are low potassic (Al_2O_3 = 10.55-14.99 wt.%), low magnesial, low titanoferous formations. They belong to highpotassic calcareous-alkaline series (Fig 2b).

Plutonic formations, which are comagmatic with volcanites of the bimodal complexes, are represented by diorites, monocites, quartz diorites, quartz monocites, granodiarites, quartz sienites, subalkalic granites, granites, subalkalic leucogranites, leucogranites. Intrusions are forming large irregularly-shaped bodies, stocks and dikes. Large massifs have complicated structure, where almost all varieties of magmatic rocks of the complex are included. As a rule, low intrusions are monorocks. The rock forming minerals of plutonic rocks are identical with volcanites and they differ only in percentage ratio of the minerals. The plutonic formations of the acid content relate to the granites of A-type, as far as granitoids, which are comagmatic to volcanites of medium-basic content, correlate with the formations of I-and S-types (Fig. 4)

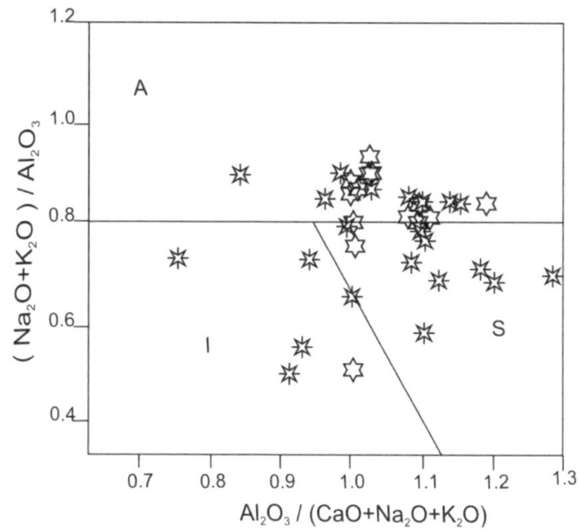

Fig. 4. A diagram for the separation A-, S-, I- types of granitoids by petrochemical indications (Chappell & White, 1974; Maeda, 1990). The conventional signs of the rocks are on Fig. 2.

A correlation analysis was made by sampling, formed out of 84 samples and including all the varieties of the rocks of the bimodal complexes of Southern and Northern frames of Western link of Mongol-Okhotsky belt (Fig. 5). Two associations of elements were singled out - SiO_2-K_2O and TiO_2-P_2O_5-FeO-CaO-Al_2O_3-MnO-MgO-Sr-Ba-Zr, which are connected with each other by negative correlation dependences. Positive correlation relation with silica is noted for Rb (r = 0.566 with r5% = 0.514) и Nb (r = 0,342 with r5% = 0.514) (Fig. 5j, 5k), fuzzy negative one for Y (r = - 0,601 with r5% = 0.514) (Fig. 5h). Thereby, a single trend of correlation reveals almost in all examined correlations of petrogenic and rare-earth elements with SiO_2, which is typical for bimodal series.

4.3 Geochemical characteristics of the rocks of the bimodal complexes

All the rocks of bimodal complexes are enriched in light rare-earth elements (La/Yb)n = 5.5-33.6 (10 - 20 values prevail), in which the accumulation of the La relative to Sm (La/Sm)n = 2.9-10.5 is significantly shown (3 - 4 values prevail), as far as the level of fractioning of heavy lanthanoids is lower (Gd/Yb)n = 0.9-3.9 (Fig. 6a). The accumulation of the rare-earth elements is happening parallelly to the manifestation of the Eu anomaly, which depth depends on the grade of plagioclase fractioning. Correspondingly Eu minimum in basic - medium rocks is weakly shown and it correlates to Eu/Eu* = 0.70-0.95 for the rocks in the Southern frames, 0.50-0.75 - for the rocks in Northern frames of the belt. It is deeper in acid formations (Eu/Eu* = 0.33-0.59 and 0.16-0.77 accordingly). Multielemental spectra are characterized by stable negative anomalies Nb, Ta and Ti for all the varieties of the rocks and very changeable anomaly Sr: for granitoids and acid volcanites it is negative and for the basic - medium rocks it is from weakly negative to almost positive (Fig 6b). The content of Ba, Rb, Th, K is marked by positive anomalies (Fig. 6b). It should be mentioned, that differences in the contents of such elements as Zr, Hf and Rb take place in the content of microelements between bimodal complexes of Northern and Southern frames.

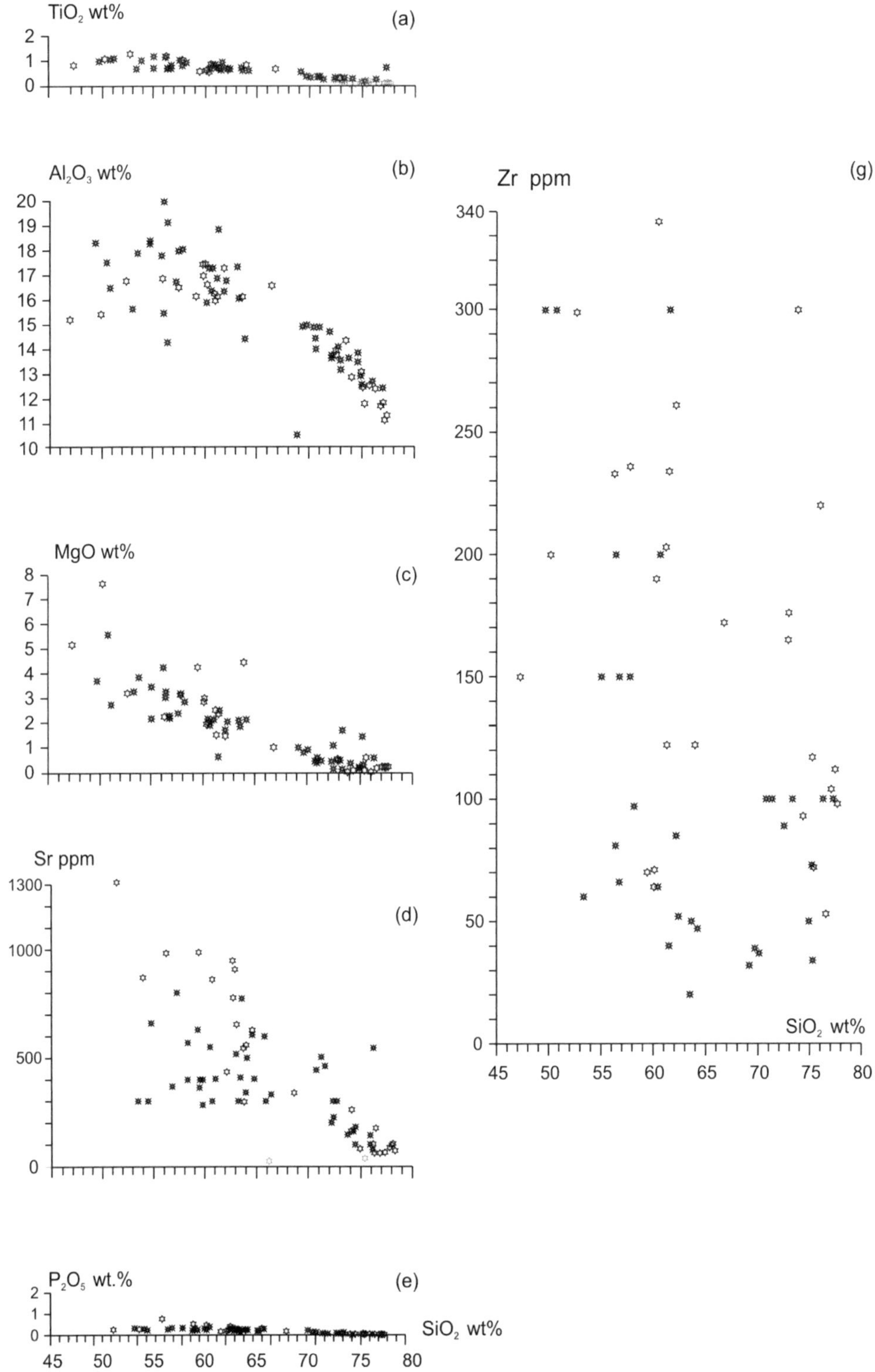

TiO_2 wt%
(a)
Al_2O_3 wt%
(b)
MgO wt%
(c)
Sr ppm
(d)
P_2O_5 wt.%
(e)
SiO_2 wt%
Zr ppm
(g)
SiO_2 wt%

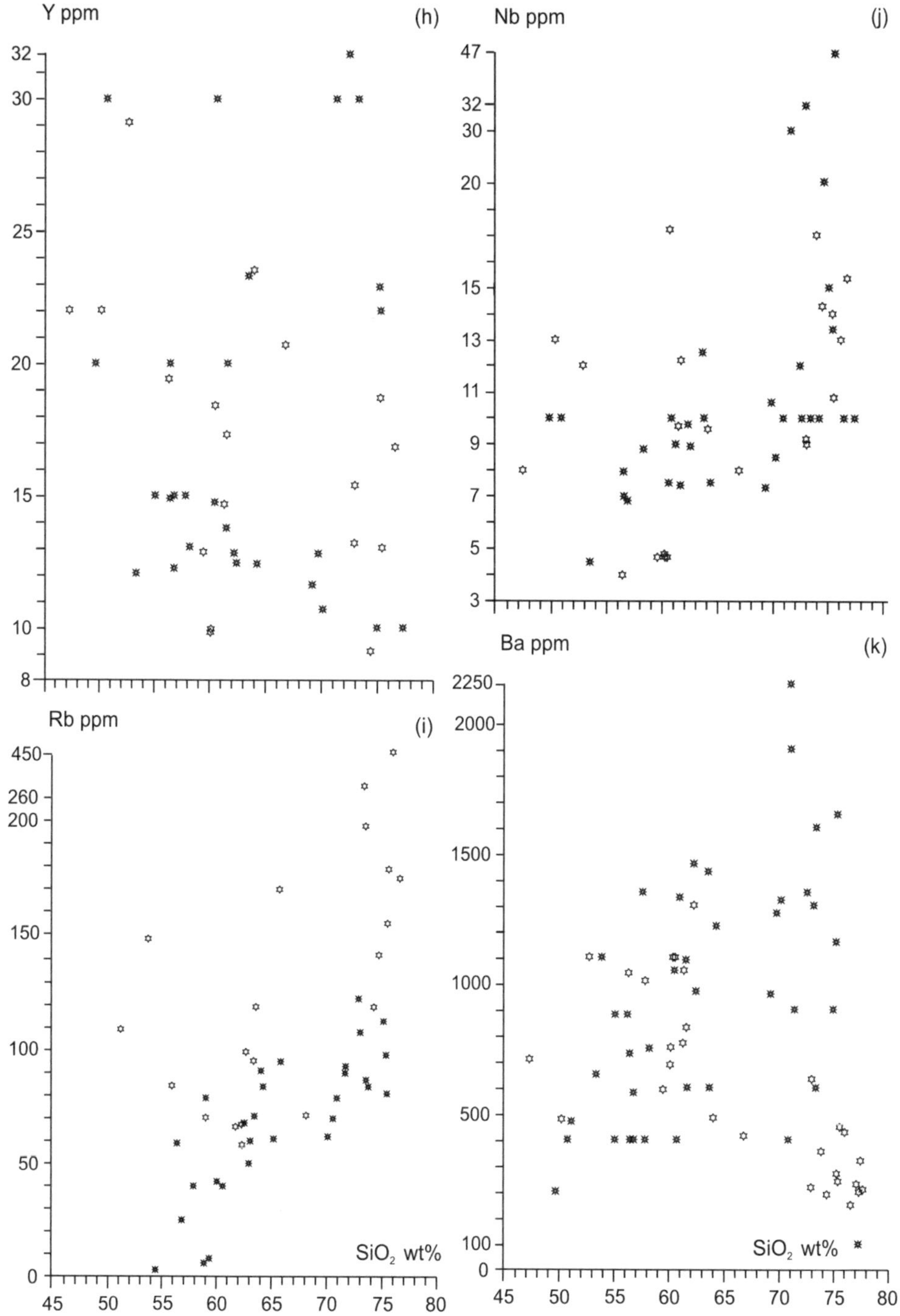

Fig. 5. The correlations of the content of petrogenic and rare elements to the content of SiO_2 in the rocks of volcano-plutonic complexes of bimodal series. The conventional signs of the rocks are on Fig. 2.

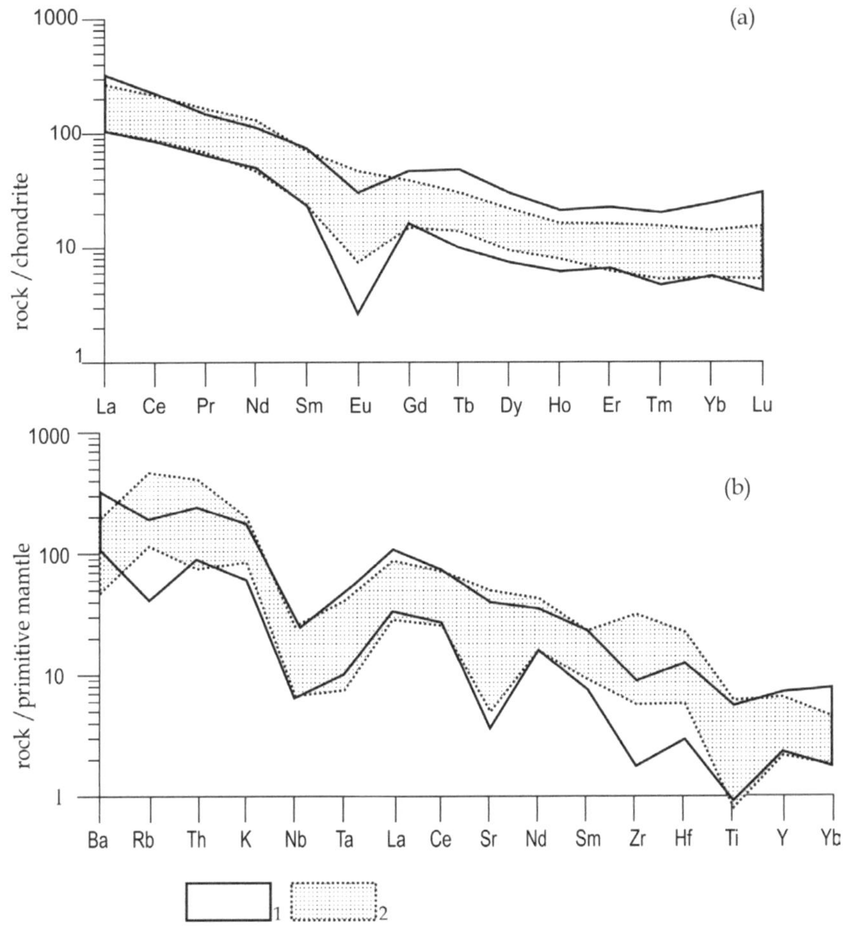

Fig. 6. Concentrations of the rear elements standardized to the composition of Chondrite (a) and primitive mantel (b). Compositions of chondrite C1 and primitive mantel are brought according to the data (Sun & McDonough, 1989). Northern frames - Bomnaksky complex (1) and Southern - Galkinsky (2).

The rocks of the examined bimodal complexes are close to some magmatic formations, were developed on the Western continuation of Mongol-Okhotsk orogenic belt and separated in Mongol-Okhotsk rifting zone, by bimodal character of development of the granitoids of A-type and by the row of geochemical characteristics (Vorontsov et al., 2007).

The manifestations of the sharp minimums Nb, Ta, Ti, and maximums in Ba, Rb, Th, K contents are typical for the rocks of the researched complexes, the same is for the formations of the rifting zone, but in the frames of the Eastern link, the alkali-salic rocks are absent. The content of highly charged elements are lower: Nb, Ta, и Zr, Hf, and by the growth of silica content, the content of Ba и Zr is mostly reducing (Fig. 5k: fig. 5i).

4.4 Isotope characteristics

The formations of the Galkinsky bimodal complex are characterized by staunch isotope structures with variations of the correlation $^{87}Sr/^{86}Sr$= 0.7057-0.7063, 0.7081-0.7084 and wide interval of the values of $\varepsilon Nd_{(T)}$ = (-0.6) – (-3.6) (Kozyrev, 2000a; Sorokin et al., 2006). The model Nd age - T_{Nd}(DM-2st) – correlates with a very narrow interval 975-1314 Ma., which might show the substantional homogeneity of substratum of melting with crustal component of late Riphean. For the formations of Bomnaksky bimodal complex, the following correlation is typical: $^{87}Sr/^{86}Sr$ =0.70592-0.70620, 0.70648-0.70773 with pretty narrow interval - $\varepsilon Nd_{(T)}$ = (-11.77) – (-12.20) (Kozyrev, 2000a; Striha & Rodionov, 2006). This might show a substantional homogeneity of substratum of melting with crustal component of early Proterozoic.

5. Discussion of the results

The substantial composition of the described bimodal volcano-plutonic complexes depends on the changes of the mineral composition of the forming rocks: the decrease of the role of dark colored minerals and growth of the role of feldspar-quartz component is mentioned in the rocks; this might be an evidence of fractional crystallization. From trachybasaltes-basaltes to trachyrhyolites a considerable decrease in Sr and Ba content is also happening. The enrichment of residual melt is mentioned against the background of a sharp decrease of the elements. On the other side, irregular, often reduced, content of MgO (to 7.67 wt.%), Ni (to 63 ppm), Co (mostly 10-20 ppm) is denying the straight connection of it's melting out of mantle peridotites (Palme & O'Neill, 2003). This fact is confirmed by low correlations of Nb/Ta (6.1-16.0, an exception is the value for trachyandesitic basaltes of Galkinsky complex even 30); as for the rocks of mantel provenance the value is close to 17.5 (Green, 1995). For the formation of the possible mantle sources, that affected the formation of the rocks of examined complexes, the behavior of the pairs of incompatible elements was analyzed. The elements have got an ability to accumulate in residual melt independently from the magma content (Fig. 7). Pairs of the residual-elements by the differentiation (anatexis) of magmas, the product of different sources, are displayed on logarithmic diagrams of their content as straight lines, crossing the sources compositions, and accumulating in residual magmas in the process of their differentiation. Analisis of the correlation of the concentrations of residual-elements' pairs (Fig. 7) shows the proximity of the bimodal series' rocks by similar correlations of model source of the island ark with the participation of enriched sources like OIB, CC and source like MORB. One can't deny the collaboration of crystallizational, which affects the accumulation of the elements.

Figure 7f shows, that almost all the varieties of the bimodal complexes' rocks are plotted in the area of high ratios Th/Ta = 9.7-41.4 (the model value of the source LAB close to 9) with relatively constant value of correlation Nb/U (Fig. 7a). An exception is the data for some granitoids, where the concentration of SiO_2 is higher than 75%. The affection of the process of magmas differentiation on the value of correlation Th/Ta can be checked with an index of magmas differentiation, denominated by the content of niobium and silica. Fig. 5 shows that the concentration of niobium is accompanied by growth of SiO_2 content. This is corresponding with the degree of differentiation of magmas and decrease of the degree of the melt of the source. The values for the middle content of CC are considerably exceeding, as the growth of the content of silica is happening at the correlation of Th/Ta; so it can be considered, that the process is more connected with the differentiation of magmas that are forming the bimodal complexes, but with that we should not except the role of the mixture of such sources like MORB, IAB, CC.

We should notice the correlation Nb/U (Fig. 7a), which value basicly fits the size of field of undersubductional source of LAB. The increase of the content of Nb and SiO_2 is accompanied by an insignificant reduction of the value of Eu/Eu* (from 0.86-0.95 to 0.16-0.23).

This demonstrates that there was differentiation with the growth of silica acidity with the participation of fractioning of feldspar. This is pretty characteristic of subductional calk-alkali magmatic associations. It can be stated, that the formations of the bimodal complexes are a result of upper continental crust assimilation. As it was already mentioned (Fig.7j), for all the varieties of the bimodal complexes' rocks a straight correlation Nb and SiO_2 is stated. For the formations with the content of SiO_2 < 57 wt.% participation of the sources IAB - M-MORB - E-MORB is possible with the growth of the content of Nb from IAB to E-MORB. Therefore the growth of Nb occurs by relatively weak changing content of silica, which might confirm the affection of this source on the magma's structure, but not the affection of the magma's differentiations. By the growth of the content of SiO_2 (57-64 wt.%), where the straight correlation between Nb and SiO_2 is more apparent, the participation of the magmas' differentiation and assimilation of the continental crust and their magmas might be proposed. Further, with the growth of the content of SiO_2 (>64 wt.%) and Nb (to 47 ppm), when the straight relation between Nb and SiO_2 significantly moves away from the composition of CC, forming a hyperbola, the basic part in forming the acid rocks of the bimodal complexes, more likely, belongs to differentiation of magmas.

If we take a look at geochemical diagrams with the shown values of different characteristics of the sources and mantle processes according to data (Sun & McDonough, 1989; Hofman & Jochum, 1996; Niu & Batiza, 1997 et al.). The formations of the examined complexes are pretty distant from the deplicated source of mantle on those diagrams (correlations Th_n/Ba_n-Th_n/La_n; Sr_n/Ce_n-Nb_n/La_n; Nb/Th-Zr/Nb; La/Yb-Th/Ta etc.) and points of their content are concentrated close to the values of enriched mantle (EM). By the data of correlations Ti/Y (380-162 and 225-43); Lu/Hf (0.04-0.10 and 0.05-0.32); Sm_n/Yb_n – the basic part of the values fits the interval 3.5-5 and 2-4 in basic-medium and acid varieties correspondingly. A conclusion can be made there. A component of the bimodal complexes of medium-basic composition was formed at more significant depth and might be less enriched in the crustal material. It is confirmed by the analysis of the correlations of Rb/Sr: 0.01-0.32 for medium-basic rocks and 0.14-3.98 for acid types. It is known, that these correlations for the rocks of similar composition of continental crust and granite-metamorphic stratum are 0.02 and 0.32 (Taylor & McLennan, 1985).

It is stated in the diagram Ba/K - $^{87}Sr/^{86}Sr$ (Fig. 8) that the inclusion in the magmatic process of the continental sedimentary materials starts already with the formation of the rocks of middle-basic content. It is present, although the figurative dots are significantly removed aside of trend of contamination, at the formation of acid varieties. All this allows us to assume, that by the formation of acid varieties the process of differentiation of magmatic materials has had an important meaning. And figurative dots of mostly basic volcanites (Fig. 8) are removed to the trend of an enriched source.

The demonstration of the rock's characteristics on the diagrams, helping to identify the conditions of melting in mantle source (Brandshaw et al., 1993; Gill, 1981 et al.), showed that initial magmas formed by the melt of granet-spinel lherzolite (degree of the melt is in limits 1- 5%), containing not more than 4% of granet. Geochemical characteristics also allow us to analyze the role of the processes of partly melting of mantle substratum and fractional

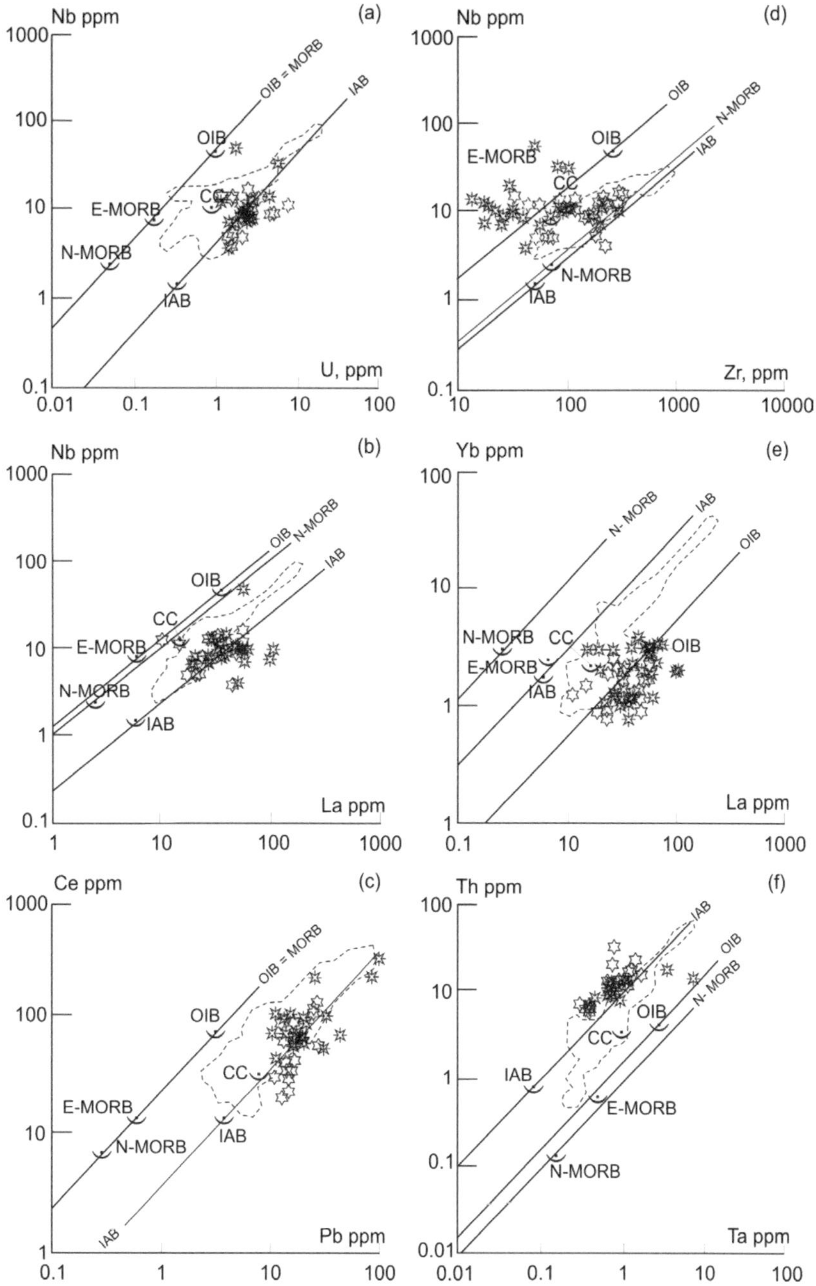

Fig. 7. The variations of the concentrations of incompatible microelements in the rocks of volcano-plutonic bimodal complexes of the framing of Mongol-Okhotsky orogenic belt. The model compositions of basalts of middle-oceanic mountain ridges normal (N-MORB) and enriched (E-MORB), of ocean islands (OIB) and island arcs (IAB), continental crust (CC). The lines of even correlations for the appropriate sources and correlations for the Paleozoic bimodal associations of Western link of Mongol-Okhotsk orogenic belt (dotted line) from the work of (Kovalenko et al., 2010). The conventional signs of the rocks are on Fig. 2.

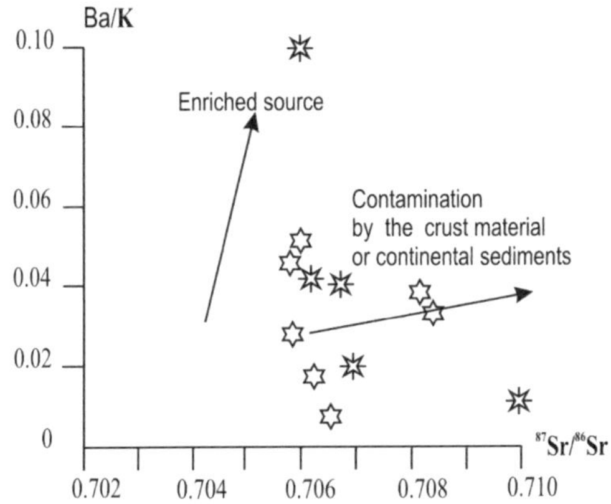

Fig. 8. The location of the content of the rocks of bimodal volcano-plutonic complexes is on the diagram $^{87}Sr/^{86}Sr$ - Ba/K (Pouclet, 1995). The conventional signs of the rocks are on Fig. 2.

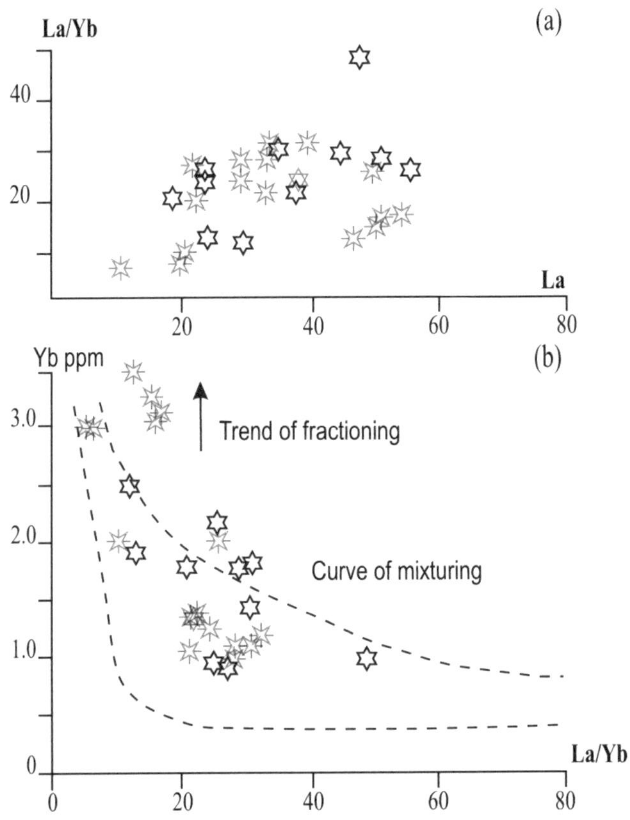

Fig. 9. The location of the compositions of the rocks of basic - middle content of bimodal complexes on the diagrams: a) La - La /Yb, б) La/Yb - Yb. Trend of fractioning and lines of mixing by (Shaw et al., 2003). The conventional signs of the rocks are on Fig. 2.

crystallization of primary magmas. It is stated that by the fractioning of abyssal magmas the content of La significantly varies, and the correlation of La/Yb is changing relatively weak.

This is reflecting the location of figurative points of volcanites of basic-medium content of bimodal complexes on the diagram of correlation of the elements (Fig. 9a).
On the correlational diagram of La/Yb - Yb (Fig. 9b) they are located in the area of trend of melting mixture, that were formed by the partial melting both sphinel and granet - lherzolites, that contain mica (Shaw et al., 2003). But a part of figurative points and Galkinsky and Bomnaksky complexes are strongly attached to a trend of fractioning. About the participation of granet and spinel peridotites in the formation of the rocks of the examined bimodal complexes says the dependence of normalized to chondrite (by Sun & McDonough, 1989) correlation of (Tb/Yb)n from the value of correlation K/Nb (Wang et al., 2002). For Galkinsky complex the value of (Tb/Yb)n = 2.08-2.66, and for Bomnaksky - (Tb/Yb)n = 1.66-1.78; 1.8-2.88. It can be proposed, that the formation of the rocks of Galkinsky and complex occurrend in the zone of stability of granet. The formations of Bomnaksky complex were formed both in a zone of stability of granet and in the less depths, where the melts are equilibrium to spinel pallial protolites. Probably, this is an explanation of the differences of geochemical characteristics of the examined bimodal complexes.
Isotope data of volcanites and plutonic formations coordinate with isotope characteristics of containing than blocks of the earth's crust. Thus in the Southern frames they lay over the Northern edge of Amur continent, in it's composition the formations of Riphean folding with corresponding crust sources: T_{Nd}(DM-2st) = 975-1314 Ma. And in the Northern frames the rocks of the bimodal complex are overlying the formations of Stanovoy terrane (Southern framing of Siberian platform). Early Proterozoic formations are widely developed there. The characteristics of the crust source of the bimodal complex (T_{Nd}(DM-2st) = 1901-1937 Ma) correlate to the time period.
According to the primary isotope compositions Sr and Nd in the formation of the rocks of the examined complex, the affection of the mantle can not be expected. The mantle is enriched in radiogenic ^{87}Sr - EM-II (Fig. 10) for the Southern framing and EM-I - for the Northern framing of Mongol-Okhotsk belt. More likely, the differences of the composition of the sources are connected with the differences of the composition of the foundation, in which frames of the complexes were formed.
To take a good look at the possible affection of the plume on the magmatic late Mesozoic process in the region we use a calculation (Fitton et al., 1997), this determines the probability of the presence of the plume source: $\Delta Nb = 1.74+\log(Nb/Y) - 1.92*\log(Zr/Y)$. According to the shown formula, for some basic - medium varieties of the complexes positive values are typical - ΔNb (+1.548 до +0.049) or very close to them ΔNb = -0.013, this might testify in favor of the proposal about primary formation of the rocks under the affection of the plume source. Moreover the identifying factors for the basic rocks of the mantle plumes are their moderate enrichment (Fig. 6a); values of the relations (La/Sm)n>1.8 и (Ce/Yb)n>7 (Schilling et al., 1983; Le Roe et al., 1983). Except singular values, these correlations for medium - basic rocks are (La/Sm)n = 2.9-5.8; (Ce/Yb)n = 8.0-27.0. The correlation Nb/Y - Zr/Y (Fig. 10) is also about the affection of the plume source.
The fields of the model composition of basalts: oceanic islands (OIB), medium-oceanic ridges normal (N-MORB), island arc (ARC), oceanic plateau (OPB). Magmatic sources: primitive mantle (PM), abyssal depleted mantle (DEP), low abyssal depleted mantel (DE), recycled component (REC), enriched component (EN), upper crust (UC), enriched mantle with high Rb/Sr (EMII), enriched mantle with high Nd/Sm (EMI), enriched mantle with high U/PB. Arrow signs reflect the effect of the volumetric melting (F) and subductional enrichment (SUB).

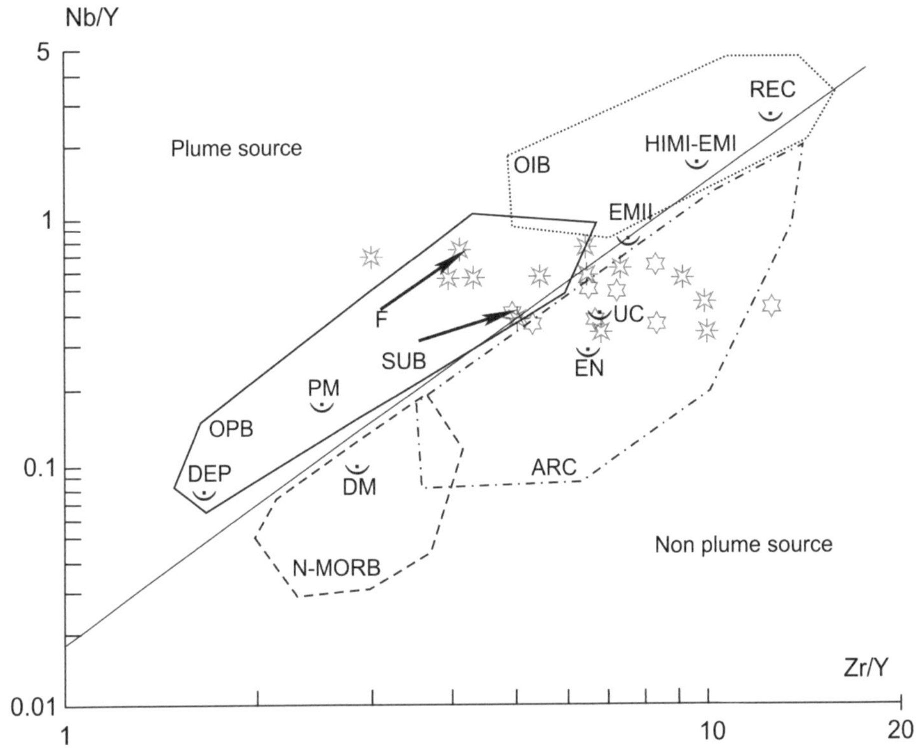

Fig. 10. The location of the rocks of the basic - medium composition of bimodal volcano-plutonic complexes on the diagram in the coordinates of pair correlations of Zr/Y - Nb/Y in ratio to typical basalts and magmatic sources by (Condi, 2005). The conventional signs of the rocks are on Fig. 2.

Late Mesozoic - Cainozoic concentric-zonal structure - Aldan-Zeisky plume was separated by the data of geophysical research, spatial parameters and peculiarities of the spreading of the viscous subcrustal layer (Petrischevsky & Khanchuk, 2008). The authors consider that the central part of the plume was inverted (droped) and continues to cave not long time ago (early Cretaceous). Location of the central part of Aldan-Zeysky plume practically is a contour of the location of the bimodal complexes in the frames of Eastern link of Mongol-Okhotsk orogenic belt (Fig. 1).

It was already mentioned about the resemblance of a few geochemical characteristics of the rocks of the researched complexes with such characteristics of the formations of the late Paleozoic - Mesozoic Central-Asian rifting systems. This fact illustrates Fig. 7, where the figurative points of the rocks of the bimodal complexes of the Eastern flank of Mongol-Okhotsky orogenic belt or partially cover them.

On the diagram I_0^{Sr} - $\varepsilon Nd_{(T)}$ (Fig. 11) the figurative points of Galkinsky complex get into the frames of the fields of the compositions of Early - Late Mesozoic intraplate magmatic formations of Central-Asia are translocated to the area EM-II. The figurative points of Bomnaksky complex (Northern framing) are translocated to the area of an enriched source EM-I.

To this differences relay: the presence of normal alkalinity in the composition of the examined complexes of the volcanites. Also, there are relatively low concentrations of the highly charged elements: Nb, Ta, Zr, Hf.

Such characteristics make them closer to their products of subductional origin. Pretty low statistics of isotope data of the foundation rocks in the area of the development of magmatites of the bimodal complexes do not allow us to give a unique explanation to these facts. We can only make a proposal that the crust material coming into the magmas chambers was enriched with subductional component. That leaded to a partly mixture and the affection of the inherited characteristics of the foundation rocks on the formation of the bimodal complex.

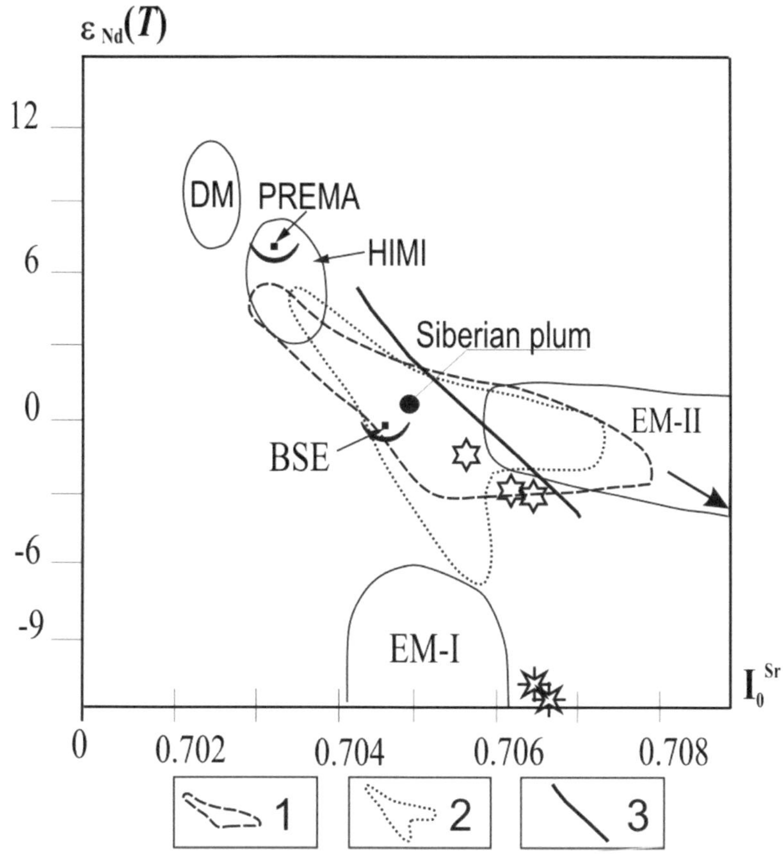

Fig. 11. Correlation of the composition of the rocks of the bimodal volcano-plutonic complexes of Eastern frames of Mongol-Okhotsk belt with the fields of intraplate magmatic rocks of Central Asia and typical sources of mantle by primary isotope compositions of Sr and Nd. The fields of rocks: 1 - Early Mesozoic and 2 - Late Paleozoic of Central Asia. 3 - trend of correlation of the composition of the basalts of the bimodal series of Early Mesozoic Mongol-Transbaikalian region by (Yarmolyuk et al., 2002). The data from the work (Types of magma, 2006) and data PREMA and BSE from (Zindler & Hart, 1986) were used to make a diagram. The conventional signs of the rocks are on Fig. 2.

But Late Mesozoic magmatism in the frames of the Eastern link of Mongol-Okhotsky belt has a row of differences from more ancient analogical magmatism of Central-Asian riftogenesis.

6. Geodynamical reconstructions

In the Mesozoic era, the examined part of the region was enveloped by the collisional processes that were caused by the convergense of North-Asian and Sino- Korean cratons and by the closure of the Eastern flank of the Mongol-Okhotsk basin (Sonenschein et al.; 1990; Parfenov et al., 1999). We should remind that the Late Mesozoic formations of bimodal series have a linear spreading along the Southern and Northern borders of Mongol-Okhotsk belt. Along the Southern border on the East their spreading is framed by the structures of Bureja-Jiamusy superterrane. Along the Northern frame - in the same direction - they are changing into younger calk-alkali formations of Okhotsk-Chukotsky orogenic belt (Fig. 1). Similar formations are not revealed on the South of Mongol-Okhotsk orogenic belt. But the bimodal complexes become widely developed on the West, in frames of Western branch of the examined belt. The bimodal complexes are one of the components of the Early Mesozoic North-Mongolian - East-Transbaikalian rifting zone. A pretty grounded geodynamical model was worked out for the zone (Yarmolyuk & Kovalenko, 2000; Yarmolyuk et al., 2002; Kovalenko et al., 2003; Vorontsov et al., 2004, Vorontsov et al., 2007). The basic point of the model is in simultaneously existing conditions of the pressure of the plume on the area that is under conditions of the collisional pressure.

On the discrimination diagram of the primary isotope characteristics of strontium and neodymium (Fig. 11) the points of the rocks of Galkinsky volcano-plutonic Late Mesozoic complex are close or match with the fields of early and Late Mesozoic intra plate magmatic formations of Central Asia. The points are superposed with the lower part of trend of correlation of basalts of Mongol-Okhotsk area's bimodal series. These correlations for the basic rocks are also brought close to the points that identify the location of the similar characteristics for Siberian plume and BSE.

The figurative points of the formations of the bimodal complexes of acid composition on the diagrams of the tectonic situations are concentrated in the field of the collisional conditions of the formation (in singular cases - intra platform) (Fig. 12a) or on the border of collisional-intra platform conditions of the formation (Fig 12b). A field of basalts of the island arc with a removal and partly location in the field of basalts of the continental rifts and traps is defined for the rocks of the basic-middle content (Fig. 12c).

The obtained data show that the rocks of the bimodal volcano-plutonic complexes that were formed in the frames of the Eastern link of Mongol-Okhotsk orogenic belt correlate with the intro continental formations in Central Asia by series of geochemical characteristics and isotope data. It was stated, that participation of a singular mantle source is possible in the formation of the formation of all late Mesozoic - early Cretaceous volcanites of Central Asia (Yarmoluk & Kovalenko, 2000; Kozlovsky et al., 2006). It is characteristic peculiarity is the high values of the correlation Zr/Hf (38-50). For the rocks of the researched complexes the value Zr/Hf correlates to 34-52 - Southern frame of the belt, 25-66 - Northern frame. This allows us to use the well-known geodynamical model, by the examination of the tectonic script of the formation of Galkinsky and Bomnaksky volcano-plutonic complexes.

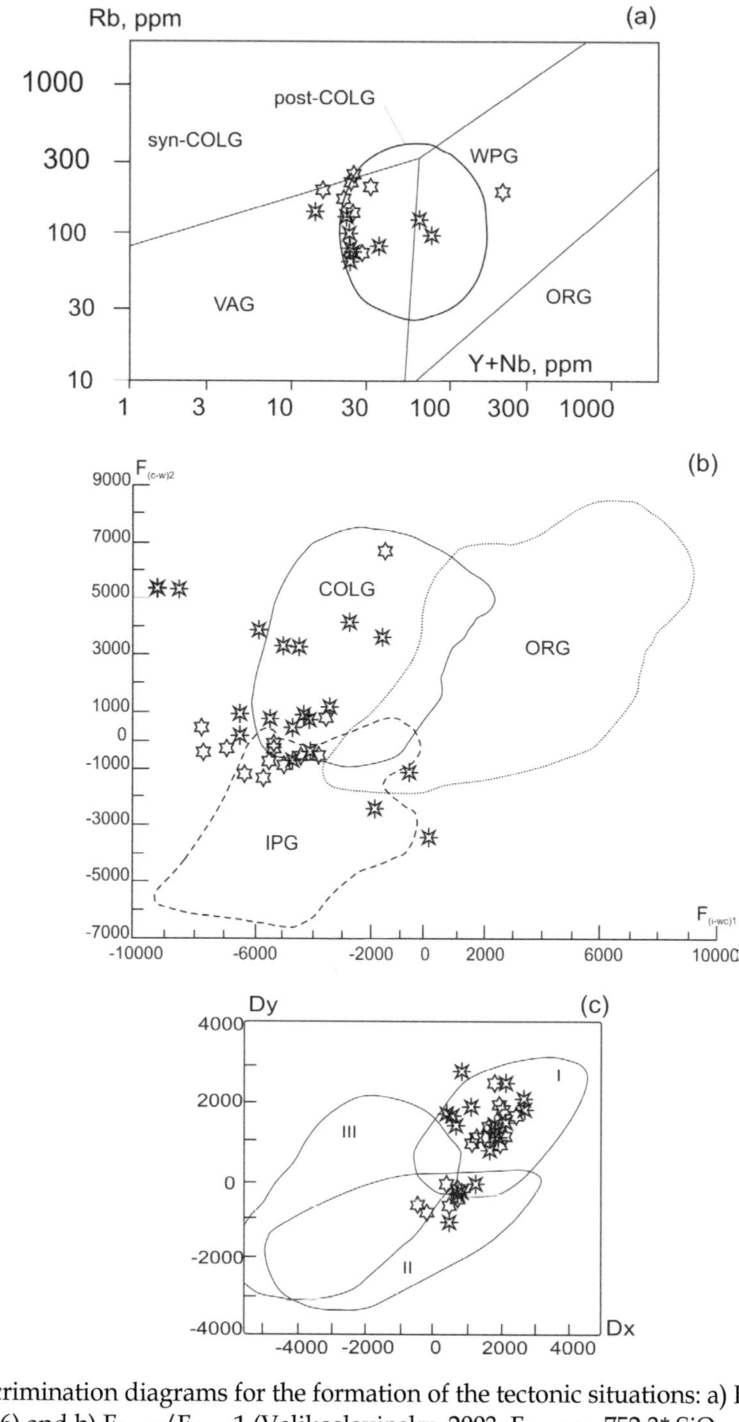

Fig. 12. Discrimination diagrams for the formation of the tectonic situations: a) Rb-Y+Nb (Pearce, 1996) and b) $F_{(c-w)2}/F_{(i-wc)}1$ (Velikoslavinsky, 2003; $F_{(c-w)2}$ = -752.3* SiO_2 – 6537.06* TiO_2 – 25.6* Al_2O_3 - 928.96* $Fe_2O_3^*$ +1928.07* MgO - 464.21* CaO - 1808.19* Na_2O - 272.16* K_2O + 8675.33* P_2O_5 + 71073.5; $F_{(i-wc)1}$ = 2432.42* SiO_2 + 7900.33* TiO_2 + 2512.12* Al_2O_3 +

1380.23* FeOt + 2616.55* MgO + 3480.51* CaO + 3045.39* Na_2O + 645.91* K_2O - 241285.5) for the salic formations; c) D_x/D_y (Velikoslavinsky & Glebovitsky, 2005; D_x = (176.94* SiO_2) - (1217.77* TiO_2) + (154.51* Al_2O_3) - (63.1* FeOt) - (15.69* MgO) + (372.43* CaO) + (104.41* Na_2O) - (19.96* K_2O) - (873.69* P_2O_5) - 11721.488; D_y = (94.39* SiO_2) - (103.3* TiO_2) + (417.98* Al_2O_3) - (55.63* FeOt) + (57.61* MgO) + (118.42* CaO) + (502.02* Na_2O) + (6.37* K_2O) + (415.31* P_2O_5) - 13724.66) for the basinrocks. The fields of the basalts: I – island arcs, II – traps, III – continental rifts. The conventional signs of the rocks are on the Fig. 2.

7. Conclusion

The bimodal volcano-plutonic complexes of subalkali – normal petrochemical series were formed along the Southern and Northern borders of the eastern flank of Mongol-Okhotsk orogenic belt in the interval 119 – 97 Ma. The duration of the formation of similar complexes (to 35 Ma) is marked for the intercontinental formations of Central Asia (Yarmolyuk et al., 2000; Yarmolyuk et al., 2002; Kovalenko et al., 2003; Voroncov et al, 2007).

Both shpinel and granet peridotites were a protolite for the basaltoids of the complexes, and geochemical characteristics of basaltoids are minimum Ta and Nb with maximum K and Pb, point at their formation in the situation of convergent borders of the plates. The enrichment of the incompatible elements is characteristic for all the varieties of the rocks of the bimodal complexes, with the decreased content of Ta, Nb, Ti and more high presence of Ba, Sr, K и Pb. As A.A. Vorontsov and coauthors notice (Voronysov et al., 2007) these peculiarities are characteristic for all late Paleozoic – early Mesozoic intra plate rocks of Central – Asian riftous system, its development is connected with the overlapping of the continental lithosphere of the plume source.

Thus, it can be stated: the formation of the bimodal complexes in the frames of the Eastern link of Mongol-Okhotsk orogenic belt accompanied the collision of North-Asian and North-Korean continents at the affection of the plume source. The result of this process was a final closure of Mongol-Okhotsk basin a the end of Early Cretaceous: 119-97 Ma.

8. References

Antonov, A. Yu. (2007). Geochemistry and petrology of Mesozoic – Cainozoic magmatic formations of Southern framing of Aldansky shield. The problems of geodynamics, *Pacific geology,* Vol.26, No.2, (March - April 2007), pp. 56-81, ISSN 1819-7140.

Antonov, A. Yu., Dril, S.I., & Bankovskaya, E.V. (2001). Rb-Sr isotope characteristic of allochthonous and autochthonous Late Mesozoic granitoids of Stanovoy ridge (Southern limitation of Aldansky shield), *Pacific geology,* Vol.20, No.4, (July - August 2001), pp. 61-75, ISSN-1819-7140.

Bogaticov O.A (1983). *Magmatic rocks.* Moskva: Nauka, 367 p.

Bogatikov, O.A. & Kovalenko V.I. (2006). *Types of magma and their sources in the history of the Earth,* Moskva: Institute of Geology of ore deposits Russian Academy of Science, ISBN 5-88918-013-4, 280 p.

Brandshaw, T.K., Hawkesworth, C.J. & Gallagher, K. (1993). Basaltic volcanism in the Southern Basin and Range: no role for a mantle plume, *Earth and Planetary Science Letter,* Vol.116, (1993), pp45-62, ISSN 0012-821X.

Chappel, B.W. & White, A.I.R. (1992). I-and S-type granites in the Lachlan Fold Belt. *Transactions of the Royal Society of Edinburgh: Earth Science,* Vol.83, (1992), pp. 1-26.

Chzhan Khun, Chzhao Chunczin, Yao Chzhen' & Cuan' Khen' (2000). Dynamic bases of Mesozoic volcanism in the Northern part of Big Khindan (China), *Pacific geology,* Vol.19, No.1, (January - February 2000), pp. 109-117, ISSN-1819-7140.

Condie K.S. (November 2004). Higt field strength element ratios in Archean basalts: a window to evolving sources of mantle plumes? In: *Litos,* (2005), Available from http: // www.ees.nmt. edu/ condie/ pubs/ Condie Archean_plumes.pdf.

Derbeko, I.M. (1998). On the issue of the separation of North-Eastern flank of Selengino-Vitimsky volcano-plutonc belt at the territory of Amur region, *Problems of genesis of magmatic and metamorphic rocks,* p. 92, Saint Petersburg, Russia, May 25-27, 1998.

Derbeko, I.M., Sorokin, A.A. & Agafonenko, S.G. (2008a). Geochemical peculiarities of acid magmatism of North-Western flank of Khingan-Okhotsk volcano-plutonic belt (Esopic and Yam-Alinskaya zones), *Pacific geology.* No.1, (January - February 2008), pp. 61-71, ISSN-1819-7140

Derbeko, I.M., Sorokin, A.A., Salnikova, E.B., Kotov, A.B., Sorokin, A.P., Yakovleva, S.Z., Fedoseenko, A.M. & Plotkina U.V. (2008b). The age of acid volcanism of Selitkanskaya zone of Khingan-Okhotsky volcano-plutonic belt (Far East), *Doklady Akademii Nauk,* Vol.418, No.2, (January 2008), pp. 221-225, ISSN 0869-5652.

Derbeko, I. M. (2004). Early cretacous intrusive and volcano-plutonic complecxes of the North-Greater Khingan zone (Amur region, Russia) and their role in AU-AG Mineralization, *Metallogeny of the Pacific Northwest: Tectonics, Magmatism and Metallogeny of active continental margins,* pp. 93-96, ISBN 5-8044-0470-9, Vladivostok, Russia, September, 10-20, 2004.

Derbeko, I.M., Vyunov, D.L., Kozyrev S.K. & Ponomarchuk, V.A. (2009) Conditions for the formation of a bimodal volcano-plutonic complex, within the southern margin on the eastern flank of the Mongol-Okhotsk orogenic belt, *International Symposium Large igneous provinces of Asia, mantle plumes and metallogen,* pp. 73-75, ISBN 978-5-94301-089-7, Novosibirsk, Russia, August 6-9, 2009.

Fitton, J.G., Saunders, A.D. Norry M.J., Hardarson,. B.S. & Taylor, R.N. (1997). Termal and chemical structure of the Iceland plume, *Earth and Planetary Science Letters,* Vol.153, pp. 197-208, ISSN- 0012-821X.

Frost, B.R., Barnes, C.G., Collins, W.J., Arculus, RJ, Ellis, DJ. & Frost, C.D. (2001). A geochemical classification for granitic rocks, *Petrology,* Vol.42, pp. 2033-2048, ISSN-0022-3530.

Geologycal map of Priamurie and neighbouring territories. Scale 1:2 500 000 (1999). *Explanatory note,* Saint Petersburg - Blagoveschensk - Harbin: Ministry of nature resources of Russian Federation, Ministry of geology and mineral resources of China, 135 p.

Gill, J.B. (1981). *Orogenic andesites and plate tectonic,* New York, ISBN 3540106669, 390 p.

Gordienko, I.V. (1987). *Paleozoic magmatism and geodynamics of Central-Asian folded belt.* Moskva: Nauka, 237 p.

Green, T.H. (1995) Significance of Nb/Ta as an indicator of geochemical processes in crust-mantle system, *Chemical Geology,* Vol.120, (1995), pp. 347-359, ISSN 0009-2541.

Gusiev, G S. & Khain, V E. (1995). About the correlations of Baikal-Vitimsky, Aldan-Stanovoy and Mongol-Okhotsky terrains (South of middle Siberea), *Geotectonic,* No.5, (September - October 1995), pp. 68–82, ISSN 0016-853X.

Hoffman, A.W. & Jochum, K.P. (1996). Source characteristics derived from very incom patible trace elements in Mauna Loa and Mauna Kea basalts, Hawaii Scientifi c Drilling Project, *Journal of Geophysical Research,* Vol.101, (1996), pp. 11831–11839, ISSN 0148-0227.

Khanchuk A.I. (2006). *Geodynamics, magmatism and metallogeny of the Russia East,* Vladivostok: Dalnauka, ISBN 5-8044-0634-5, 572 p.

Kovalenko, V.I., Yarmoluk, V.V. & Salnikova, E.B. (2003) Sources of the magmatic rocks and formation of early Mesozoic tecktono-magmatic areal of Mongol-Transbaikalian magmatic area: geological characteristic and isotope geochronology, *Petrology,* Vol.11, No.2, (Marct - April 2003), pp. 164-178, ISSN-0869-5911

Kovalenko, V.I., Kozlovsky, A.M. & Yarmoluk, V.V. (2010). Comendite containing subductional volcanic associations of Han-Bogdinsky region, Southern Mongolia: results of the geochemical research, *Petrologiya,* Vol.18, No.6, (November - Decemberl 2010), pp. 351-380, ISSN-0869-5911.

Kozak, B.P. *State geological map of RF scale 1:200 000.* Edition 2. Series Stanovaya. Sheet N-51-XYI (Urusha). Explanatory report. Saint Petersburg, 195 p.

Kozlovsky, A.M., Yarmoluk, V.V., Salnikova, E.B., Savatenkov, B.M. & Kovalenko, V.I. (2005). The age of bimodal and alkali – granites magmatism Goby-Tianshan' riftous zone, mountain ridge Test, Southern Mongolia, *Petrology.* Vol.13, No.2, (March – April 2005), pp. 218-224, ISSN-0869-5911.

Kozirev, S.K., (2000a). On the issue of the age and the sequence of the formation of Low Cretaceous volcanic complexes of Gonzhinsky ledge and it's framing (Amure region), *Conference on Correlation of Mesozoic continental formations of Far East and the Eastern part of Transbaikalian,* pp. 59-64, Chita, Russia, Oktober 23-27, 2000.

Kozirev, S.K. (2000b). On the issue of the substantial composition, age and geology-structural position of Galkinsky volcanic complex and about the possible presence of volcanic complexes of late Jurassic age on the territory of Gonzhinsky ledge and it's framing (Amur region), *Conference on Correlation of Mesozoic continental formations of Far East and Transbaicalian.* pp. 64-68, Chita, Russia, October 23-27, 2000.

Le Bas, M., Le Maitre, R.W., Streckeisen A. & Zanettin, B. (1986). A chemical classification of volcanic rocks based on the total-silica diagram, *Journal of Petrology,* Vol.27, (March 1986), pp. 745-750, ISSN-0022-3530.

Le Roex, A. P., Dick, H.J., Reid, A.M. & Erlank, A.J. (1982). Ferrobasalts from the Spiess Ridge segment of the Southwest Indian Ridge, *Earth Planetary Science Letters.,* Vol.60 (October 1982), pp. 437-451, ISSN 0012-821X.

Maeda, J. (1990). Opening of the Kuril Basin deduced from the magmatic history of Central Hokkaido, North Japan, *Tectonophys,* Vol.174, No.3-4, (March 1990), pp. 235-255, ISSN 0040-1951.

Martinuk, M.V., Riamov, S.A. & Kondratieva, V.A. (1990). *Explanatory report to the scheme of dismemberment and correlation of magmatic complexes of Khabarovsky region and Amur region.* Khabarovsk, Russia: Industrial geological organization, p.215.

Neimark, L.A. Larin, M.A. & Ovchinnikova, G.Vl. (1996). U-Pb geochronologic and Pb-isotopic evidence for the Mesozoic mineralization stage of the Archean Stanovoi Megablock, Aldan-Stanovoi Shield, *Petrology,* Vol.4, No4, (Iuly – August 1996), pp. 421-435, ISSN-0869-5911.

Niu, Y. & Batiza, R. (1997). Trace element evidence from seamounts for recycled oceanic crust in the Eastern Pacific mantle, *Earth and Planetary Science Letters,* Vol.48, (May 1997), pp. 471–483, ISSN 0012-821X.

Palme, H. & O'Neill, H.St.C. (2003). Cosmochemical estimates of mantle composition, *Treatise on geochemistry.* Elsevier Ltd., Vol.2, pp. 1-38, ISBN 0-08-043751-6

Pearce, J. (1996). Sources and settings of granitic rocks, *Episodes,* Vol.24, (1996), pp. 956-983, ISSN 0705-3797.

Pouclet, A., Lee, J.-S., Vidal, P., Cousens, B. & Bellon, H. (1995). Cretaceous to Ce.zoic volcanism in South Korea and the Sea of Japan: magmatic constrains on the opening of the back-are basin, Volcanism Associated with Extension at Consuming Plate Margins, *Geological Society Special Publication,* No.81, (1985), pp. 169-191.

Parfenov, L.M., Berezin, N.A., Khanchuk, A. I., Badarch, G., Belichenko, V.G., Bulgatov, A.N., Drill, S.I., Kirillova, G.L., Kuzmin, M.I., Nokleberg, U., Prokopiev, A.V., Timofeev, V.F., Tomurtogoo, O. & Yan', H. (2003). The model of the formation of the orogenic belts of Central and Northern-Eastern Asia, *Pacific geology,* Vol.22, No.6, (November - December 2003), pp. 7-41, ISSN-1819-7140.

Parfenov, L.M., Popeko, L.I. & Tomurtogoo, O. (1999). The problems of tectonics of Mongol-Okhotsk orogenic belt, *Pacific geology,* Vol.18, No.5, (September - October 1999), pp. 24-43, ISSN-1819-7140.

Petrischevsky, A. M. & Khanchuk, A.I. (2006) Kainozoic plume in Upper Priamurie, *Doklady Akademii Nauk, Geohpysics,* Vol.406, No.3, (March 2006), pp. 384-387, ISSN 0869-5652.

Schilling, J.-G., Zajac, M., Evans, R., Johnston, T., White, W., Devine, J. D. & Kingsley, R. (1983). Petrologic and geochemical variations along the Mid-Atlantic Ridge from 29° N to 73°, *Nort Amerrican Jyrnal. Sciences,* Vol.283. No.6, (1983), pp. 510-586.

Sharma, M., Basu, A.R. & Nesterenko, G.V. (1992). Temporal, Sr-, Nd- and Pb-isotopic variations in the Sibirian flood basalts: implications for the plume-source characteristics, *Earth and Planetary Science Letters,* Vol.113, (1992), pp. 365-381, ISSN 0012-821X.

Shaw, J.E., Baker, J.A., Menzies, M.A., Thirlwall, M.F. & Ibrahim, K.M. (2003). Petrogenesis of the largest intraplate bolcanic field on the Arabian Plate (Jordan): A mixed lithosphere-astenosphere source activated by litosperic extension, *Petrology,* Vol.44, No.9, (September 2003), pp. 1647-1679. ISSN 0022-3530.

Sonenshein, L.P., Kuzmin, M.N. & Natapov, L.M. (1990). *Tectonics of lithosphere platforms on the territory of USSR.* Vol.1. Moskva: Nedra, 328 p.

Sorokin, A.A. & Ponomarchuk, V.A. (2002) Umlekan-Ogodzha Early Cretaceous magmatic belt (North margin of the Amurian superterrane): duration of magmatism, *Geochim. Cosmochim. Acta.* Vol.66, No.15A: A728, ISSN 0016-7037.

Sorokin, A.A., Ponomarchuk, V.A., Sorokin, A.P. & Kozirev, S.K. Geochronology and correlation of Mesozoic magmatic formations of the Northern edge of Amur superterrain, *Stratigraphy. Geological correlation,* Vol.12, No.6, (November - December 2004), pp.36-52, ISSN 0869-5938.

Sorokin, A.A., Kotov, A.B., Kovach, V.P., Soropkin, A.P., Ponomarchuk, V.A., Derbeko, I.M. & Melnikova, O.V. (2006) Isotope-geochemical peculiarities and the formation of Mesozoic volcano-plutonic complexes of Upper Priamurie. *Materials of the III-ed all*

Russian conference on isotope geochronology, Vol.2, pp. 305-307, Moscow, Russia, June 6-8, 2006.

Strikha, V.E. (2006). Late Mesozoic collisional granitoids of Upper Priamurie: new geochemical data, *Geochemistry,* No.8, (August 2006), pp. 855-872, ISSN 0016-7525.

Strikha, V.E. & Rodionov, N.I. (2006). Cretaceous granet - leykogranet association of Stanovoy terrain: new geochronological, geochemical and isotope-geochemical data, *Doklady Akademii Nauk,* Vol.406, No.3, (Januar 2006), pp. 375-379.

Sun, S.S. & McDonough, W.F. (1989). Chemical and isotopic systematics of oceanic basalts: implications for mantle composition and processes, In: *Magmatism in the ocean basins.* Geological Society, London, pp. 313-345.

Taylor, S.R. & McLennan, S. M. (1985). *The continental crust: its composition and evolution.* Blackwell, Oxford, 379 p.

Velicoslavinsky, S.D. (2003). Geochemical typification of acid magmatic rocks of leading geodynamical situations, *Petrology,* Vol.26, No.2, (March - April 2003), pp. 363-380, ISSN-0869-5911.

Velicoslavinsky, S D. & Glebovicky, V.A. (2005). New discriminational diagram for classification of the island arc and continental basalts on the base of petrochemical data, *Doklady Akademii Nauk,* Vol.401, No.2, (March 2005), pp. 213-216, ISSN 0869-5652.

Voroncov, A.A., Yarmoliuk, V.V., Likhin, D.A, Dril, S.I, Tatarnikov, S.A. & Sandimirova, G. P. (2007). Sources of magmatism and geodynamics of the formation of Early Mesozoic North - Mongolian - East-Transbaikalian riftous zones, *Petrology,* Vol.15, No.1, (January - February 2007), pp. 37-60, ISSN-0869-5911.

Wang, K., Plank, T., Walker, J.D. & Smith, E.I. (2002) A mantle melting profile the Basin and Range, SW USA, Journal of Geophysical Research, Vol.107, No.B1. ISSN 0148-0227.

Yarmoluk, V.V. & Kovalenko, V.I. (1991). *Riftogenous magmatism of the active continental edges and it's ore content.* Moskva, Russia: Nauka, 263 p.

Yarmoluk, V.V. & Kovalenko, V.I. (2000). Geochemical and isotope parameters of abnormal mantel of North Asies in late Mesozoic (by data of the examination of intraplatal basalt magmatism), *Doklady Akademii Nauk,* Vol.375, No.4, (December 2000), pp. 525-530, ISSN 0869-5652.

Yarmoluk, V.V., Kovalenko. V.I. & Kuzmin, M.I. (2000) North-Asian superplume in Phanerozoic: magmatism and depth geodynamic, *Geotectonics,* No.5, (September - November 2000), pp. 3-29, ISSN 0016-8521.

Yarmoluk, V.V., Kovalenko, B.I., Salnikova, E.B., Budnikov, S.V., Kovach, V.P., Kotov, A.B. & Ponomarchuk, VA. (2002). Tectono-magmatic zonzlity, sources of magmatic rocks and geodynamics of early Mesozoic Mongol-Transbaikalian area, *Geotectonics,* No.4, (July - Avgust 2002), pp. 42-63, ISSN 0016-8521.

Zindler, A. & Haris, S. (1986). Chemical geodynamics, In: *Annual review of earth and planetary sciences.* Vol.14 (A87-13190 03-46). Palo Alto, CA, Annual Reviews, Inc., (1986), pp. 493-571.

6

Hotspot Concept: The French Polynesia Complexity

Claudia Adam
CGE/Univ. Evora
Portugal

1. Introduction

At the surface of the Earth, volcanism is found in several tectonic contexts. It is largely concentrated at the plate margins: at divergent plate boundaries, mid-oceanic ridges, where new tectonic plates are created, and at convergent margins, subduction zones, where the lithospheric plates dive into the mantle. In the interior of the oceanic plates, we find however linear volcanic chains, composed of several volcanoes aligned along the direction of the plate motion. Their origin has been attributed to the drifting of the lithospheric plates over a fixed, hot mantle upwelling, deeply rooted in the mantle. Since several years, this concept is debated and the existence of the plumes themselves is questioned.

Here we focus on French Polynesia, a region characterized by a great concentration of volcanism and situated on the South Pacific Superswell, a wide area associated with numerous geophysical anomalies, including anomalously shallow seafloor considering its age, a dip in the geoid, and a mantle characterized by slow seismic velocities. 14% of the active volcanism is concentrated in an area covering less than 5% of the globe. A wide range of volcanic features should be noted: en echelon ridges, isolated seamounts and chains of midplate volcanoes. The characteristics of these chains often depart from the classical definition of hotspots. In particular, the broad depth anomalies surrounding the chains, called swells, display peculiar morphologies. These characteristics are however well recovered by a numerical model based on highly resolved seismic tomography model, describing the first 240 km of the upper mantle. This demonstrates that a direct link exists between the surface observations and mantle flows. However, even if the dynamics of the shallowest part of the mantle is sufficient to explain the surface observations, the existence of the secondary plumes at the origin of the hotspot chains, cannot be accounted for without involving a deeper component: the South Pacific superplume. This latter displays a complex signature in tomography models where it appears as broad low velocity anomalies throughout the lower mantle up to 1000 km, depth at which they split into narrower and more localized anomalies, a few hundred kilometers in diameter. Two of these narrow upwelling are associated with hotspots - the Society and Macdonald ones -, whereas the upwellings at the origin of the other chains seem to be restricted to the upper mantle. The pattern pointed out by the tomography is well retried by analogical experiments where two layers of miscible fluids are superimposed in a tank heated from below and cooled from above. In some conditions, long-lived thermochemical domes that oscillates vertically are produced. Experimentally, secondary plumes are observed at the top of the rising domes.

2. Hotspot concept

Hotspot chains are chains of midplate volcanoes, surrounded by wide shallow regions, called swells. The chains are composed of several volcanoes, aligned along the direction of the tectonic plate motion. The age of volcanism increases linearly along the chain, and active volcanism is often found at one extremity of the chain. One of the most classical example of such chains is Hawaii. This pattern, partially observed hundreds of years ago, has fascinated people and several explanations for its origin have been proposed.

2.1 First hypothesis on the origin of linear volcanic chains: ancient legends

The notion of volcanism migration is even older than the plate tectonics theory. According to Hawaiian legends, the fuming Pele goddess get angry with her sister after a terrible quarrel and went south-east, building in her way Diamond head on the Oahu island, Haleakala on the Maui island and the Kilauea on Hawaii, where she is living now - the actual active extremity of the chain. Another explanation, points out to the Namazu giant carp, leaving beneath Japan, which would be responsible for the Mount Fuji eruption when it shifts position. An extrapolation of this Japanese legend, (Holden and Vogt, 1977), makes Namazu swimming in the mantle, leaving behind it a buoyant trail of tholeiitic bubbles, rising ponderously, and creating chains of midplate volcanoes (Fig. 1).

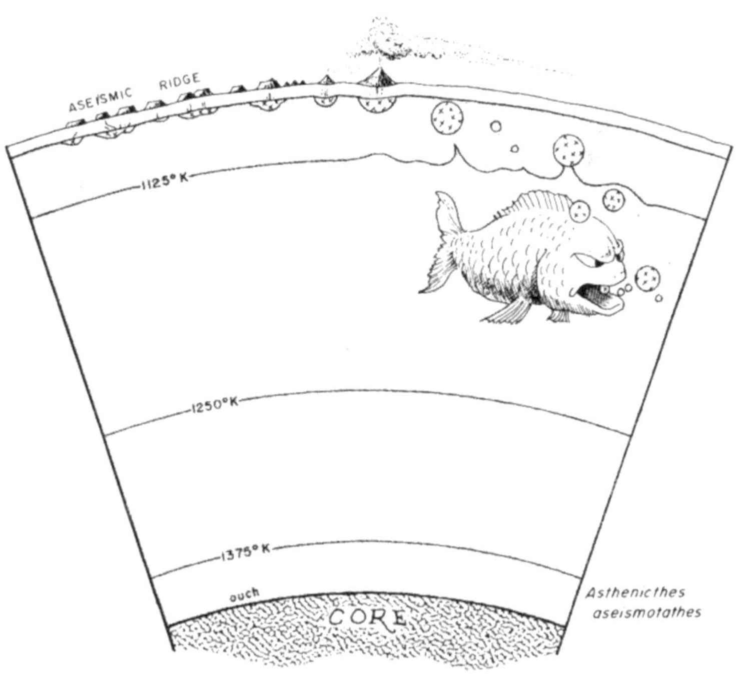

Fig. 1. Alternative to mantle plume theory (based on ancient Japanese legend), from Holden and Vogt (1977).

2.2 The hotpot concept: Mantle plumes and plate tectonics

More recently, the hotspot concept emerged at the same time that the plate tectonics theory (Wilson, 1963). As hotspot chains (Fig. 2) are linear volcanic alignments, parallel to the direction of the plate motion, and displaying regular volcanism age progression, with active volcanism emplaced at one extremity of the chain, Morgan (1971, 1972) proposes that the origin of these tracks may be due to deep mantle plumes. He imagines the plumes as vertical conduits through which hot mantle flows upward. The hotspot tracks are then due to the drifting of a plate over a stationary mantle source. The plate displacement pushes the old volcanoes away from the source as young volcanoes are formed above the source (Fig. 2).

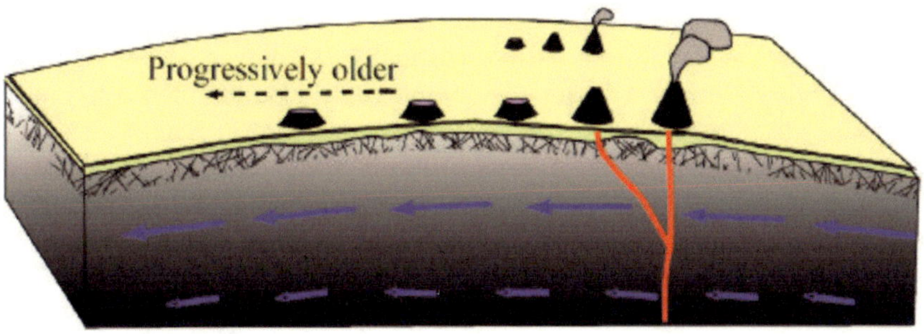

Fig. 2. The hotspot concept: the hotspot tracks are due to the drifting of a plate over a stationary mantle source. The plate displacement pushes the old volcanoes away from the source as young volcanoes are formed above the source- from Clouard (2001)

The plume itself would be characterized by a mushroom-shaped head and a thin, long stem. When the head of the plume reaches the lithosphere, it produces a massive volcanic event, and creates traps (or oceanic plateaus) at the surface. This episode is followed by the interaction of the stem with the lithosphere, which produces the linear volcanic chain.

2.3 The plume debate

The simple model of the interaction of a hot mantle upwelling, deeply rooted in the mantle, and the overriding lithosphere has since then been challenged, and the existence of plume is now questioned.

The characteristics which fuels the more energetically the 'plume debate' is the depth at which they initiate (Clouard & Bonneville, 2001; Sleep, 1990; Anderson, 2000). From the fluid dynamics point of view, a plume can only initiate from instabilities out of a thermal boundary layer. Morgan (1971, 1972) proposes first that plumes initiate in the lower mantle, but later has an idea of a second type of hotspot (Morgan et al. 1978). Other authors invokes superficial sources which use the weakness zones of the lithosphere to express themselves at the surface (Turcotte & Oxburgh, 1973; Anderson, 1975; Foulger et al., 2005; Foulger, 2010; Anderson, 2010). This last hypothesis, also called theory of Plate Tectonics Processes (PTP), points to a passive mantle. The volcanism emplacement would then be controlled by the stress field in the plate, and the magnitude of volcanism by the fertility of the underlying shallow mantle (Foulger et al., 2005; Foulger, 2010; Anderson, 2010). Higher mantle temperatures are not required in this case, and only the shallow part of the mantle is involved in the volcanism emplacement.

Geochemist have tried to answer this open question. Rocks from hotspot volcanoes (OIB: Ocean Island Basalts) appear indeed enriched in noble gazes when compared to Mid Oceanic Ridge Basalts (MORB). As mid-oceanic ridges sample a shallow part of the mantle, plumes must tap deep reservoirs, enriched in noble gazes, which remained isolated from intermixing. However, recent studies indicate that there is no need to invoke deep isolated reservoirs to account for the geochemical signatures of OIB (Anderson, 1998; Allègre, 2002). In then appears that "geochemistry will not deliver the silver bullet for proving or disproving plumes" (Hofmann & Hart, 2007), neither discriminate the depth at which mantle upwellings initiate.

If the plumes at the origin of hotspots are hot mantle upwellings, they mantle beneath midplate chains should be characterized by anomalously low shear velocities (Vs). However, most of the available tomography models lack resolution at the scale of the plume, and therefore stems are not easily identifiable. Moreover, a low velocity anomaly does not unswervingly imply a hotter region, but could also be interpreted in terms of a chemical anomaly (Karato, 2008).

The morphology of the volcanic chain itself bring information about the plume dynamics. According to the classical definition, first proposed by Morgan (1971, 1972), there must be a flood basalt near the oldest extremity of the chain, and the linear chain should display a long and monotonous age progression. The swell surrounding the chain is a direct consequence of the buoyant plume upwelling, and therefore, is also commonly used as the parameter to quantify the hotspot strength (Sleep, 1990; Courtillot et al., 2003; Vidal & Bonneville, 2004, Adam et al., 2005).

By analyzing the previously described criteria, Courtillot et al. (2003) show that three kind of plumes may coexist, each of them corresponding to a boundary between the CMB and the seafloor: those which initiate at the CMB (primary plumes or Wilson-Morgan), the secondary plumes initiating at the transition zone (also called secondary hot spots) and the "Andersonian" plumes that may be due to a passive response to forms of lithospheric breakup (Anderson, 2010; Foulger et al., 2005). According to their analysis, the primary (or Wilson-Morgan) hotspots in the Pacific may be Hawaii, Louisville and Eastern, and the secondary ones Caroline, Macdonald, Pitcairn, Samoa and Tahiti (see Fig. 3.) Let us now see in a more practical way which are the characteristics of hotspot chains, by considering what is happening in French Polynesia.

3. French polynesia region

3.1 Description of the french polynesia volcanism

This study focuses on the French Polynesia (Fig. 3, 4), a region particularly interesting for the study of intraplate volcanism. There is indeed a great concentration of volcanism: 14% of the active volcanism is concentrated in an area covering less than 5% of the globe. The young volcanoes (Fig. 4) in the region are the Macdonald (Norris & Johnson, 1969) and Arago (Bonneville et al., 2002) seamounts in the Austral archipelago, Mehetia and Tehetia (Talandier and Okal, 1984) in the Society Islands, and Adams seamount (Stöffers et al., 1990) in the Pitcairn-Gambier alignment. A wide range of volcanic features should be noted: en echelon ridges, chains of midplate volcanoes and isolated seamounts. The five main chains in the region are the Society, the Marquesas, the Tuamotu, Pitcairn-Gambier and the Cook-Austral (Fig. 4). Their characteristics often depart from the classical definition of a hotspot chains, which accounts for the simple interaction of a mantle upwelling with the overriding lithosphere. The age progression in the volcanic chains is often short (0–4.2 Ma for the Society Islands (Duncan & McDougall, 1976) and 0.5–6.0 Ma for the Marquesas (Desonie et al., 1993; Diraison, 1991)) and the orientation of the chains do not systematically correspond to the motion of the oceanic plate, like in the Marquesas where the chain is rotated 30°

clockwise from the direction of the absolute plate motion (Desonie et al., 1993). The lithosphere seems to exert a considerable influence on the location of the volcanism, like in the Austral Islands where two periods of linear volcanism, separated by 10 m.y., are superimposed on the same volcanic edifices (Bonneville et al., 2002). Let us see now, in more details, the characteristics of the main chains.

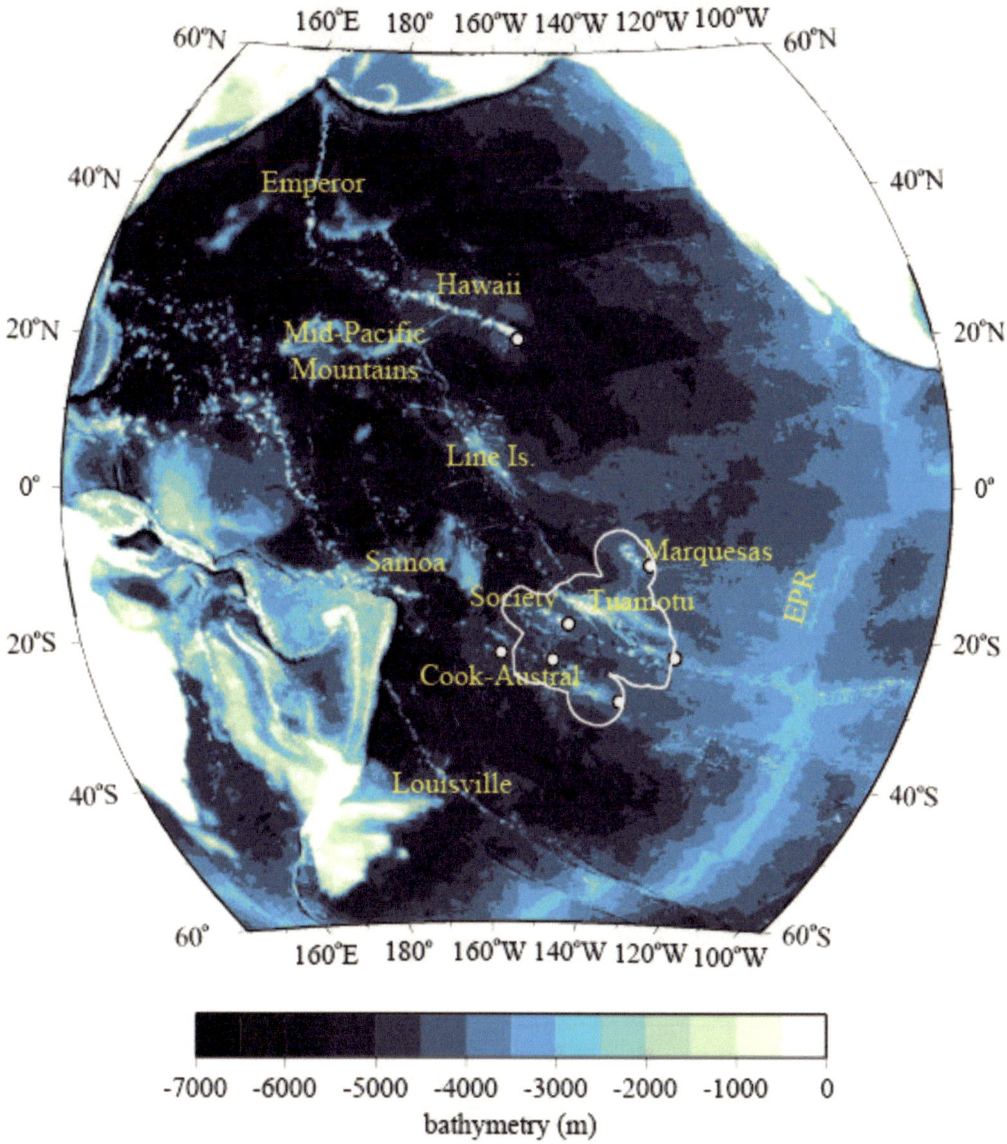

Fig. 3. Bathymetry of the Pacific, from Smith and Sandwell (1997). The names of the main volcanic chains are reported in yellow. The French Polynesia region is delimitated by the white contour.

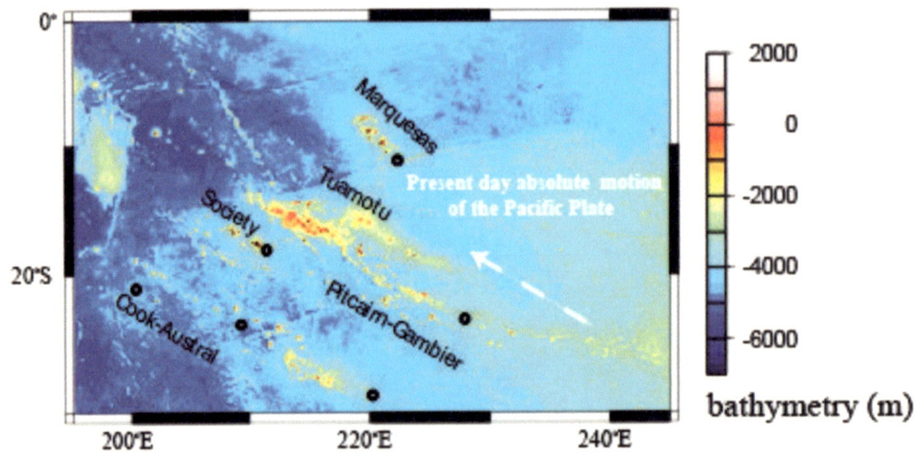

Fig. 4. Bathymetry of the French Polynesia region, obtained with Smith and Sandwell's [1997] grid. The names of the volcanic chains are reported. The disks represent the emplacement of young volcanism.

3.1.1 Society

The Society chain (Fig. 5) is the only classical case of hotspot in our study area. The Society islands are situated between latitudes 16°S and 19°S and longitudes 147°W and 153°W on a seafloor displaying ages between 65 and 95 Ma. They stretch along a 200-km-wide and 500-km-long band orientated in the direction of the present-day Pacific plate motion: N115 ± 15°. The age progression is uniform from the youngest submarine volcano, Mehetia [0.264 Ma (Duncan & McDougall, 1976)], situated at the southeast extremity, to the oldest dated island, Maupiti [4.8 Ma (White & Duncan, 1996)]. The topographic anomaly associated with this alignment (Fig. 5b) stretches along the volcanic chain. Its maximal amplitude, 980 m, is reached 30 km northwest of Tahiti, and is not correlated with any volcanic structure. For this volcanic chain, the swell description corresponds to the one previously reported for hot spot swells: it stretches along the volcanic chain and subsides along the direction of the plate motion; the swell's maximum is located roughly 200 kilometers downstream from the active volcanism.

3.1.2 Cook-Austral

The Cook-Austral chains extend from the island of Aitutaki (140°W, 29°S) to the active submarine volcano Macdonald (160°W, 19°S) in a band more than 2200 km long and 240 km wide (Fig. 6). The chains are composed of several dozens of seamounts, 11 islands and 2 atolls. Although oriented roughly in the direction of present Pacific plate motion (N115°), the spatial and temporal pattern of both the aerial and submarine volcanoes is rather complex. The age of the oceanic crust along the chain ranges from about 39 to 84 Ma (Mayes et al., 1990).

The particular geometry and morphology of the chains suggested two distinct but parallel alignments. McNutt et al. (1997) determined the existence of two additional chains of volcanoes (Taukina and Ngatemato) near the active Macdonald seamount at the southeast end of the chain, 20-34 Ma older than the Macdonald volcanism. Both recent and old ages (Turner & Jarrard, 1982; Barsczus et al., 1994) recorded on Aitutaki and Rurutu islands from

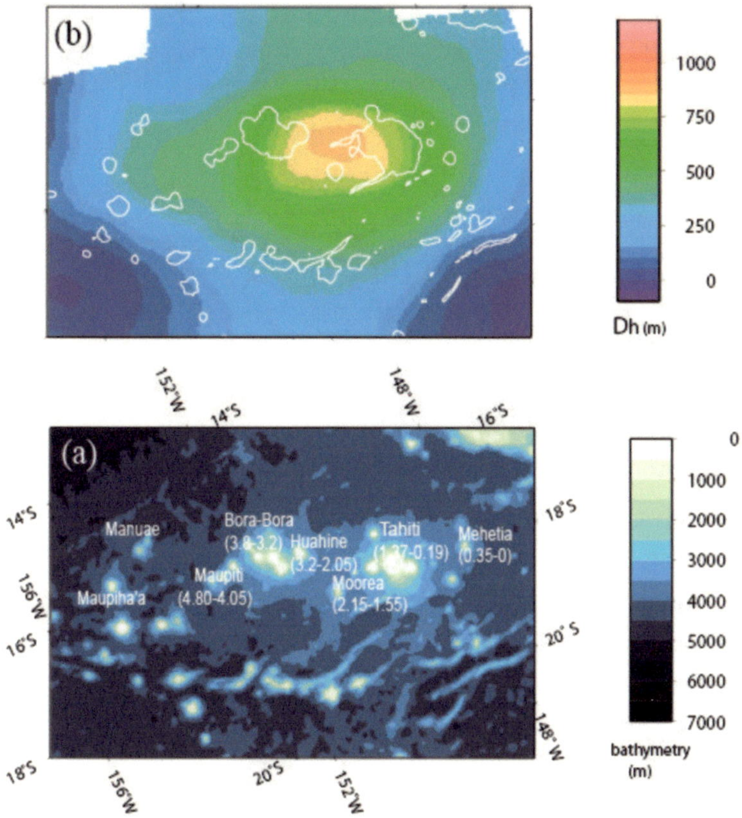

Fig. 5. Society volcanic chain. (a) Bathymetry. The volcanism ages are in Ma. (b) Hot spot swell. Gray line is the 3000 m isobath. The maps are projected along the direction of the present Pacific plate motion (N115°). Figure extracted from Adam et al. (2005)

basaltic samples require the existence of two other hot spots (Rarotonga and Rurutu). Bonneville et al. (2002) attribute the most recent volcanic stage of Rurutu to the Arago Seamount, a very shallow seamount located 120 km east-southward of Rurutu and sampled during the ZEPOLYF2 cruise. Up to 6 distinct hot spot tracks have been identified so far (Bonneville et al., 2006): from northwest to southeast, Rarotonga, "old" and "young" Rurutu, Macdonald, Taukina and Ngatemato of which three are probably still active: Rarotonga and Macdonald (which are named after the emplacement of the active volcanism) and young Rurutu, now active at the Arago Seamount volcano.

These three active tracks are characterized by coherent age progression and distinct geochemical signatures (Bonneville et al., 2006). Concerning Rarotonga, neither the observed age (1 Ma), nor isotopic analysis allows us to connect it with the above mentioned tracks. It seems to be an isolated hot spot. The same observations can be made for Aitutaki, an atoll situated less than 250 km north of Rarotonga and dated at 1.2 Ma (Turner & Jarrard, 1982). The young and old Rurutu tracks overlap each other. Sometimes this occurs on the same volcano, as in Rurutu and Arago. These two tracks occur in the same area as an older phase of volcanism which created the Lotus guyot (54.8 Ma) and the ZEP2-1 seamount (55.8 Ma). To the south, the volcanism which is now active at the Macdonald, and which is responsible for the more recent stage at Marotiri (4 Ma) and ZEP2-19 (8.9 Ma) and for the creation of

Rapa (5.1 Ma) and Mangaia (20 Ma), also overlaps an older stage which built Ra (29 Ma), Marotiri (32 Ma), ZEP2-19 (20-33 Ma) and the Neilson Bank (40 Ma). The existence of several stages of loading on the same volcanic edifice seems to be the rule rather than the exception in this region.

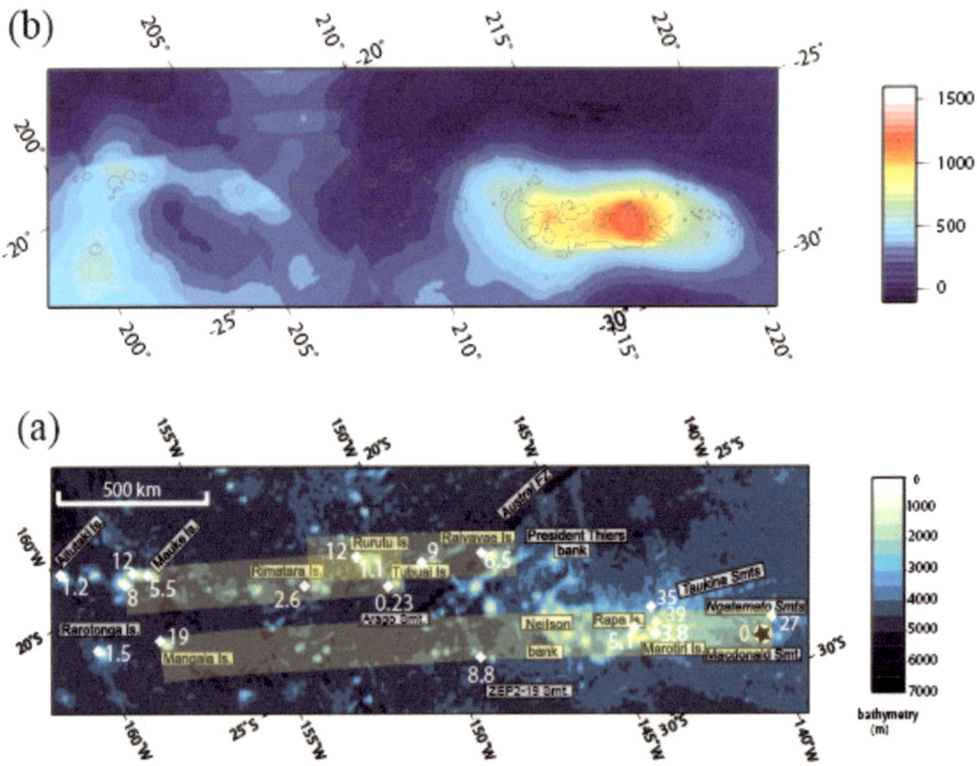

Fig. 6. Cook-Austral volcanic chain. (a) Bathymetry projected along the direction of the present Pacific plate motion (N115°). The light green ribbons represent the "old Rurutu", "young Rurutu," and Macdonald hot spot tracks as proposed by Bonneville et al. (2002). Note that volcanoes are often not located exactly on the middle of a track, which could indicate the importance of lithospheric control rather than a change in the location of the magmatic source. The volcanism ages are in Ma. (b) Hot spot swell. Gray line is the 3000 m isobath. Figure extracted from Adam et al. (2005)

Two positive depth anomalies are observed - one centered on Rarotonga Island and the other one linked to the southwestern branch with Macdonald seamount -, whereas the active Arago seamount is associated with a bathymetric low (Adam et al., 2010). The swell associated with Rarotonga has a circular shape and its maximal amplitude reaches 600 m. The Macdonald swell has an irregular shape: a part of the swell stretches along the axis connecting Macdonald and Rapa; another part is shifted northwest and is not correlated with any recent volcanic structure. The spatial length scale of the swell is about 1250 km. The maximal amplitude reaches 1220 m and is located 375 km from the active volcanism (Macdonald), almost twice the distance observed for the Society.

3.1.3 Marquesas

The Marquesas chain seems to be a classical hot spot, with a regular age progression (Duncan & McDougall, 1974; Diraison, 1991; Brousse et al., 1990) from a seamount southeast of Fatu Iva, which is only a few hundred thousand years old (Desonie et al., 1993), to the Eiao atoll, situated northwest of the chain, displaying an 5.3 Ma age. The direction of this volcanic chain varies according to the authors: McNutt et al. (1989) report a N140–146°E direction, whereas Brousse et al. (1990) prefer a N160°-170°E direction. In all cases, this direction differs from those of the Pacific plate motion.

The analysis of seismic velocities points out a crustal thickening of several kilometers (Wolfe et al., 1994; Caress and Chayes, 1995). McNutt and Bonneville (2000) show that the swell associated with this alignment is mostly due to underplating. The swell spreads along the chain axis. Its maximal amplitude (640 m) is reached on the main axis between Nuku Hiva and Hiva Oa, 275 km away from the most recent volcanism. It has an irregular shape since its width is almost constant all along the volcanic alignment. When the swell is due to a classical plume-lithosphere interaction, it is more important near the youngest part of the chain. For the Marquesas, this difference confirms the hypothesis of underplating at its origin.

3.1.4 Tuamotu

Situated between the Marquesas and Austral Fracture Zones, the Tuamotu volcanic chain has the characteristics of both island chains and oceanic plateaus. The sixty atolls composing the Tuamotu fall under two parallel alignments orientated N115°E, thus suggesting a hot spot origin (Morgan, 1972; Okal & Cazenave, 1985). They are superimposed on a large plateau capped with sediments, limestone and basalt layers. The origin of the plateau still remains controversial. Ito et al. (1995) propose a scenario according to which the northern segment of the Pacific-Farallon spreading center propagates southward in an inner pseudofault and a failed rift, and northward in an outer pseudofault. The southern discontinuities would enclose a block of lithosphere transferred from the Farallon plate to the Pacific plate. This zone of discontinuity focuses volcanism along the Tuamotu Plateau, channels the magma uplift and is then responsible for the plateau shape and morphology.

3.2 The south Pacific superswell

The term "superswell" was first used by McNutt & Fischer (1987) to describe a broad area of the Pacific corresponding roughly to French Polynesia (Fig. 3). This region is characterized by numerous geophysical and geochemical anomalies. There is, as previously discussed, a high concentration of volcanism; the seafloor is anomalously shallow considering its age; there is dip in the geoid; the mantle beneath the superswell is characterized by slow seismic velocities.

3.2.1 Depth anomaly

The term 'superswell' itself comes from the fact that the seafloor in French Polynesia is unusually shallow compared to other seafloors of the same age. This area subsides less rapidly away from the East Pacific Rise than any thermal subsidence model of the oceanic lithosphere predicts. Mammerickx & Herron (1980) and Cochran (1986) use the GEBCO bathymetry chart to point out a 15 million km^2 elevated area in the South Pacific, to the west of Easter Island. Using SYNBAPS bathymetry, Van Wykhouse (1973) and McNutt & Fischer (1987) find a broad area of the Pacific where the seafloor is 250 to 750 m too shallow and name it the South Pacific Superswell. This depth anomaly is challenged by Levitt & Sandwell (1996). They propose that the Superswell could be an artifact of the poor sampling and gridding of ETOPO5, which in oceanic regions corresponds to the SYNBAPS database. Subsequent studies using original ship depth soundings (McNutt et al., 1996; Sichoix et al., 1998) confirm that the seafloor is indeed 1 km shallower.

The precise description of the location and extent of the superswell remained approximate for a while. One talked indeed about "a region in the South Pacific, to the west of Easter Island" (Mammerickx & Herron, 1980) or about "a broad area of French Polynesia" (McNutt, 1998). The design of filtering methods, especially adapted for bathymetry (Hillier& Watts, 2004; Adam et al., 2005) allow a precise quantification of the seafloor uplift in this area. In Fig. 7 we show the depth anomaly found through the MiFil method (Adam et al., 2005, Adam & Bonneville 2005). It extends between latitudes 10°N and 30°S and longitudes 130°Wand 160°W and has a maximal amplitude of 680 m on a seafloor displaying ages between 30 and 115 Ma. It is obviously not a simple swell as one thought, but a hemispheric shaped feature, composed of two branches. The southern branch corresponds to the location of French Polynesia. The northern branch, described by Adam & Bonneville (2005), is not clearly correlated with volcanic features. The only volcanic chain over this area is the Line Islands, which is a fossil alignment displaying ages between 35.5 and 93.4 Ma (Schlanger et al., 1984).

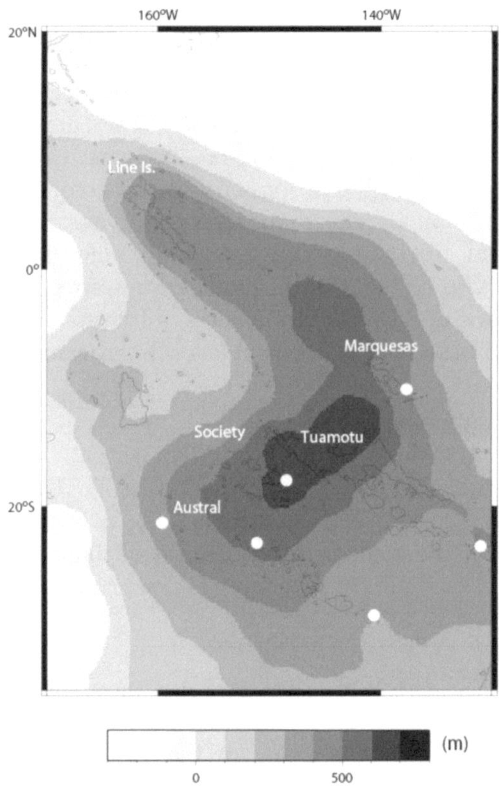

Fig. 7. Depth anomaly associated with the South Pacific Superswell, from Adam & Bonneville (2005) , computed through the MiFil method (Adam et al., 2005). The white disks represent the emplacement of recent volcanism.

3.2.2 Geoid anomaly

The Superswell area, is also associated with a geoid anomaly. The geoid is an equipotential surface which corresponds to the mean surface of oceans. Departures from this reference bring information on the mass repartition at the interior of the Earth. The correlation between the depth anonaly associated with the Superswell and the geoid is still debated since it is not very clear which long wavelength gravitational field best represents the Superswell. Hager (1984) and Watts et al. (1985), report a geoid high over the Superswell (for spherical harmonic degree and order, of n= 2-10 and n>10 respectively), while McNutt & Judge(1990), McNutt (1998) and Adam & Bonneville (2005) suggest this area is correlated with a geoid low (n > 4, 7 < n < 12, n=6 respectively).

The influence of choice of the considered degrees and orders illustrated in Fig. 8. This parameter is rather important since it helps constraining the phenomena at the origin of the observed anomalies. The association of swells with a geoid high may be explained by isostatic compensation (Crough,1978, 1983) where the rise of the seafloor is supported by a density deficiency within the lithosphere. A geoid low however, requires dynamic support. The only model which can explain altogether the observed anomalies is a convective mantle where the low-velocity zone is located immediately below the lithosphere (McNutt, 1998).

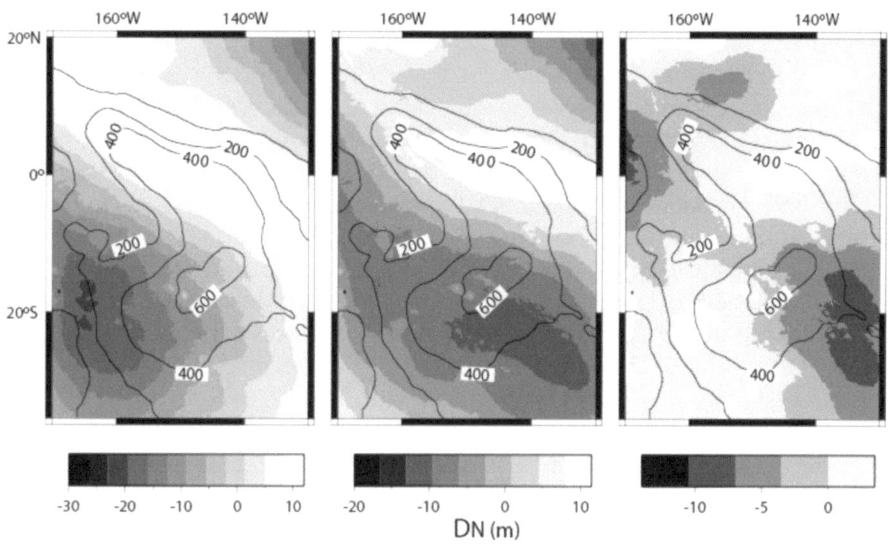

Fig. 8. Geoid anomaly over the superswell: (a) $\lambda > 8000$ km, (b) $\lambda > 6700$ km, and (c) $\lambda > 5700$ km. Lines show isovalues of the depth anomaly. Figure extracted from Adam & Bonneville (2005).

The phenomena at the origin of depth and geoid anomalies are constrained by dynamic models, which try to reproduce the convection occurring in the mantle. Some of them use as an input tomography models, which will be discussed in the following.

3.2.3 Anomalous mantle

In order to 'look' at the deep mantle, seismic tomography is the most appropriate tool. Seismic tomography is a technique to image a heterogeneous structure of the Earth's interior using vast amount of seismic data, of which principle is same as that of a medical CT scan.

Studies generally report that the mantle beneath the superswell is characterized by slow seismic velocities (McNutt & Judge, 1990; Su et al., 1992; Montagner, 2002), but the mantle structure remained poorly resolved for a long time, due to the sparse coverage by seismic stations. This resulted in blurred and even inconsistent images among different studies: some studies have described a whole mantle-scale broad plume (Zhao, 2004), others have proposed a continuous plume conduit generated from a broad plume near the core-mantle boundary (Ritsema & van Heijst, 2000; Montelli et al., 2006; Takeuchi, 2007, 2009), and still others have suggested plumes that do not continue throughout the mantle (Mégnin & Romanowicz, 2000).

The recent deployment of two new networks of broad band seismic stations, emplaced on the seafloor (9 BBOBS stations, (Suetsugu et al., 2005)) and on the islands (10 PLUME stations (Barruol et al., 2002)) has lately rectified this. The emplacement of these stations is represented in Fig. 11. From the data obtained through these new stations, several models have been developed. The tomography model provided by Isse et al., (2006) is based on Rayleigh waves inversion and provides the most accurate view of the shallowest part of the mantle (depths < 240 km). In its first version, it did however not include the PLUME data. These latter are included by Suetsugu et al. (2009), who determine the mantle structure of the upper mantle down to 400 km by inverting the dispersion for the S wave velocity. The authors also update the map of the MTZ thickness by applying the receiver function analysis to the PLUME data. To analyze the lower mantle structure, Suetsugu et al. (2009) design a P wave travel time tomography. Tanaka et al.(2009a, 2009b) also obtain a P and S wave traveltime tomography and a global P wave traveltime tomography (2009a) from the BBOBS and PLUME data. The main results are summarized in the composite Fig. 9.

There is no uniform anomaly throughout the upper mantle, and the average S velocity profile beneath the South Pacific superswell is close to that of other oceanic regions. On shorter scales, there are slow (of about 2-3%), localized anomalies, which could represent narrow plumes in the upper mantle. They appear to be deep-rooted beneath Society, Macdonald and Pitcairn, and superficial beneath the Marquesas (<150 km). These anomalies are not necessarily vertical. The one in the vicinity of the Macdonald hot spot, is located 200–300 km northwest of the hot spot.

The mantle transition zone (MTZ) is defined as a layer bounded by two seismic discontinuities: the 410 km and 660 km discontinuities, which represent mineral phase transition, and provide indirect information regarding the temperature. The average MTZ thickness beneath the South Pacific is close to the global average, and values beneath most of the stations are nearly normal, with some notable exceptions. The MTZ is indeed thinned near the Society, the Macdonald and the Pitcairn hot spots, suggesting that hot plumes may be ascending from the lower mantle.

Low velocity anomalies are found throughout the lower mantle, while the lateral dimension of the anomalies changes drastically at a depth of 1000 km (Fig. 9). Below 1000 km, slow anomalies of 1% are as large as 3000 km* 3000 km. They seem continuous down to the core–mantle boundary. On the other hand, above 1000 km, the slow anomalies split into narrower and more localized anomalies, a few hundred kilometers in diameter.

These new images suggest that the narrow slow anomalies observed in the first 400 km of the upper mantle may be connected to the top surface of the superplume. These narrow anomalies are smaller than the spatial resolution of the lower mantle model. The distribution of the MTZ thickness, however, supports the presence of these narrow anomalies. In the case of Society and Macdonald, these anomalies also coincide with the slower anomalies in the upper mantle, which suggests that they represent continuous plumes that ascend from the top surface of the superplume. The Marquesas hot spot seem to be restricted to the upper mantle, as well as Pitcairn, which is characterized by slow upper mantle and a thin MTZ , but with no slow anomalies at the top of the lower mantle.

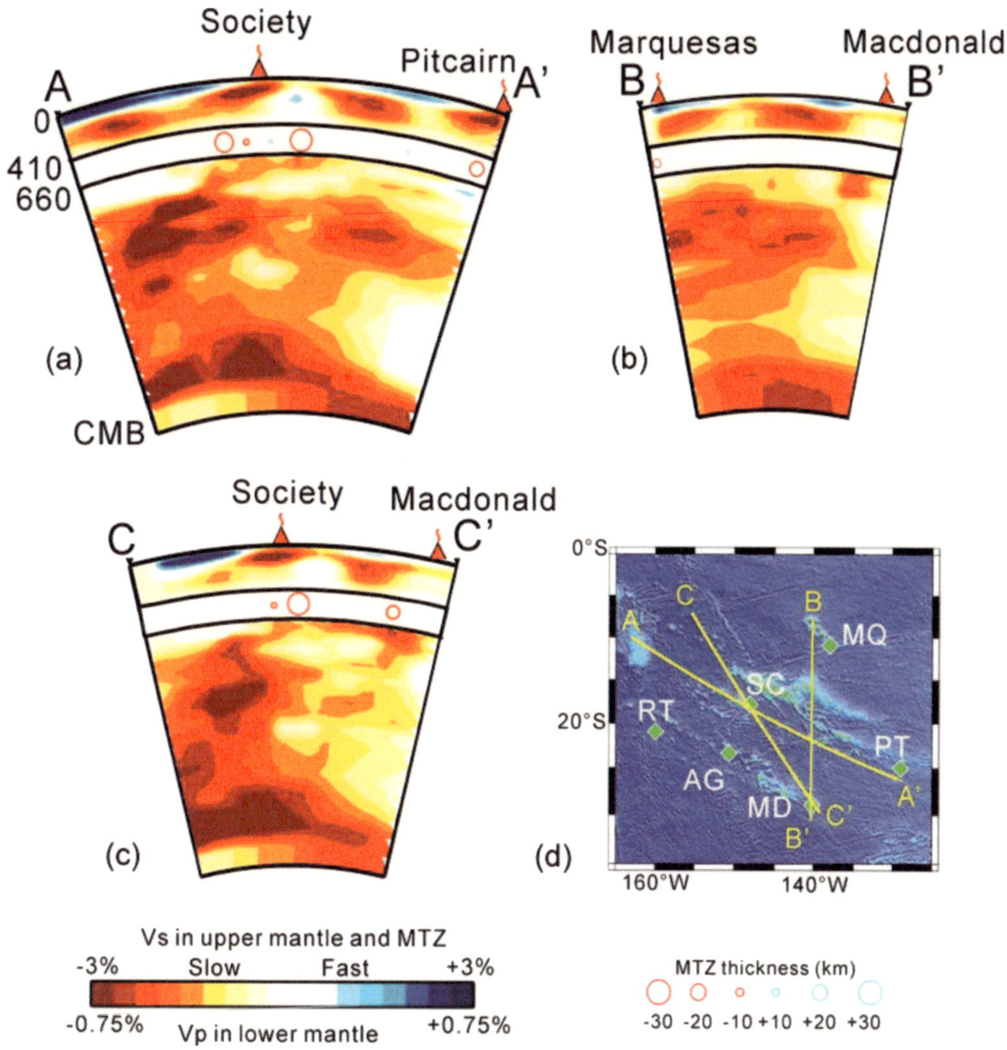

Fig. 9. Composite tomography from Suetsugu et al. 2009 (a–c) Cross sections of seismic structure in the entire mantle. S wave velocities are shown for the upper mantle (0–410 km). P wave velocities are shown for the lower mantle (660–2900 km). Velocity scales are ±3% in the upper mantle and ±0.75% in the lower mantle. (d) Positions of the profiles. Circles plotted in the MTZ (410–660 km) are the locations of the MTZ thickness estimated near the profiles within 2.5°, where red and blue represent thicker and thinner than the global average, respectively, and the size is proportional to the deviation from the average. Green diamonds in are active hot spots.

4. Models, interpretation

The phenomenon connecting the surface observations (volcanic chains, swells, the geophysical anomalies associated with the South Pacific Superswell) and the mantle structure are still debated. The dynamics of mantle plumes and Superplumes, and the origin of their associated swells, remain indeed some of the most controversial topics in geodynamics. In the following, we will provide a detailed description of Adam et al. (2010) study, which focuses on the dynamics of the shallowest part of the mantle under the French Polynesia region.

4.1 Dynamics of the French Polynesia plumes: Results of numerical simulations based on a highly resolved tomography model

Adam et al. (2010) use a new regional seismic tomography model obtained through the inversion of Rayleigh waves (see section 3.3.3), which high resolution allows obtaining information at the scale of plumes. They construct a numerical model of the mantle flow beneath the French Polynesia region. We will describe at first the numerical model, then compare the outputs of the numerical modeling to the surface observations.

4.1.1 Model

Convection model

Adam's et al. (2010) study is a local study, where the computation domain extends between latitudes 0 and 32°S, longitudes 174 and 232°E, and depths 0–240 km. Following the work of Yoshida (2008), the authors used the finite volume method for the discretization of the basic equations.
The non-dimensionalized expressions of the mass and momentum equation are:

$$\nabla \cdot v = 0 \qquad (1)$$

$$-\nabla p + \nabla \cdot \left\{ \eta \left(\nabla v + \nabla v^{tr} \right) \right\} + R_{ai} \delta \rho \, e_r \qquad (2)$$

where v is the velocity vector, p the dynamic pressure, η the viscosity, $\delta\rho$ the density anomaly, Ra_i the instantaneous Rayleigh number, and e_r, the unit vector in the radial direction. The superscript tr indicates the tensor transpose. The impermeable and shear stress-free conditions are adopted on the top (0 km depth) and bottom (240 km depth) surface boundaries. The flows across lateral boundaries are taken to be symmetric.
The resulting normal stress acting on the top surface boundary (σ_{rr}) is

$$\sigma_{rr} = -p + 2\eta \frac{\partial v_r}{\partial r} \qquad (3)$$

where v_r is the radial velocity. The dynamic topography δh is obtained from the normal stress through the equation

$$\delta h = \frac{\sigma_{rr} - \langle \sigma_{rr} \rangle}{\Delta \rho g} \qquad (4)$$

where $\langle \sigma_{rr} \rangle$ is the averaged σ_{rr} over the top surface boundary, $\delta\rho$ the density contrast between the mantle and sea water densities and g, the gravity acceleration.

Model parameters

Ratio of density anomaly to seismic velocity anomaly, $R_{\rho/vs}$

The density to velocity heterogeneity ratio $R_{\rho/vs}$ used to convert the seismic waves velocity anomalies into density anomalies, is a primordial parameter of the model since it allows to make geodynamical inferences from a tomography model. In the range of viscosity profiles tested by Adam et al. (2010), the density to velocity heterogeneity ratio, $R_{\rho/vs}$, allowing the best fit of the observations is situated between 0.14 and 0.24. These values are quite close to the value inferred from mineral physics ($R_{\rho/vs}$=0.2, Karato, (2008)). The $R_{\rho/vs}$ is slightly bigger than the former estimation found through similar geodynamic models ($R_{\rho/vs}$=0.1 to 0.15 for Forte and Mitrovica (2001), $R_{\rho/vs}$= -0.01 to 0.07 according to Simmons (2007)), certainly because these previous studies never focused on such shallow depths. The results presented hereafter use a density to velocity heterogeneity ratio $R_{\rho/vs}$ of 0.17.

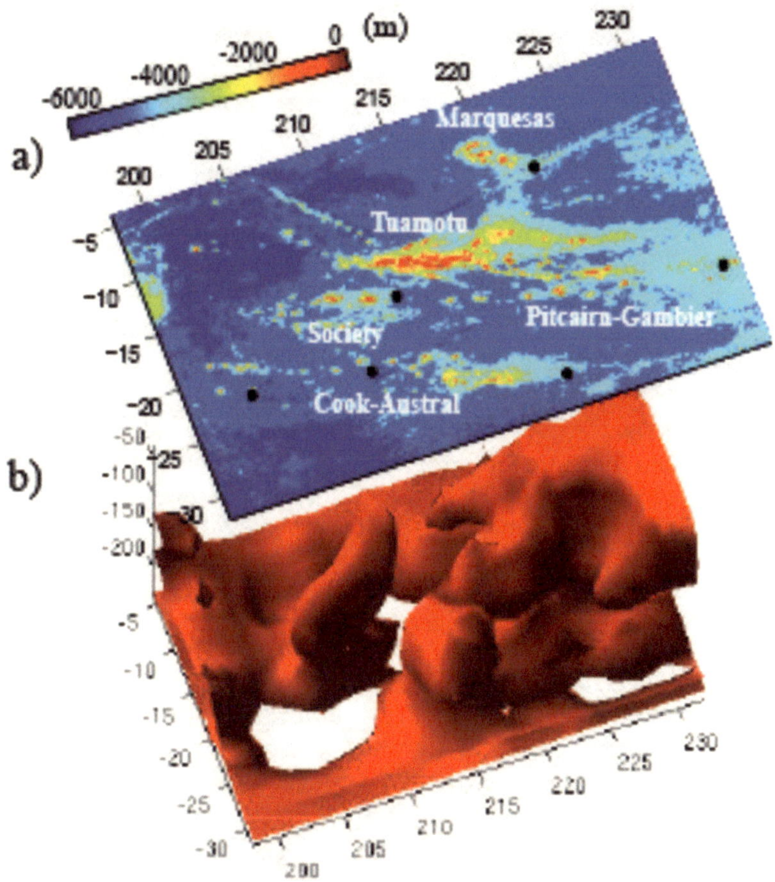

Fig. 10. French Polynesia region - from Adam et al. (2010). a) Bathymetry and names of the hotspot chains. The black disks represent the young volcanoes. b) View of the mantle density anomalies deduced from the tomography model. The red iso-surface represents the -30 kgm-3 density anomaly.

Viscosity profile

Adam et al. (2010) also tested the effects of vertical variation of the viscosity (see their Supplementary Online Material). The introduced viscosity profile describes a highly viscous lithosphere overlying a low-viscosity asthenosphere. Note that while the absolute velocity of mantle flow is affected by the choice of the relative viscosity of each layer, the dynamic topography does not depend on its absolute viscosity. We present here the results obtained with viscosity of 10^{23} Pas between depths 0 and 30 km and viscosity of 10^{20} Pas between depths 30 and 240 km.

Influence of deeper heterogeneities

In the present study Adam et al. (2010) limited the depth range from 0 to 240 km, which covers the lithosphere and asthenosphere. To examine the effects of the density heterogeneities situated deeper than 240 km, the authors also constructed a composite tomography model in which they impose in the upper part (0-240 km depths) the local Rayleigh wave tomography model (Isse et al., 2006) and in the lower part (260-660 km) the S20RTS model (Ritsema & van Heist, 2000). They then compute the whole upper mantle convection (0-660 km) by varying the viscosity below the asthenosphere. The viscosity profile above 240 km remains the same as previously described (highly viscous lithosphere overlying a low-viscosity asthenosphere). When the viscosity below the asthenosphere is greater than in the asthenosphere, say, $\eta=10^{21}$-10^{22} Pas between depths 240 and 660 km, the convection is very similar to the convection obtained in the present study (depths 0-240 km), because the viscosity increase at the base of the asthenosphere (240 km depth) reduces the flow velocity created in the lower part of the mantle. The increase of viscosity below the asthenosphere is a reasonable assumption, which has been supported by many previous studies (Dziewonski & Anderson, 1981). They therefore conclude that the convection computed from the upper 240 km of the mantle is representing well the actual convection in this depth range.

4.1.2 Hotspot swells and dynamic topography

The dynamic topography computed by Adam et al. (2010) is presented in Fig. 11. The authors notice a good overall correlation between the observed (Fig. 11 b) and the modelled swells (Fig. 11 c), in spite the fact that they have been obtained from totally independent data (bathymetry and seismic tomograms respectively). Previous studies (Crough, 1978; Sleep, 1990) show that midplate chains are generally associated with bathymetric highs. The emplacement, wavelength and amplitude of the French Polynesia swells are well recovered by Adam et al. (2010) model, indicating the authors reproduce well the actual mantle flow. The swell over the Society chain, the only classical hotspot chain in the study area has indeed the same characteristics that the dynamic topography. The swell over Pitcairn-Gambier and the circular swell associated with Rarotonga, an isolated active volcano, are also well retrieved.

As discussed in section 3.1.2, the situation is quite puzzling for the Macdonald chain, since previous studies demonstrate that most of the volcanism there may be produced by non-hotspot processes (McNutt et al., 1997; Jordahl 2004). The fact that the swells are well modelled by the dynamic topography, demonstrates that the buoyant ascent of the plume plays a major role. In section 3.1, we have seen that the Marquesas and the Tuamotu swells probably have shallow origins, respectively crustal underplating (McNutt & Bonneville, 2000) and a large plateau capped with sediments, limestone and basalt layers (Ito et al.,

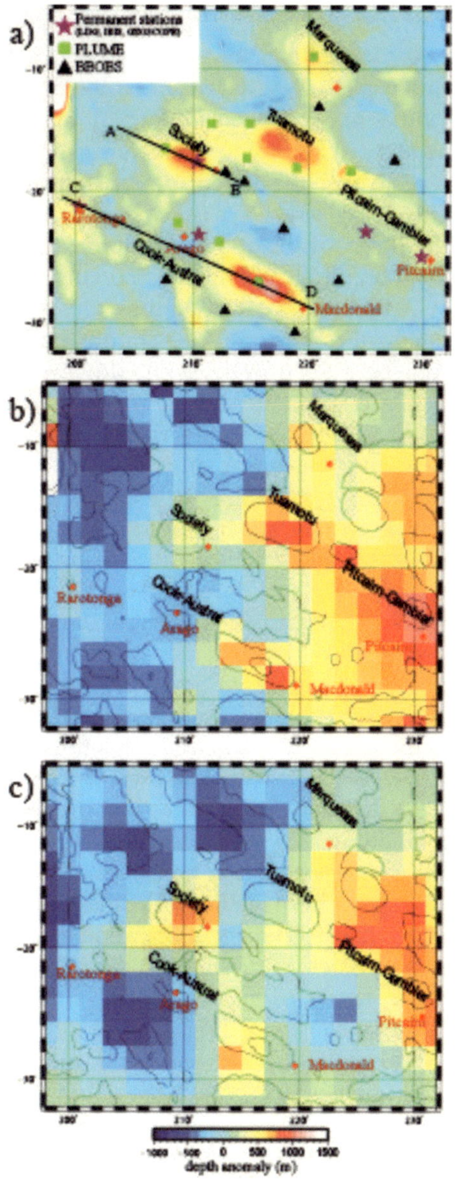

Fig. 11. French Polynesia swells- from Adam et al. (2010): a) The colour map represents the observed swells, determined by the same method that Adam et al. (2005). The red disks represent the location of active volcanism. The seismic station emplacement is shown by the magenta crosses (permanent stations), the green squares (PLUME stations) and the red circles (BBOBS). The AB and CD profiles are used to make depth cross sections (see Fig. 12 and discussion in the text). b) Observed swells with age correction and 2°*2° sampling. The black lines represent the isocontours of the original swells displayed in panel a. c) Dynamic topography obtained through a computation using $R_{\rho/vs}$=0.17 and a viscosity profile which describes a highly viscous lithosphere (10^{23} Pas between depths 0 and 30 km) overlying a low-viscosity asthenosphere (10^{20} Pas between depths 30 and 240 km).

1995). It is then not surprising that these chains are not associated with any important dynamic topography. The new modelling however, points out a distinct area of localized upwelling, east of the Tuamotu plateau, which may be the first evidence that the present-day mantle dynamics contributes to the observed depth anomaly.

The model also recovers the observed bathymetric lows such as the one associated with the active Arago volcano, there again with the correct emplacement, wavelength and amplitude. This pattern is actually surprising since active volcanism is generally associated with bathymetric high, created by the buoyant uplift of a plume. Latter, we will check in more details what is happening, by looking at the convection occurring beneath this region.

The convection modelled by Adam et al. (2010) is represented in Fig. 12. The pattern is quite complex. Volcanism emplacement is not due to a simple vertical ascent of a plume, but rather to a complex interaction between upwelling and downwelling flows. The only case in the study area which corresponds to the definition originally proposed (Morgan, 1968), is the Society. On the depth cross section along the AB profile (Fig. 12 a), we can see indeed that the buoyant source (negative density anomaly) located under this chain creates a vertical upwelling reaching the lithosphere directly beneath the active volcanism. For all the other chains, the convection pattern is more complex. It is not merely created by deep buoyant sources but mainly controlled by the surrounding mantle flows. Therefore, the inferred mantle convection involves lateral flows, tilted conduits and the buoyant source is generally far from the active volcanism. However, Adam et al. (2010) find that the emplacement of each active volcano can be explained by the modelled flows.

The Cook-Austral chain has often been taken as an example to argue against the plume theory (McNutt at al., 1997; Jordahl 2004), since volcanic stages overlap on this chain (Bonneville at al., 2006) and the latest volcanic emplacement is apparently controlled by the stresses left in the lithosphere by previous loadings (McNutt at al., 1997). It is then interesting to check the convection under this chain. On the depth cross section along the CD profile (Fig. 12b), a large low density body, deeper than 50 km, creates an upwelling beneath the observed swell maximum, but far from the active volcanism occurring at Macdonald. At the base of the lithosphere the flow becomes horizontal and streams towards the Macdonald. This explains the volcanism emplacement. Even if this latter may be facilitated by structural discontinuities of the lithosphere, as previously suggested (McNutt at al., 1997; Jordahl 2004), the present result demonstrates that the buoyant ascent of the plume plays a major role in the volcanism loading.

The bathymetric low observed near the Arago active volcano is also puzzling. We can see that the mantle under this volcano is indeed characterized by densities slightly higher than the surrounding mantle (Fig. 12 b). This accounts for the observed and modelled bathymetric lows but does explain the volcanism emplacement. One would expect a hot buoyant mantle upwelling. How can active volcanism be associated with cooler mantle? Maybe through the downwelling current this configuration creates. This downwelling current may produce lateral tensile stresses and then cracks near the base of the lithosphere. The magma will simply follow this lithospheric discontinuities and erupt as pressure decrease along the way.

The low density buoyant source associated with the upwelling creating Rarotonga, is situated far from this volcano, at depths greater than 140 km and more than 500 km west of the island. The mantle in the immediate vicinity is indeed slightly cooler. The upwelling reaches the lithosphere through a tilted conduit, and occurs surprisingly through a mantle which is not especially hotter than the surroundings. The buoyancy created by the deep source (140-240 km depths), should be strong enough to counterbalance the adjacent mantle buoyancy.

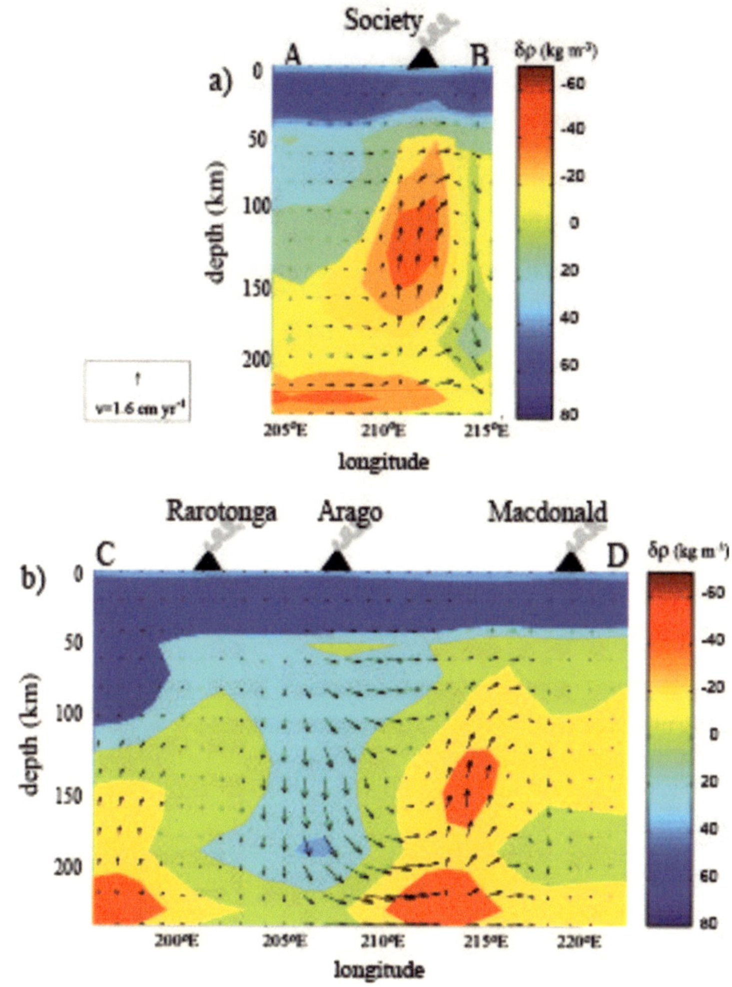

Fig. 12. Depth cross section along the AB (a) and CD (b) profiles displayed in Figure 11 - from Adam et al. (2010). The colour map represents the density anomalies and the arrows the convection driven by them. The schematic volcanoes represent the active volcanism emplacement.

4.1.3 Buoyancy fluxes

Quantitatively, the measure of the plume strength is given by the buoyancy flux, which measures the flux of material from the mantle. The buoyancy flux can be computed through two independent ways, one based on swell morphology (B_{swells}), and the other on the mantle flow (B_{dyn}) (Davies, 1988, Sleep 1990). B_{swells} has been obtained through the formula

$$B_{swells} = WE(\rho_m - \rho_w)V_l \tag{6}$$

where E is swell's amplitude, W its lateral extent, $(\rho_m\text{-}\rho_w)$ the density contrast between the mantle and the seawater and V_l the Pacific plate velocity

B_{dyn} is defined at each depth as:

$$B_{dyn} = -\int_s (v_r \delta\rho)\, ds \tag{7}$$

where v_r is the radial velocity obtained by the dynamic model and $d\rho$ the input density anomaly. The integration is carried out over the horizontal cross-sectional area of plumes. We assumed that regions with $v_r > 0$ ($v_r < 0$) are those of upwelling (downwelling) plumes.

The values obtained by Adam et al. (2010) are represented in Fig. 13. We can see that the results obtained through the two approaches are very consistent. Indeed, both of the B estimations give the same lineup of the hotspot strength. From the strongest to the weakest

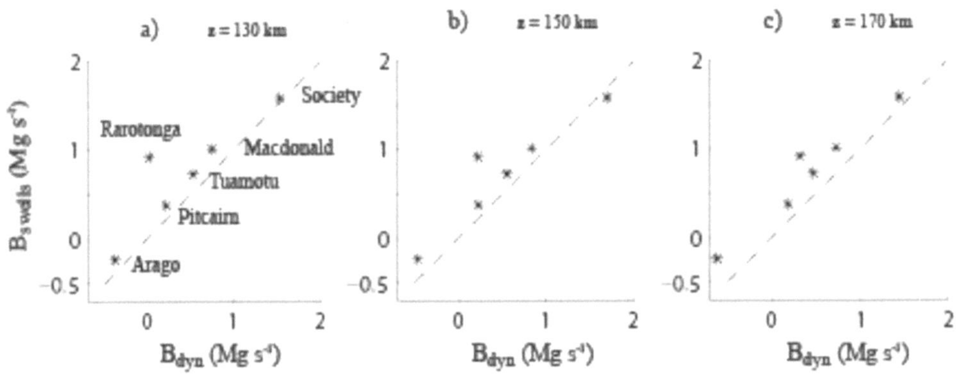

Fig. 13. Buoyancy fluxes from Adam et al. (2010). The values of the buoyancy fluxes obtained from the observed swells (B_{swells}) are represented as a function of the buoyancy fluxes obtained through the dynamic computation (B_{dyn}) for various depths (130 km, 150 km, and 170 km). We represent the values obtained through a computation using $R_{\rho/vs}$ =0.17 and a viscosity profile which describes a highly viscous lithosphere 10^{23} Pas between depths 0 and 30 km) overlying a low-viscosity asthenosphere (10^{20} Pas between depths 30 and 240 km). The dashed line represent the B_{dyn}=B_{swells} curve.

we find the Society, Macdonald, Rarotonga, Tuamotu, Pitcairn and Arago. The Society and Macdonald ones are the only "strong" plumes, with B >1Mgs^{-1} (Courtillot et al., 2003). The fact that the buoyancy fluxes values found through the two independent approaches are consistent, implies that the heat transported by mantle plumes can be accurately evaluated from a careful estimation of the swell morphology. This is important for constraining the role that plumes play into the total heat flow on Earth. Adam et al. (2010) find that the total buoyancy flux B_{total} of the five hotspots (except Marquesas and Arago) is 4.7 Mgs^{-1}. Taking the thermal expansivity α=2.0×10^{-5} K^{-1} and the specific heat cp=1250 Jkg^{-1}K^{-1}, the total heat flow is estimated as Q=$B_{total}\, Cp/\alpha$=0.29 TW (Davies, 1999), which accounts for 9 % of the total plume heat flow, 3.4 TW (Sleep, 1990), and for around 1% of the total heat flow out of the Earth's mantle, 36 TW (Davies, 1999)

5. Discussion

French Polynesia is a complex region, characterized by numerous geophysical anomalies. There is indeed a high concentration of volcanism: 14% of the active volcanism is concentrated in an area covering less than 5% of the globe. There is a wide range of volcanic features: en echelon ridges, isolated seamounts, and five chains of midplate volcanoes.

Their characteristics often depart from the classical definition of a hotspot chains: the age progression in the volcanic chains is often short, the orientation of the chains do not systematically correspond to the motion of the oceanic plate, and the lithosphere seems to exert a considerable influence on the location of the volcanism, like in the Cook-Austral Austral Islands, where several periods of linear volcanism are superimposed.

The morphology of the swells associated with these chains is also peculiar and does not correspond to the classical definition of a hotspot swell (Crough, 1978; Sleep, 1990). Adam et al. (2010) show that they are well correlated to the dynamic topography modelled from a regional, highly resolved seismic tomography model, which concerns only the first 240 km of the upper mantle. This demonstrates that a direct link exists between the surface observations and mantle flows. However, according to the classical definition, plumes should be deep-rooted buoyant mantle upwellings. We have seen that the convection occurring in the shallowest part of the mantle (0–240 km depths) is sufficient to explain the active volcanism and the observed swells, while the roots of some hotspots are probably located at greater depths, as indicated by a recent tomography model (Suetsugu et al., 2009). The buoyancy created by these deeper sources is apparently not required to explain the surface observations.

The snapshot of the mantle provided by the tomography models display a complex pattern. Low velocity anomalies are found throughout the lower mantle, but their dimension changes drastically at a depth of 1000 km. The broad anomaly extending from the CMB, splits indeed into narrower and more localized anomalies a few hundred kilometers in diameter. Narrow anomalies are observed in the upper mantle. The Society and Macdonald ones seem to be connected to the lower mantle, whereas the other are restricted to the upper mantle.

We have shown that the dynamics of the narrow plumes concerns the shallowest part of the mantle (depths 0-240 km). What about the broad deeper anomaly, called superplume? Through the model described in section 4.1.1, the dynamics of this superplume has also been studied (Adam, Yoshida & Suetsugu, personal communication). The effects of the upper and lower mantle have been tested separately. The tomography models are the S20RTS model (Ritsema & van Heist, 2000) for the upper mantle and the Obayashi & Fukao model (2000) for the lower mantle. The preliminary results displayed in Fig. 14 show that the depth anomaly associated with the South Pacific Superswell can only be accounted for when considering the dynamics of the lower mantle. Indeed, the dynamic topography computed from the upper mantle (lef panel Fig. 14) has a faint amplitude and the wavelength is much shorter than the observations. The wavelength and emplacement of the observed depth anomaly is well recovered when modeling the dynamics of the lower mantle (right panel Fig. 14). This points out to the fact that the Superswell is a phenomenon involving mainly the lower mantle.

What is the connection between all these observed features? In a traditional view, hot upwellings in the lower mantle are stuck below the ''660' discontinuity'. Suetsugu's et al. (2009) study reveals however that the top of the superplume is located at a depth of 1000 km and that there are no broad slow anomalies immediately beneath the ''660. Moreover, the morphology of the superplume imaged by Suetsugu's et al. (2009) does not correspond to the image of a thermal plume, as defined on the basis of laboratory and numerical experiments, which often show that purely thermal plumes are characterized by a broad head and a narrow tail. This points out to a thermo-chemical origin. The size of the Superswell also indicate such an origin. Models studying the dynamics of boundary layers, show that the maximum diameter reached by a plume that has only a thermal origin is in the range of a few hundred kilometers (Bercovici & Kelly, 1997; Schubert et al., 2004). If other phenomena such as compositional effect are considered, this might stabilize the boundary layer and allow it to thicken toward the superplume dimension.

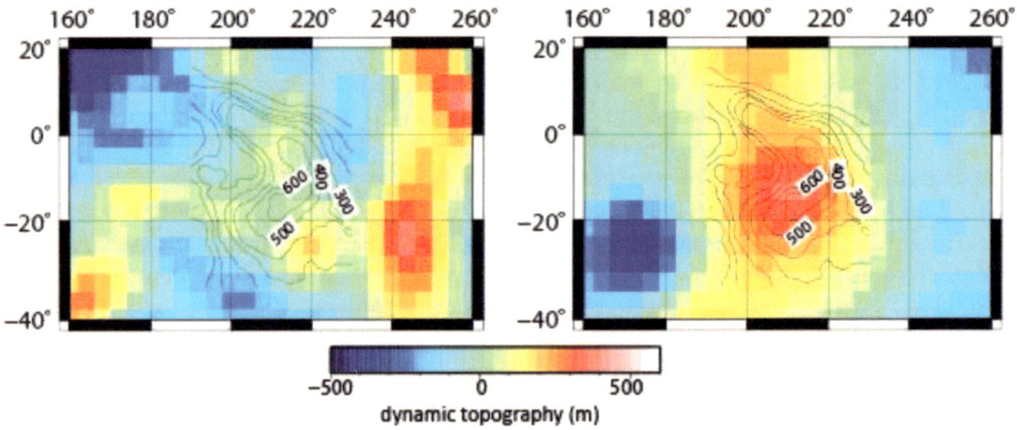

Fig. 14. Dynamics of the South Pacific Superswell. The left and right panels represent the dynamic topography computed for the upper and lower mantle respectively (Adam, Yoshida & Suetsugu, personal communication). The black isocontours represent the depth anomaly associated to the Superswell (Adam & Bonneville, 2005).

Courtillot et al. (2003) argue out that two types of upwellings can occur in the lower mantle: the primary plumes and the superplumes. The type of upwelling depend on the local buoyancy ratio (i.e., the ratio of the stabilizing chemical density anomaly to the destabilizing thermal density anomaly). These results are based on analogical experiments where two layers of miscible fluids with different kinematic viscosities, densities and depths, superimposed in a tank heated from below and cooled from above (Davaille, 1999; Le Bars & Davaille, 2004; Kumagai et al., 2007). For high buoyancy ratio, long-lived thermochemical plumes are produced, whereas for intermediate buoyancy ratios, large thermal domes develop. In this case, the lower (respectively upper) fluid is heated (respectively cooled), becomes lighter (respectively heavier) and rises up (respectively sinks) until the thermal effects are cancelled by the chemical density anomaly. This large domes oscillates vertically. Experimentally, secondary plumes are observed at the top of the rising domes (arrows in Fig. 15), where the ratio of the chemical density anomaly to the thermal density anomaly is higher than in the rest of the mantle.

The morphology obtained through these experiments is very similar to the geometry of the superplume and narrow plumes imaged by Suetsugu et al. (2009). The Society and the Macdonald seem to be still connected with the superplume. The other observed narrow plumes may represent (remnant) secondary plumes that have detached from the superplume. Such a dome structure has also been found through numerical simulations (McNamara & Zhong, 2004; Farnetani & Samuel, 2005; Ogawa, 2007). Analogic experiments also shown that secondary plumes are generated sporadically and do not last for a long time (Davaille, 1999; Ogawa, 2007). This may explain the short age progressions of the Polynesian hot spots (Duncan & MacDougall, 1976; Clouard & Bonneville, 2005). Moreover, if the emphasized secondary plumes entrain chemically distinct superplume materials, from different location and in different proportions, this would explain the enriched mantle signature found in the South Pacific OIB, and the different geochemical signatures that have been observed (Bonneville et al., 2006). If the experimentally observed vertical oscillation of the large dome are actually representative of the superplume behavior, this would also explain the present-day relative quietness of the volcanic activity and deep location of the superplume. Vertical oscillation will indeed induce pulsation in the volcanic activity, which was increased in the Cretaceous, when it created large oceanic plateaus.

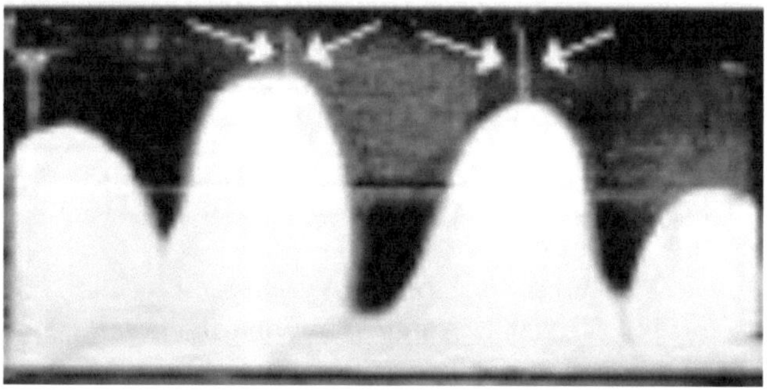

Fig. 15. Convection driven by two layers of miscible fluids submitted to a temperature gradient. For intermediate values of the stabilizing chemical density anomaly to the destabilizing thermal density anomaly, large domes develop and oscillate vertically through the whole layer. Thin plumes rise from the upper surfaces of the domes (white arrows) (Davaille, 1999).

The volcanism pattern and mantle dynamics under the French Polynesia region are quite complex. Magma uplift obviously does not occur through deep vertical conduits as first proposed (Morgan, 1971; 1972), but rather through narrow secondary plume, probably initiating at top of a large oscillating dome, the superplume. The volcanism emplacement is not a passive phenomenon, only due to discontinuities of the lithosphere. Mantle dynamics is required, even if it involves only the shallowest part of the mantle (0 - 240 km depths). Anderson (2010) uses the results of Adam et al. (2010) on the dynamics of secondary plumes, to argue that hotspots are shallow phenomena, involving only the shallow asthenosphere. Even if the dynamics of this later is sufficient to explain the surface observations, the narrow secondary plumes may have not existed without the superplume, which appears to be related with the lower mantle dynamics.

6. References

Adam, C., Vidal V. , & Bonneville A. (2005). MiFil: a method to characterize hotspot swells with application to the South Central Pacific. *Geochem. Geophys. Geosyst., 6, Q01003, doi:10.1029/2004GC000814.*

Adam, C. & Bonneville A. (2005). Extent of the South Pacific Superswell, *J. Geophys. Res.*, 110, B09408, *doi:10.1029/2004JB00346.*

Adam, C., Yoshida M., Isse T., Suetsugu D., Fukao Y., & Barruol G. (2010). South Pacific hotspot swells dynamically supported by mantle flows. *Geophys. Res. Lett.* VOL. 37, L08302, doi:10.1029/2010GL042534.

Allègre C.J. (2002). The evolution of mantle mixing, *Philos. Trans. R. Soc. London* 360, 1-21.

Anderson, D. (1975). Chemical plumes in the mantle. *Tectonophysics*, 86, 1593-1600.
Anderson, D. (1998).The Helium Paradoxes, *Proc. Natl. Acad. Sci. USA* 95, 4822-4827.
Anderson, D. (2000). The thermal state of the upper mantle: no role for mantle plumes. *Geophys. Res. Lett.*, 27, 1593-1600.
Anderson, D. (2010). Hawaii, Boundary Layers and Ambient Mantle-Geophysical Constraints.. *J. of Petrology,* doi:10.1093/petrology/egq068
Barruol, G., Bosh, D., Clouard, V., Debayle, E., Doin, M.P., Fontaine, F., Godard, M., Masson, F., Tommasi, A. & Thoraval C. (2002). PLUME investigates the South Pacific Superswell. *Eos, Trans. Am. Geophys. Un.*, 83, 511-514.
Barsczus, H., Guille, G., Maury, R., Chauvel, C., & Guillou, H. (1994). Two magmatic sources at Rurutu Island (Austral Islands, French Polynesia) and the Austral Hotline. In *Eos trans.*, 75, 323.
Bercovici, D., & Kelly, A. (1997). The non-linear initiation of diapirs and plume heads. *Phys. Earth Planet. Inter.*, 101, 119–130.
Bonneville, A., Lesuave, R., Audin, L., Clouard, V., Dosso, L., Gillot, P., Hildenbrandt, A., Janey, P., & Jordhal, K. (2002). Arago seamount: the missing Hot Spot found in the Austral islands. *Geology,* 1023-1026.
Bonneville, A., Dosso, L., & Hildenbrand, A. (2006), Temporal evolution and geochemical variability of the South Pacific superplume activity, *Earth Planet. Sci. Lett.*, 244, 251–269.
Brousse, R., Barsczus, H., Bellon, H., Cantagrel, J., Diraison, C., Guillou, H., & Léotot, C.(1990). Les Marquises (Polynésie française); volcanologie, géochronologie, discussion d'un modèle de point chaud. Bull. Soc. Géol. France, 6, 933-949.
Caress, D., & Chayes, D. (1995). New software for processing data from side-scancapable multibeam sonars. In *Oceans'95: Challenges of our changing global environment*, 2, 997-1000, Washington DC, NY, USA: MTS-IEEE.
Clouard V. (2001). Etude géodynamique et structurale du volcanisme de la Polynésie française de 84 Ma à l'actuel, Thèse de l'Université de la Polynésie Française, 214 p.
Clouard, V., & Bonneville, A. (2001). How many Pacific hotspots are fed by deep mantle plumes? Geology, 695-698.
Clouard, V., &. Bonneville, A. (2005), Ages of seamounts, islands and plateaus on the Pacific plate, in *Plates, Plumes, and Paradigms*, edited by G. R. Foulger et al., Geol. Soc. Am. Bull., 388, 71–90, doi:10.1130/2005.2388(06).
Cochran, J. (1986). Variations in subsidence rates along intermediate and fast spreading mid-ocean ridges. *Geophys. J. R. Astr. Soc.*, 87, 421-454.
Courtillot, V., Davaille, A., Besse, J., & Stock, J. (2003). Three distinct types of hotspots in the Earth's mantle. *Earth Planet. Sci. Lett.*, 205, 295-308.
Crough, S. (1978). Thermal origin of midplate hotspot swells. *Geophys. J. R. Astron. Soc.* 55, 451-469 .
Crough, S. (1983). Hotspot swells. *Ann. Rev. Earth Planet. Sci.*, 11, 165-193.
Davaille, A. (1999). Simultaneous generation of hotspots and superswells by convection in a heterogeneous planetary mantle. *Nature*, 402, 756-760.
Davies, G. (1988). Ocean bathymetry and mantle convection. 1. Large scale flow and hotspots, *J. Geophys. Res.*, 93(B9), 10,467–10,480.

Davies, G. (1999), *Dynamic Earth: Plates, Plumes and Mantle Convection*, Cambridge Univ. Press, Cambridge, U.K.

Desonie, D., Duncan, R., & Natland, J. (1993). Temporal and geochemical variabilityc of volcanic products of the Marquesas hotspot. *J. Geophys. Res.*, 98, 17649 17665.

Diraison, C. (1991). Le volcanisme aérien des archipels polynésiens de la Société, des Marquises et des Australes-Cook. Unpublished doctoral dissertation, Univ. de Bretagne Occidentale. (413 p.)

Duncan, R., & McDougall, I. (1974). Migration of volcanism with time in the Marquesas Islands, French Polynesia. *Earth Planet. Sci. Lett.*, *21* (4), 414–420.

Duncan, R., & McDougall, I. (1976). Linear volcanism in French Polynesia. *J. Volcanol. Geotherm. Res.*, 197-227.

Dziewonski, A.M. & Anderson D.L. (1981). Preliminary reference Earth model, *Phys. Earth Planet. Inter.*, 25, 297-356.

Farnetani, C. G., and H. Samuel (2005), Beyond the thermal plume paradigm, *Geophys. Res. Lett.*, 32, L07311, doi:10.1029/2005GL022360.

Forte, A. & Mitrovica J. (2001). Deep-mantle high-viscosity flow and thermochemical structure inferred from seismic and geodynamic data. *Nature*. 410, 1049-1056.

Foulger, G.R., Natland, J.H., Presnall, D.C., & D.L. Anderson, D.L. (2005), *Plates, Plumes, and Paradigms*, edited by G. R. Foulger et al, Geol. Soc. Am. Special Volume 388, pp. 881.

Foulger, G.R. (2010), Plates vs Plumes: A Geological Controversy, *Wiley-Blackwell*, ISBN 978-1-4443-3679-5, pp. 328.

Hager, B. (1984). Subducted slabs and the geoid: Constraints on mantle rheology and flow, *J. Geophys. Res.*, 89, 6003- 6015.

Hillier, J., & Watts, A. (2004). "Plate-like" subsidence of the East Pacific Rise- South Pacific Superswell system, *J. Geophys. Res.*, 109, B10102, doi:10.1029/2004JB003041.

Hofmann, A. W., & Hart, S.R. (2007), Another nail in which coffin? *Science*, 315, 39–40.

Holden J.C. & Vogt P.R . (1977). Graphic Solution to Problems of Plumacy, *EOS Trans., AGU*, 56 573-580.

Isse, T., D. Suetsugu D., H. Shiobara H., Sugioka, H., Kanazawa, T., & Fukao, Y. (2006). Shear wave speed structure beneath the South Pacific superswell using broadband data from ocean floor and islands, *Geophys. Res. Lett.*,33, L16303, doi:10.1029/ 2006GL026872.

Ito, G., McNutt, M., & Gibson, R. (1995). Crustal structure of the Tuamotu Plateau, 15°S, implications for its origin. *J. Geophys. Res.*, 100, 8097-8114.

Jordahl, K., McNutt, M., & Caress, D. (2004), Multiple episodes of volcanism in the Southern Austral Islands: Flexural constraints from bathymetry, seismic reflection, and gravity data, *J. Geophys. Res.*, 109, B06103, doi:10.1029/2003JB002885.

Karato, S.I. (2008), Deformation of Earth Materials: An Introduction to the Rheology of Solid Earth, Cambridge Univ. Press, New York, 2008.

Kumagai, I., Davaille, A., & Kurita, K. (2007). On the fate of thermally buoyant mantle plumes at density interfaces, *Earth Planet. Sci. Lett.*, 254, 180 - 193, doi:10.1016/ j.epsl.2006.11.029.

Le Bars, M., and A. Davaille (2004), Whole layer convection in a heterogeneous planetary mantle, *J. Geophys. Res.*, 109, B03403, doi:10.1029/2003JB002617.

Levitt, D., & Sandwell, D. (1996). Modal depth anomalies from multibeam bathymetry: Is there a South Pacific Superswell? *Earth Planet. Sci.Lett.*, 139, 1-16.

Mammerickx, J., & Herron, E. (1980). Evidence for two fossil spreading ridges in the southeast Pacific, *Geol. Soc. Am. Bull.*, 91, 263- 271.

Mayes, C., Lawver, L., & Sandwell, D. (1990). Tectonic history and new isochron chart of the South Pacific. *J. Geophys. Res.*, *95* (B6), 8543-8567.

McNamara, A. K., and S. Zhong (2004). Thermochemical structures within a spherical mantle: Superplumes or piles? *J. Geophys. Res.*, 109, B07402, doi:10.1029/2003JB002847.

McNutt, M., & Fischer, K. (1987). The south Pacific Superswell. *Am. Geophys. Union Geophys. Monogr*, 43, 25-34.

McNutt, M. (1988), Thermal and mechanical properties of the Cape Verde Rise, *J. Geophys. Res.*, 93, 2784–2794.

McNutt, M., Fischer, K., Kruse, S., & Natland, J. (1989). The origin of the Marquesas fracture zone ridge and its implication for the nature of hot spots. *J. Geophys. Res.*, 91, 381-393.

McNutt, M., & Judge, A. (1990). The Superswell and mantle dynamics beneath the south Pacific. *Science*, 248, 969-975.

McNutt, M., Sichoix, L., & Bonneville, A. (1996). Modal depths from shipboard bathymetry: There is a South Pacific Superswell, *Geophys. Res. Lett.*, 23, 3397- 3400.

McNutt, M., Caress, D., Reynolds, J., Jordahl, K., & Duncan, R. (1997). Failure of plume theory to explain midplate volcanism in the Southern Austral Islands, *Nature*, 389, 479–482.

McNutt, M. (1998). Superswells. *Rev. Geophys.*, 36, 211-244.

McNutt, M., & Bonneville A. (2000), A shallow, chemical origin for the Marquesas swell, *Geochem. Geophys. Geosyst.*, 1, 1014, doi:10.1029/1999GC000028.

McNutt, M. (2006), Another nail in the plume coffin?, *Science*, 313, 1394–1395.

Mégnin, C., & Romanowicz, B. (2000). The three-dimensional shear velocity structure of the mantle from the inversion of body, surface and higher-mode waveforms, *Geophys. J. Int.*,143, 709–728, doi:10.1046/j.1365-246X.2000.00298.x.

Montagner, J.-P. (2002). Upper mantle low anisotropy channels below Pacific Plate *Earth Planet. Sci. Lett.*, 632, 1-12.

Montelli, R., Nolet, G., Dahlen, F.A. & Masters, G. (2006), A catalogue of deep mantle plumes: New results from finite frequency tomography, *Geochem. Geophys. Geosyst.*, 7, Q11007, doi:10.1029/2006GC001248.

Morgan, W. J. (1968)., Rises, trenches, great faults, and crustal blocks, *J. Geophys. Res.*, 73(6), 1959–1982.

Morgan, W. (1971). Convection plumes in the lower mantle. *Nature*, 230, 42-43.

Morgan, W. (1972). Plate motion and deep mantle convection. *Geol. Soc. Am. Man.*, 132, 7-22.

Morgan, W., Rodriguez, & Darwin. (1978). A second type of Hotspot Island. *J. Geophys. Res.*, 83, 5355-5360.

Norris, A., & Johnson, R. (1969). Submarine volcanic eruption recently located in the Pacific by SOFAR hydrophones. *J. Geophys. Res.*, *74*, 650–664.

Obayashi, M., & Fukao, Y. (2001). Whole Mantle Tomography with an Automatic Block Parameterization. *SSJ meeting, B08.*

Ogawa, M. (2007). Superplumes, plates, and mantle magmatism in two-dimensional numerical models. *J. Geophys. Res.*,112, B06404, doi:10.1029/2006JB004533. Parmentier, E., Turcotte, D., & Torrance, K. (1975). Numerical experiments on the structure of mantle plumes. *J. Geophys. Res.*, 80 (32), 4417-4424.

Okal, E., & Cazenave, A. (1985), A model for the plate tectonic evolution of the east central Pacific based on Seasat investigations. Earth Planet. Sci. Lett., 72, 99-116.

Ritsema, J., & van Heijst, H.J. (2000)., Seismic imaging of structural heterogeneity in Earth's mantle: Evidence for large-scale mantle flow, *Sci. Prog.*, 83, 243–259.

Schlanger, S., Garcia, M., Keating, B., Naughton, J., Sager, W., Haggerty, J. Philipotts, J., & Duncan, R. (1984). Geology and geochronology of the Line Islands, *J. Geophys. Res.*, 89, 11,216- 11,272.

Schubert, G., G. Masters, P. Olson, and P. Tackley (2004), Superplumes or plume clusters?, *Phys. Earth Planet. Inter.*, 146, 147–162.

Sichoix, L., Bonneville, A., & McNutt, M. (1998)., The seafloor swells and superswell in French Polynesia, *J. Geophys. Res.*, 103, 27,123–27,133.

Simmons, N. (2007). Thermochemical structure and dynamics of the African superplume. *Geophys. Res. Lett.* 34, L02301,doi:10.1029/2006GL028009.

Sleep, N. (1990). Hotspots and mantle plumes: some phenomenology. *J. Geophys. Res.*, 95 (B5), 6715{6736.

Smith, W., & Sandwell, D. (1997)., Global sea floor topography from satellite altimetry and ship depth soundings. *Science*, 277, 1956–1962.

Stöffers P. & the Scientific Party of cruise SO-65 of F.S.Sonne (1990). Active Pitcairn hotspot found. *Mar. Geol.*, 95, 51-55.

Su, W., Woodward, R., et Dziewonski, A. (1992). Joint inversions of travel time and waveform data for the 3-D models of the Earth up to degree 12. *Eos Trans.*, 73, 201-202.

Suetsugu, D., Shiobara, H., Sugioka, H., Barruol, G., Schindele, F., Reymond, D., Bonneville, A., Debayle, E.,. Isse, T., Kanazawa T. & Fukao Y. (2005). Probing South Pacific mantle plumes with ocean bottom seismographs. *Eos Trans. Am. Geophys. Un.*, 86, 429-435.

Suetsugu, D., Isse, T., Tanaka, S., Obayashi, M., Shiobara, H., Sugioka, H., Kanazawa, T.,. Fukao, Y., Barruol, G., & D. Reymond., (2009), South Pacific mantle plumes imaged by seismic observation on islands and seafloor, *Geochem. Geophys. Geosyst.*, 10, Q11014, doi:10.1029/2009GC002533. D.L. Turcotte,D.L. & E.R. Oxburgh,E.R. (1973). Mid-plate tectonics. *Nature* 244, 337-339.

Takeuchi, N. (2007). Whole mantle SH-velocity model constrained by waveform inversion based on three-dimensional Born kernels. *Geophys. J. Int.*, 169(3), 1153–1163, doi:10.1111/j.1365-246X.2007.03405.x.

Takeuchi, N. (2009). A low-velocity conduit throughout the mantle in the robust component of a tomographic model. *Geophys. Res. Lett.*, 36, L07306, doi:10.1029/2009GL037590.

Talandier, J., & Okal, E. (1984). The volcano seismic swarms of 1981-1983 in the Tahiti-Mehetia area, French Polynesia. , *J. Geophys. Res.*, *89*, 11216–11234.

Tanaka, S., Obayashi, M., Suetsugu, D., Shiobara, H., Sugioka, H., Yoshimitsu, J., Kanazawa, T., Fukao, Y. &. Barruol, G. (2009a). P wave tomography of the mantle beneath the South Pacific Superswell revealed by joint ocean floor and island broadband seismic experiments. *Phys. Earth Planet. Inter.*, 172, 268–277, doi:10.1016/j.pepi.2008.10.016.

Tanaka, S., Suetsugu, D., Shiobara, H., Sugioka, H., Kanazawa, T., Fukao, Y., Barruol, G., &. Reymond, D. (2009b). On the vertical extent of the large low shear velocity province beneath the South Pacific Superswell, *Geophys. Res. Lett.*, 36, L07305, doi:10.1029/2009GL037568

Turcotte, D.L. & E.R. Oxburgh, E.R. (1973). Mid-plate tectonics. *Nature* 244, 337-339.

Turner, D., & Jarrard, R. (1982). K-Ar dating of the Cook-Austral island chain : A test of the hot-spot hypothesis. *J. Volcanol. Geotherm. Res., 12*, 187–220.

Van Wykhouse, R. (1973). SYNBAPS, *Tech. Rep.* TR-233, Natl. Oceanogr. Off., Washington, D. C.

Vidal, V., & Bonneville, A. (2004). Variations of the Hawaiian hot spot activity revealed by variations in the magma production rate. *J. Geophys. Res.* 109, B03104. doi:10.1029/2003JB002559.

Watts, A., Parsons B., & Roufosse M. (1985). The relationship between gravity and bathymetry in the Pacific Ocean, *Geophys. J. R. Astron. Soc.*, 83, 263- 298.

White, W., & Duncan, R. (1996). Geochemistry and geochronology of the Society Islands : New evidence for deep mantle recycling. *Am. Geophys. Union Geophys. Monogr, 95*, 183–206.

Wilson, J. T. (1963). A possible origin of the Hawaiian Islands. *Canadian J. Phys.*, 41,863-870.

Wolfe, C., McNutt, M., & Detrick, R. (1994). The Marquesas archipelagic apron: Seismic stratigraphy and implications for volcano growth, mass wasting, and crustal underplating. *J. Geophys. Res.*, 99 (B7), 13591-13608.

Yoshida, M., (2008), Core-mantle boundary topography estimated from numerical simulations of instantaneous mantle flow, *Geochem. Geophys. Geosyst.*, 9, Q07002, doi:10.1029/2008GC002008.

Zhao, D. (2004). Global tomographic images of mantle plumes and suducting slabs: insight into deep earth dynamics. *Earth Planet. Sci. Lett.*, 146, 3-34.

7

Magmatectonic Zonation of Italy: A Tool to Understanding Mediterranean Geodynamics

Giusy Lavecchia[1] and Keith Bell[2]
[1]*Laboratory of Geodynamics and Seismogenesis, Earth Science Department, Gabriele d'Annunzio University, Chieti,*
[2]*Department of Earth Sciences, Carleton University, Ottawa, Ontario,*
[1]*Italy*
[2]*Canada*

1. Introduction

The Cenozoic magmatic activity of Italy is characterized, from the Alps to the Aeolian Islands, by an abundance of SiO_2-undersaturated potassic to ultra-potassic rock-types (leucite-phonolites, leucitites, kamafugites and lamproites). Rocks of sodic character, mainly sub-alkaline transitional and alkaline basaltic in composition plus some isolated lamprophyres, are situated along the western and southern side of the Tyrrhenian basin (Sardinia, Ustica, Sicily Channel, Etna) and eastward of it, within the Padan-Adriatic-Hyblean foreland (Veneto, la Queglia-Pietre Nere and Hyblean plateau). Although the sodic and potassic products belong to the same magmatotectonic domain, i.e. the Mediterranean "wide-rift system" and its shoulders, they have been attributed in the literature to contrasting geodynamic environments, one anorogenic and intra-plate, and the other orogenic and subduction-related (see Lustrino and Wilson, 2007 and references therein).
There is widespread acceptance that the sodic magmatism is intra-plate in character, but in the case of the potassic suites there is no unanimity (Lustrino et al., 2011 and references therein). Models for the origin of the potassic rocks centre around the nature of the metasomatic components which might result from dehydration or decarbonation of a slab as it is subducted into the mantle or from the upwelling of deep mantle melts and/or fluids associated with hot spots/plumes. Some papers address the problem from a geochemical point of view comparing the radiogenic isotopic signatures, especially Sr, Nd and Pb, with the composition of well-known, world-wide, mantle components (Bell et al., 2006 among many others). Another category of papers address the problem from a tectonic-structural point of view, discussing the geological and seismological proof for or against subduction (Lavecchia and Creati, 2006 and references therein).
Here, we provide an overview of the Paleocene to Present Italian igneous rocks in relationship to their tectonic setting at the time of their emplacement, and then evaluate possible alternative geodynamic scenarios for their origin, mainly based on integrated geophysical/geological and geochemical parameters.

2. The Mediterranean deformation history

The present tectonic setting of the Western and Central Mediterranean "wide-rift" system (e.g. Ligurian-Provençal and Tyrrhenian basins) and of the surrounding, outward-verging, Apennine-Maghrebian thrust belt (Fig. 1) results from a long history of deformation characterized by the alternation and/or the overprinting of contractional and extensional tectonic phases. Such a history, developed during the last 35 Ma, started while the Alpine-Betic orogeny was undergoing its final collisional stage (Faccenna et al., 2004).

Beginning from Late Cretaceous times, the Alpine Tethys ocean progressively closed with generation of the Alpine orogenic belt, presently exposed in the Alps, northwest Corsica, Calabria and in the Betics (southeast Spain) (Stampfli and Borel, 2004). By the end of the Paleocene (~55 Ma ago), the oceanic and/or ultra-thinned continental Tethyan lithosphere had been completely underthrust beneath the African continental plate. Shortening, however, along the Alpine-Betic belt continued at least to the end of the late Eocene. The Alpine collisional activity occurred together with a localized intra-continental extensional phase, which generated the Cenozoic "narrow-rift" system of eastern and central Europe (Fig.1) (Ziegler, 1992). With the beginning of the early Oligocene (~35 Ma ago), extension started to dominate again over compression. A regional, east-dipping, extensional fault system developed along the south-westward prolongation of the Central Europe Rift System, along the western border of the Western Mediterranean basin, between Corsica-Sardinia and the Provençal region (Fig. 1).

During early Miocene times (from ~22 to ~16 Ma ago), the Corsica-Sardinia block, which belonged to the European continental lithosphere, started to undergo a counter-clockwise rotation around a pivot point situated more or less north of present-day Corsica (star in Fig. 2), which led to the opening of Ligurian-Provençal basin. After a period of tectonic quiescence of a few million years, in middle Miocene times (~13 Ma ago) the extensional process started again to the east of the Corsica-Sardinia block. The Adriatic foreland rotated counter-clockwise giving birth to the progressive opening of the Tyrrhenian basin (Fig. 1a). The extensional process was mainly achieved through easterly-dipping, low-angle, and antithetic, high-angle, normal faulting which produced an extensive stretching and thinning of the crust and mantle lithosphere, with localised areas of mantle unroofing and/or the accretion of new oceanic crust (Fig. 1b). The extensional process progressively migrated eastward and is now active and seismogenic along the axis of peninsular Italy (Lavecchia, 1988; Chiarabba et al., 2005). Both the Ligurian-Provençal and the Tyrrhenian extensional phases were characterized by the development of contemporary coeval and co-axial outward-verging fold-and-thrust structures that nucleated at the outer border of the extended regions (Carminati et al., 1998; Finetti et al., 2001).

3. Magmatic phases

Each of the tectonic phases listed above was characterized by distinctive magmatic activity (Figs. 1, 2 and 3). The Late Cretaceous to Paleocene Alpine orogenic phase was substantially amagmatic, accompanied by isolated lamprophyric activity, mainly consisting of dyke swarms situated on both the African and European continental sides of the closing Tethys (Vichi et al., 2005; Stoppa, 2008 and references therein). Known Italian occurrences are located in southern Tuscany (~110-90 Ma), in the south-eastern Alps (Calceranica and Corvara in Badia, 70-68 Ma), in south-eastern Sardinia (Nuraxi Figus 62-60 Ma), in the

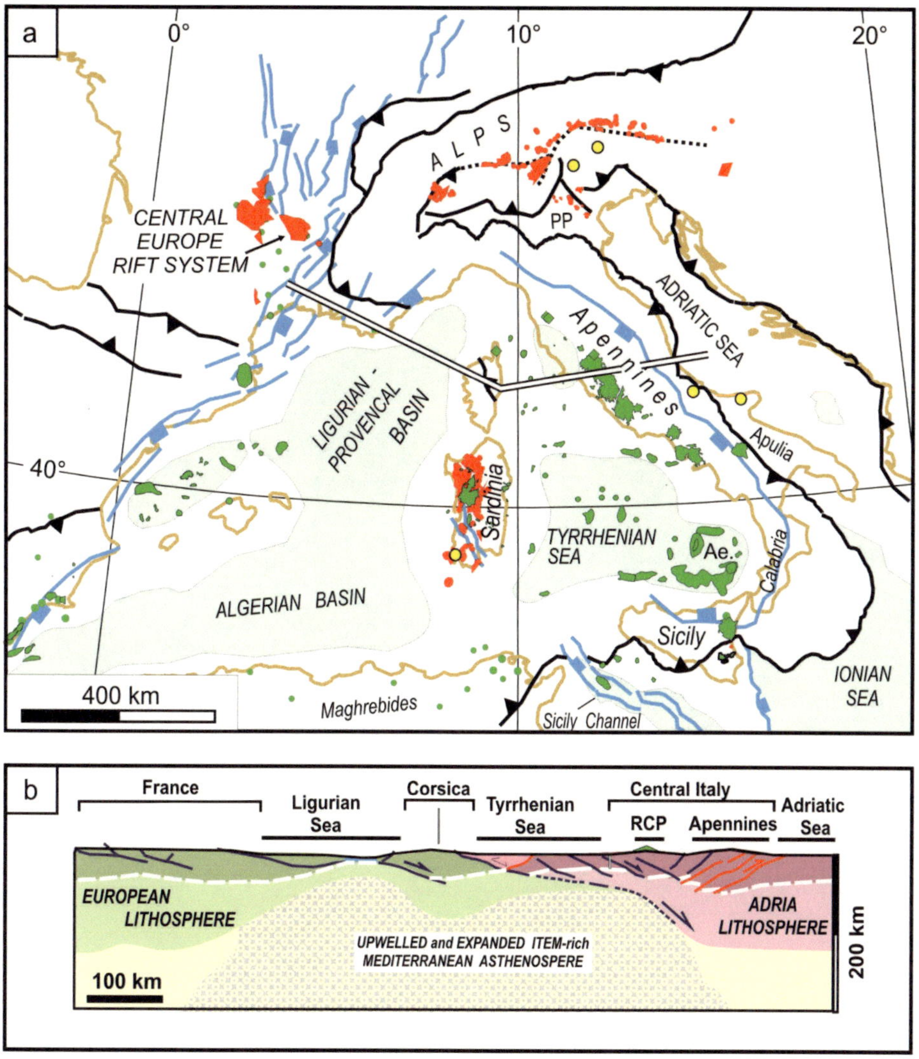

Fig. 1. a) The map shows the Western (e.g. Ligurian-Provençal) and Central (e.g. Tyrrhenian) Mediterranean extensional basins and surrounding regions, with associated volcanic occurrences and major tectonic structures. Key: A = Aeolian volcanic arc; PP = Padan Plain; blue lines = normal and normal oblique faults; black lines with triangles = major thrust fronts; green areas and spots = volcanic occurrences younger than 25 Ma; red areas = volcanic occurrences aged between 25 and 45 Ma; yellow spots = Italian volcanic occurrences older than 45 Ma. The light green basinal areas refers to regions with oceanic and ultra-thinned continental crust (Moho depth ≤ ~20 km).
b) The section shows a schematic view of the crustal and lithospheric structure of the Ligurian and Tyrrhenian "wide-rift" basins; the trace of the section is given in the map. The European and the Adriatic lithospheric domains are distinguished with different colours; the Moho discontinuity (white dashed line) is from Locardi and Nicolich (1988); the lithosphere thickness is from Suhadolc *et al.* (1990); the adopted structural style is from Lavecchia and Stoppa (1989) and Lavecchia et al. (2003).

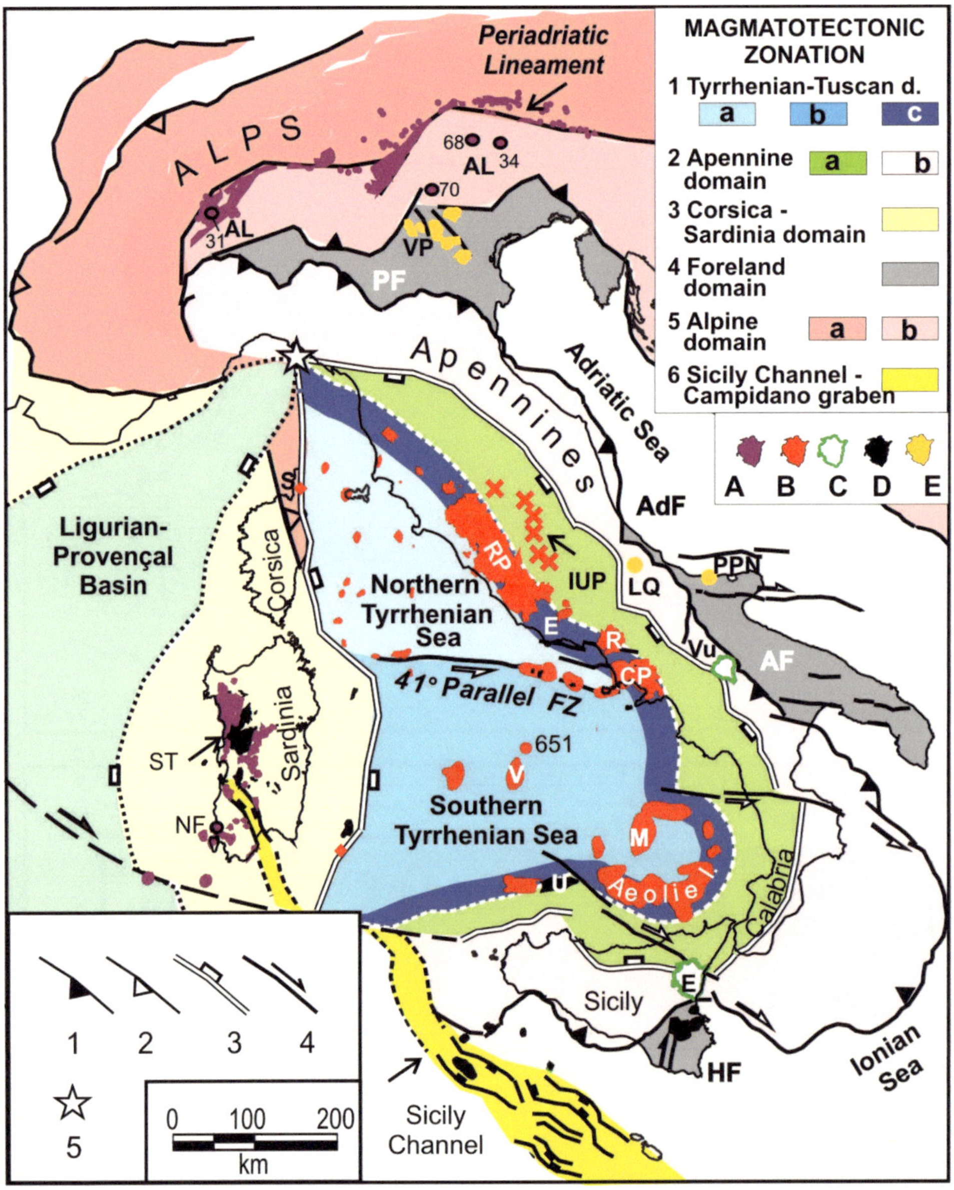

Fig. 2. Major "Magmatotectonic Domains" defined on the basis of both the compositional character and age of the igneous occurrences and the thickness and tectonic setting of the crust and the lithosphere at the time of the volcanic emplacement (after Lavecchia and Stoppa, 1996; Bell et al., submitted). The features of the various domains numbered 1, 2, 3, 4, 5 and 6 in the legend, and their subdivision in sub-domains (a, b , c), are described within the main text. The occurrences labelled A, B, C, D and E at the base of the legend in the upper right side of the figure are distinguished with different colours based on their age, composition and major features in terms of radiogenic isotopic arrays. Labels A, B and C refer to centres lying within the FOZO-ITEM isotopic array in Fig. 6; labels D and E refer to centres lying within other isotopic arrays in the same figure (possible FOZO-EM1 and

FOZO-HIMU) Key: A = late Eocene-early Miocene calc-alkaline magmatism widely distributed along the Periadriatic Lineament and the Sardinian Trough (ST) plus individual lamprophyric occurrences, with their average ages in Ma, in the eastern and western Alps (AL = Alpine lamprophyres) and in south-western Sardinia (NF = Nuraxi Figus); B = Plio-Quaternary Tuscan-Tyrrhenian and intra-Apennine magmatism, from sub-alkaline transitional in the Tyrrhenian Sea (V = Vavilov, M = Marsili, 651 = ODP drilling site), to prevailingly calc-alkaline, K-alkaline in the Aeolian islands, to HK-alkaline and ultra-alkaline in peninsular Italy (RP = Roman Province, E = Ernici; R = Roccamonfina, CP = Campanian Province which includes Vesuvius and the Phlegrean Field; IUP = Intramontane Ultra-alkaline Province); C = volcanoes lying in between the outer extensional and the outer compressional fronts (Vu = Vulture, Etna = E); D = late Miocene to Quaternary Na-alkali basalts of Sardinia, the southern Tyrrhenian border (U = Ustica), the Sicily Channel and the Hyblean plateau; E = Late Paleocene-Oligocene Na-alkaline to ultra-alkaline occurrences (VP = Veneto Province, LQ = La Queglia, PPN = Punta delle Pietre Nere) lying within the foreland domain (PF = Padan Foreland, AdF = Adriatic Foreland, AF = Apulia Foreland, HF = Hyblean foreland). The legend in the lower left corner refers to: 1 = active thrust front; 2 = Alpine thrust front; 3 = outer limit of active extension; 4 = major transcurrent fault systems; 5 = pivot point during the Mio-Pliocene opening of the triangular shaped Ligurian-Provençal and Tyrrhenian basins.

Gargano region (Punta delle Pietre Nere, 62-58 Ma) and in Abruzzo (La Queglia, 62-54 Ma) (Figs. 2 and 3).

The prevailingly Eocene Alpine collisional phase was characterized mostly by calc-alkaline and K-alkaline magmatism concentrated along the Periadriatic Lineament that extends across the entire Alpine belt in an approximate E-W direction (Macera et al., 2008) (Fig. 2). Such activity occurred in middle Eocene - late Oligocene times (~42 to 24 Ma), climaxing in the early Oligocene (~34 to 28 Ma). Isolated lamprophyric activity also occurred in the south-eastern (Val Fiscalina, 34 Ma) and in the south-western Alps (Sezia-Lanzo, Combin and Biellese, ~29 to 33 Ma) (Fig. 3). In addition, extensive Paleocene-early Oligocene volcanism of primarily basaltic composition occurred in the Veneto foreland region of south-eastern Alps (e.g. Macera et al., 2003).

The Oligocene-early Miocene Western Mediterranean extensional phase was characterized by prevalent calc-alkaline magmatic activity mainly developed between ~38 and 15 Ma, with the peak of activity taking place between ~22 and 18 Ma. Such a climax was more or less contemporary with the maximum opening of the Ligure-Provençal basin, from ~21 to 16 Ma. The magmatic products extensively outcrop within the Sardinian Trough, which extends for nearly 220 km along the western side of Sardinia (Figs. 1 and 2), and partially along the French coast (Provençe, ~34 to 16 Ma) (Savelli, 2002; Cherchi et al., 2008). It is interesting to note that the early Oligocene magmatic activity within the Sardinian Trough is contemporary with the peak of activity along the Periadriatic Lineament.

The middle Miocene-Quaternary volcanism in the Tyrrhenian Sea is characterizes by a wide range of products, whose distribution, age and petrology neatly fits with the progressive, eastward-migrating process of crustal extension and lithospheric stretching. At any given site, the magmatic activity post-dates the beginning of extensional activity by up to 2-3 Ma (Lavecchia and Stoppa, 1996). Details about the Tyrrhenian and circum-Tyrrhenian magmatic activity are given below.

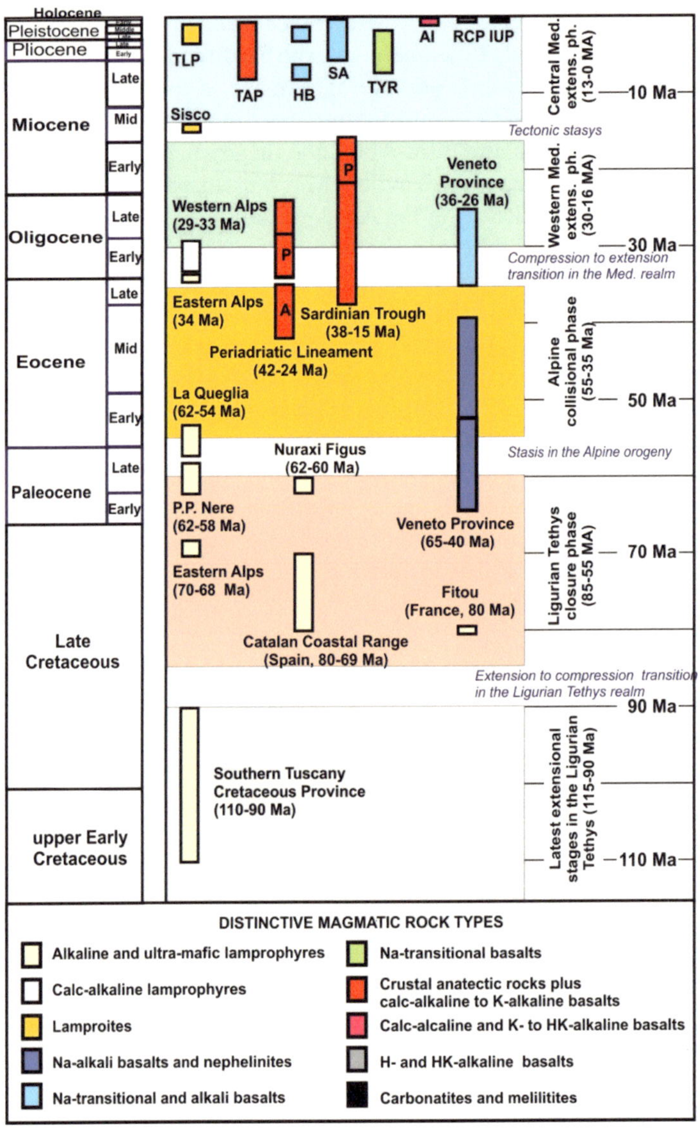

Fig. 3. Schematic chronostratigraphic chart with major tectonic and magmatic events in Italy since Late Cretaceous times. Key: A = Adamello batholith; AI = Aeolian Islands; HB = Hyblean plateau; IUP = Intramontane Ultra-alkaline Province; P = peak of magmatic activity along the Periadriatic Lineament and within the Sardinian Trough; P.P. Nere = Punta delle Pietre Nere; RCP = Roman-Campanian Province; SA = Sardinia; TAP = Tuscan Anatectic Province; TLP = Tuscan Lamproitic Province; TYR = Tyrrhenian basin.

4. Magmatotectonic domains

We define the term 'Magmatotectonic domain' as a lithospheric-scale structural domain which is homogeneous from the geometric-kinematic point of view, and closely associated with one or more well-defined igneous province. An igneous province, as considered here, consists of specific igneous associations, relatively discrete in time and space, characterized by distinctive major, minor and trace elements, as well as isotopic compositions. In Italy, the close spatial relationship between the surface distribution of the igneous provinces and well-defined, structural domains indicate that the magmatic activity is strongly controlled by the tectonics of the lithosphere. We define six major magmatotectonic domains. These are schematically shown in Figure 2.

1. The Tyrrhenian-Tuscan domain, that consists of the basinal area and its eastern onshore shoulder in Tuscany, Latium and Campania, underwent progressive eastward extension starting in late Miocene times along the eastern side of the Corsica-Sardinia block. This process progressively migrated eastward, with present activity occurring along the axis of the Apennine mountain chain (Lavecchia, 1988). The domain mainly lies within the boundary of the 50 km depth contour line that corresponds to the lithosphere-asthenosphere boundary (Panza and Suhadolc, 1990). The lithospheric stretching is mainly achieved through low-angle, east-dipping normal faulting and antithetic, high-angle faults (Fig. 1b). The consequential crustal thinning and mantle upwelling is associated with thermal highs with nodes situated mainly in the Latium and southern Tuscan regions and in the southern Tyrrhenian Sea (e.g. Rehault et al., 1987). Based on the geometry of the lithospheric structure and on the character of the magmatic occurrences, three major sub-domains can be identified, one (sub-domain 1a) northward of a nearly E-W lithospheric discontinuity known as "41° Parallel Fault Zone", one south of it (sub-domain 1b) and another running along the eastern Tyrrhenian side, from Tuscany to the Aeolian insular arc (sub-domain 1c). These sub-domains are shown in Figure 2.

   Sub-domain 1a consists of the northern Tyrrhenian Sea and of the Tuscan onshore region; it is characterized by thinned crust (20-25 km) and lithosphere (~50 km) and is typically marked by the Tuscan Magmatic Province. This province consists of distinct magmatic associations: felsic rocks of crustal anatectic origin together with subordinate, sub-alkaline basalts, late Tortonian to early Pleistocene in age (Conticelli and Peccerillo, 1992; Serri et al., 1993), and rare lamproitic rocks of Pliocene-Pleistocene age (Orciatico-Montecatini 4.1 Ma, Torre Alfina 1.3 Ma). An isolated lamproitic outcrop of middle Miocene age (~14.5 Ma) is situated at Sisco, on the north-western Corsican coast. In southern Tuscany, alkali-lamprophyric rocks of Early-Late Cretaceous age are also found (Faraone and Stoppa, 1990).

   Sub-domain 1b consists of the southern and south-eastern Tyrrhenian Sea, which is characterized by thinned to ultra-thinned continental crust (~20 to 5 km) and lithosphere (~50 to 30 km), the extension being largely achieved by top-to-the-east extensional, low-angle, normal faults. A mantle core complex of peridotitic rocks, overlain by a thin volcanic layer and by Pliocene sediments, characterizes the Vavilov basin (Mascle et al., 1991). The volcanic rocks within this sub-domain mainly consist of Na-transitional basalts and range in age from late Miocene to early Pleistocene. Alkali basalts occur at the Magnaghi and Vavilov seamounts and at Ustica and Prometeo islands. At the south-eastern side of this sub-domain the Marsili basin, coinciding with ultra-thinned crust (~10-15 km), is surrounded by the Aeolian insular arc. The latter is emplaced on thin crust (~20 km thick) and belongs to sub-domain 1c.

   Sub-domain 1c consists of the southern Latium and Campania onshore region, located in a transitional position between the Tyrrhenian rift basin and the Apennine Mountain belt. Northward of the "41° Parallel Fault Zone" (Fig. 2), it is typically marked by the magmatism of the Roman-Campanian Province (RCP), which ranges in age from middle Pleistocene to Present. In this province, which mainly consists of large volcanoes with giant calderas, the most abundant rock types are leucite tephrites and leucitites, belonging to the so-called HK-series (K_2O/Na_2O-2 >>1), (Appleton, 1972). They are also commonly associated with leucite-free, silica-saturated rocks ($K_2O/Na_20 \cong 1$) belonging to the so-called K-series. Also belonging to this sub-domain is the Aeolian Insular arc. The latter consists of calc-alkaline to K-rich rocks with an increasing potassic character; HK-rocks occur at Vulcano, Vulcanello and Stromboli (Trua et al., 2004; Francalanci et al., 2004).

2. The Apennine domain, which surrounds the Tyrrhenian basin to the east and the south, has undergone outward-verging compression since late Miocene times. The compression is still active. Seismic activity occurs along the Padan-Adriatic and the Calabrian-Sicilian thrust fronts (Lavecchia et al., 2003, 2007). Since late Pliocene times, the intra-Apennine compressional structure domain has undergone normal and normal oblique extension, which is still active and responsible for large crustal extensional earthquakes (Chiarabba et al., 2005). This domain, characterized by thickened crust (up to 40-45 km) and unthinned underlying mantle lithosphere (~100-110 km) (Fig. 4) is virtually amagmatic. Few exceptions include the small monogenic volcanoes of the Intramontane Ultra-alkaline Province (IUP) of central Italy (Stoppa and Woolley, 1997; Bailey and Collier, 2000; Lavecchia et al., 2006) and two isolated large volcanoes, Mt. Vulture and Mt. Etna. The IUP consists of a number of small monogenetic kamafugitic (kalsilite foidites or kalsilite olivine melilitites) and/or carbonatitic centres of middle-late Pleistocene age (0.74 to 0.13 Ma) which lie within a narrow area (less than 20 km wide) extending NNW-SSE for a length of ~110 km, nearly 50 km eastward from the centers of the Roman Province. The known occurrences are sited in Umbria (San Venanzo, Polino, Collefabbri), in Latium (Cupaello) and in Abruzzo (Oricola, Grotta del Cervo). The beginning of the IUP activity post-dated the onset of the extensional tectonics (middle Pliocene times) by ~3.0 Ma. The end of the IUP activity occurred while the intra-Apennine extensional regime was still active.

   Mt. Vulture is located in Lucania (southern Italy), nearly 100 km eastward from the Campanian Province, and lies between the active extensional and compressional thrust fronts, close to the boundary between the southern Apennine east-verging thrusts and the Adriatic foreland terrains (Fig. 2). The Mt. Vulture igneous rocks are trachyphonolite, phono-tephrite and melilitites. A swarm of carbonatitic maar-diatremes ~0.1 Ma years old are present in the Vulture area along the Ofanto line (Stoppa and Principe, 1998; Stoppa et al., 2009). The Etna volcano, in Sicily, also lies within the same structural position as Mt. Vulture, on the frontal thrust of the Apennine-Maghrebian chain. Etna is the largest active volcano in Europe, and its basaltic composition marks the compositional difference from the foiditic Mt. Vulture volcano.

3. The Corsica-Sardinia domain mainly coincides with the stable, on-shore and partially offshore, areas situated at the footwall of the east-dipping, normal fault system correlated with the Tyrrhenian opening (Fig. 1b). The domain, characterized by 25 to 30 km thick continental crust, is associated with widespread Na-alkaline basaltic volcanism, early Pliocene to Quaternary in age, occurring in eastern Sardinia and its offshore areas, as well as in the NW-SE striking Campidano graben (south-western

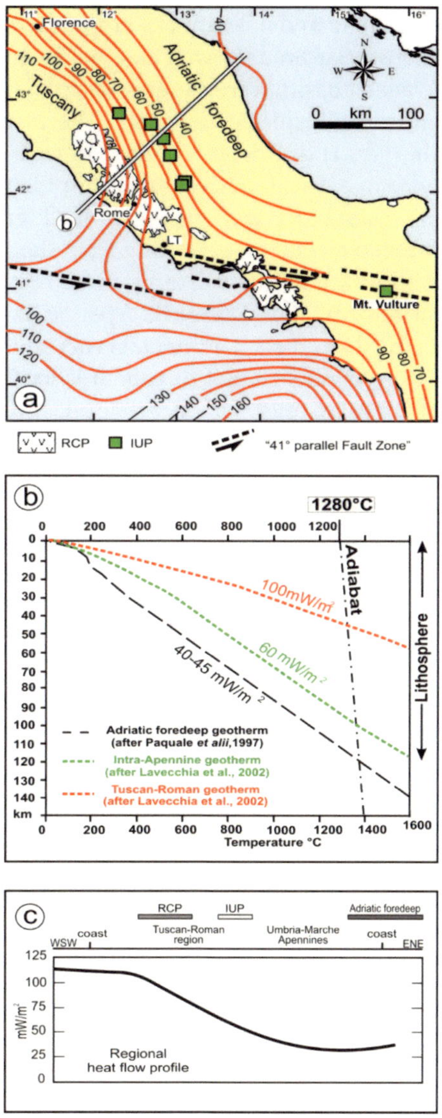

Fig. 4. a) Regional heat flow density map (mWm^{-2}) of central-southern Italy and Tyrrhenian sea (after Mongelli et al., 1989; Pasquale et al. , 1997).

b) Temperature-depth profiles for the Roman-Campanian Province (RCP) and the Intramontane Ultra-alkaline Province (IUP) (from Lavecchia et al., 2002). The geotherms have been calculated from inversion of the regional pattern of surface heat flow, assuming steady-state conditions. Thermal parameters used: mantle heat flow = $30mWm^{-2}$; upper crust thermal conductivity $k_{uc} = 2.7\ Wm^{-1}K^{-1}$; lower crust thermal conductivity $K_{lc} = 2.1\ Wm^{-1}K^{-1}$; mantle thermal conductivity $k_m = 2.6\ Wm^{-1}K^{-1}$; near surface exponential radiogenic production $H_s = 2.0*10^{-7}\ Wm^{-3}$; constant radiogenic production $H_c = 2.5*10^{-6}\ Wm^{-3}$; thickness of the layer with constant radiogenic heat production $h_c = 17$ km; characteristic length $h_r = 10$km.

c) Regional heat flow profile across the RCP and the IUP; the trace of the section is marked on the map.

Sardinia). From early Oligocene to the beginning of middle Miocene (~30 to 15 Ma), the Sardinian Trough in western Sardinia, as well as the Provençal region on the east side of the Western Mediterranean basin, were the sites of crustal anatectic and sub-alkaline magmatism, tholeiitic to calc-alkaline in character. The magmatic activity culminated between ~22 and 18 Ma (Fig. 1). It post-dated the formation, in Aquitanian times, the Sardinian trough-rift system and was concomitant with the phase of counter-clockwise rotation of the Sardinia-Corsica block and opening of the Liguria-Provençal basin.

4. The Adriatic-Pelagian foreland domain, extending from the Padan Plain to the Adriatic Sea, Apulia and the Hyblean Mountains in south-eastern Sicily (Fig. 3), consists of part of the African plate unaffected by the Alpine and Apennine compressional deformation. It is characterized by 25 to 30 km thick continental crust and lies at the footwall of both the south Alpine and the Apennine frontal thrusts. It is locally affected by discrete and localized deformational zones, mainly with strike-slip kinematics and characterized by a number of isolated magmatic occurrences. The Veneto region (south-eastern Alps), west of the Schio-Vicenza line, was the site of mafic alkaline magmatic activity from Paleocene to middle Eocene times and of prevailing Na-transitional activity during late Eocene-early Oligocene times (Figs. 2 and 3). Alkali basalts, basanites and transitional basalts are the commonest rock types (Macera et al., 2003). During late Paleocene times the Gargano-Abruzzi foreland region was characterized by two lamprophyric occurrences, possibly situated along the same major lithospheric strike-slip fault zone extending across the Gargano area of the Apulia foreland (Punta delle Pietre Nere, ~62-58 Ma) and the Abruzzi region (La Queglia, ~58-54 Ma). From late Miocene to Pleistocene times, three cycles of magmatic activity, ranging in composition from tholeiitic to nephelinitic, occurred within the Hyblean foreland, in south-eastern Sicily (8-6 Ma, 3-2 Ma, 1.5-1. 2 Ma) (Savelli, 2002). Intraplate volcanic activity with Na-alkaline affinity had also occurred in the same area from the Late Triassic to the Late Cretaceous (Beccaluva et al., 1998).
5. The Alpine deformational domain resulted from south-verging compression associated with the Alpine Tethys closure during Late Cretaceous-early Paleocene times and from double-verging compression during the Neogene collisional phase between the African and European plates. Both crust and mantle lithosphere are thickened with values of nearly 50 km and 130 km, respectively (Stampfli and Borel, 2004). The region does not show magmatic activity associated with the main Alpine Tethys closure phase, but instead is characterized by an intense, widespread, calc-alkaline activity along the Periadriatic Lineament during early Oligocene (~30-32 Ma) times. Such activity was substantially coeval with the lamprophyric activity in the south-eastern Alps (Val Fiscalina, BZ, ~34 Ma) and in the south-western Alps (Sezia-Lanzo, Combin and Biellese, ~29 to 33 Ma). It was also coeval with the onset of the calc-alkaline activity in the Sardinian Trough.
6. The Sicily Channel domain consists of a "narrow-rift" system, which extends north-westerly across the Pelagian Sea, south-westward of Sicily, with a possibly prolongation along the Campidano Graben in Sardinia. It is associated with moderately thin (~20-25 km) continental crust and characterized by normal and normal-oblique tectonic activity which dates back to the late Miocene (Beccaluva et al., 2004). The associated magmatism began in the early Pliocene and was still active in Holocene times. Most representative rock types are Na-alkaline and sub-alkaline transitional basalts, as well as peralkaline rhyolites at Pantelleria (e.g. Rotolo et al., 2006).

5. Depth and structural setting of the magmatogenic sources

No matter which of the above igneous provinces we consider, there is no general *consensus* about the composition and the depth of magma equilibration. Among the many sources proposed for the various magmatic occurrences are non-metasomatized and metasomatized lithosphere, metasomatized asthenosphere, and mesosphere. The problem is further complicated by the definition of lithosphere and asthenosphere, and whether the basis of their definition is rheological or chemical. In the case of the Quaternary igneous provinces, assuming that the present Italian lithosphere is similar to that during the Pleistocene, it is possible to speculate not only on the depth of equilibration of the magmas, based on compositional considerations, but also on the location of the sources themselves.

Following Lavecchia and Stoppa (1996), the Plio-Quaternary Tyrrhenian and peri-Tyrrhenian magmatic products may be schematically divided into two magmatogenetic groups, whose parental magmas equilibrated at different depths. The first included the Tyrrhenian transitional and sub-alkaline transitional basalts, the calc-alkaline to potassium series of the Aeolian Islands and part of the Roman-Campanian Province, as well as of the Na-series typical of Etna, the Sicily Channel and the Campidano Graben in Sardinia. These products are mainly SiO_2-saturated and are attributed to parental melts derived from a non-metasomatized, depleted mantle source. These can be related to liquids equilibrated at pressures <22 kb (i.e. at a depth less than ~70 km) within a relatively homogeneous lherzolite containing variable amounts of spinel (Olafsson and Eggler, 1983; Peccerillo and Manetti, 1985). Both garnet and spinel-peridotites have been proposed for the Sicily Channel products (Rotolo et al., 2006).

The HK-series of the Roman-Campanian Province and the ultra-alkaline products of the IUP, which are strongly SiO_2-undersaturated, highly potassic to ultra-potassic in composition with high $^{87}Sr/^{86}Sr$ ratios, are assumed to have been generated from a radiogenic and metasomatized carbonate/phlogopite-bearing peridotite (Lavecchia and Stoppa, 1996; Bailey and Collier, 2000). These melts have been related to liquids equilibrated at pressures between 22 and 24 kb i.e. at a depth of 70-80 km (Peccerillo and Manetti, 1985), and those of the IUP to pressures in the range of 28-30 kb (Cundari and Ferguson, 1994), i.e. depths of 85-100 km. Direct evidence for the mantle source composition for the IUP group is given by mantle nodules found within the volcanic rocks that commonly consist of phlogopite-clinopyroxenite, phlogopite-wherlite, spinel-wherlite and phlogopitite (Conticelli and Peccerillo, 1992; Stoppa and Lupini, 1993; Rosatelli et al, 2007). A depth of ~75 km has been estimated for the mantle xenoliths sampled at Monticchio at Mt. Vulture (Jones et al., 2000).

Assuming a value of about 1280°C for the potential temperature (TP) at the lithosphere-asthenosphere boundary (LAB) (Cundari and Ferguson, 1994; Cella et al., 1998; Federico and Pauselli, 1998) and considering the regional heat flow values of 60 and 100 mW/m^2 which characterize the IUP and RCP areas, respectively (Pasquale *et al.*, 1997), Lavecchia et al. (2002) calculated the corresponding steady-state geotherms and the lithosphere thermal thickness (Fig. 4). The adiabatic curve corresponding to a TP of 1280°C intersects the RCP and the IUP geotherms at depths of ~45-50 and ~85-90 km, respectively. The same adiabatic curve intersects the geotherm of the Adriatic foreland area (heat flow of 45-50 mW/m^2, Pasquale et al., 1997) at a depth of ~115 km and the geotherm of an average continental lithosphere (heat flow of 60 mW/m^2, Zhou, 1996) at a depth of ~110 km. On the basis of these findings, parental melts could have equilibrated within the lithosphere, close to the LAB, or within the uppermost asthenosphere (Fig. 5) The difference in the regional geothermal gradient between the RCP and the IUP essentially corresponds to a sharp LAB deepening in central Italy, moving from the thinned Tuscan lithosphere to the unthinned Adriatic lithosphere (Fig. 5). The lithospheric step is located beneath the transition area between the RCP and the IUP. The surface position of the RCP and the IUP corresponds to the surface projection of the upper and lower LAB-hinge zones, respectively. Beneath the IUP, a significant amount of melt generation would have been precluded by the almost unthinned lithosphere; beneath the RCP, it would have been allowed by the thinned lithosphere and related high stretching-factor. The different amount of melts may also be considered responsible for the different volcanic styles of the two provinces: giant calderas in the RCP and diatremes in the IUP; the latter is probably due to the small melt volume, fluidisation and gas exolution during upward migration. In both provinces, the volcanic activity post-dates, by some millions of years, the beginning of the horizontal, extensional tectonics and occurred only after the onset in middle Pleistocene times of prevailing vertical tectonics with a regional uplift. The surface distribution in central Italy of the IUP and RCP occurrences along narrow NNW-SSE bands corresponds to the surface projection of the lower and upper LAB-hinge zones (Fig. 5). This suggests that the sharp lithospheric step and

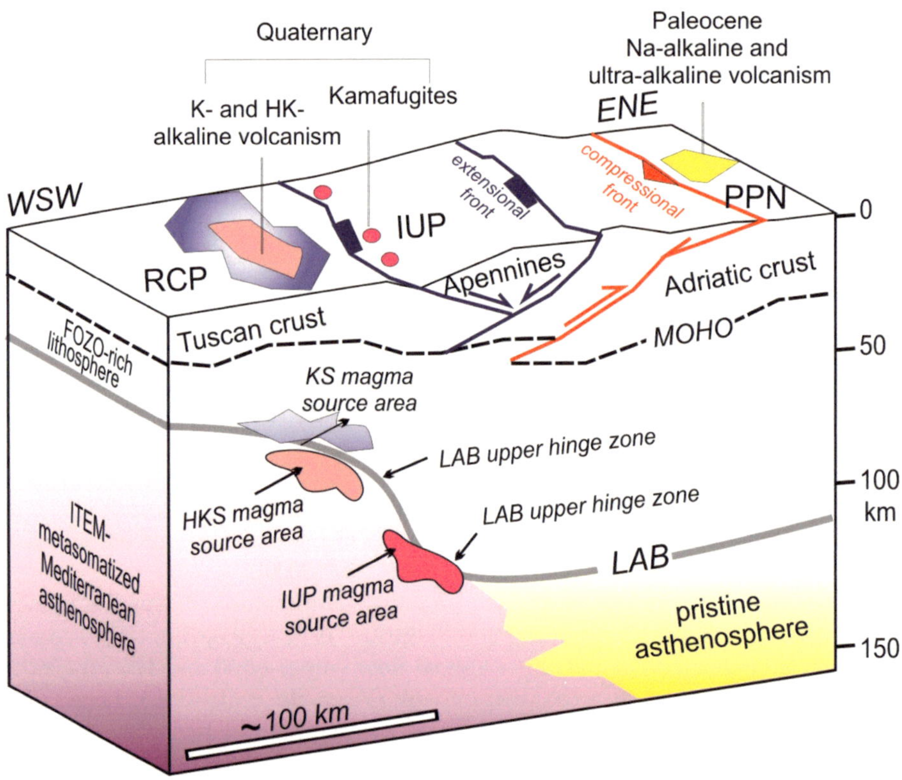

Fig. 5. Schematic block diagram across central Italy illustrating the present-day crustal and lithospheric structure and the relationships between the surface distribution of the various igneous provinces and magma types and the inferred mantle location of their parental melts (after Lavecchia and Stoppa, 1996; Lavecchia et al., 2003). Key: RCP = Roman-Campanian Province; IUP = Intramontane Ultra-Alkaline Province; PPN = Punta delle Pietre Nere. Note the inferred ITEM isotopic character of the upwelled Tyrrhenian asthenosphere.

lateral density contrast between the thinned Tyrrhenian and unthinned Adriatic lithosphere may have controlled the uprising of melts from asthenosphere reservoirs into relatively narrow bands across the lithosphere. In the case of Somma-Vesuvius and Vulture, the focusing of the volatiles from the mantle reservoir into one well-defined site might have been controlled by the intersection between the nearly E-W "41° Parallel Fault Zone" (see Figs. 2 and 4a) and the upper and lower LAB-hinge zone, respectively.
A source region close to the base of the lithosphere with normal temperatures is also proposed by Beccaluva et al. (2007) for the Paleogene Veneto province and for the other magmas of the Adria and Hyblean foreland domains. They considered that most of the magmas were equilibrated within the spinel-peridotite lithosphere mantle, from sources ranging in depth from about 30 to 100 km and with concomitant decrease in the degree of partial melting from (25% to 3%) from quartz-normative tholeites to nephelinites. Two kinds of mantle mineralogy were identified: lherzolite-bearing amphibole with phlogopite for the tholeiites to basanites, and similar sources, but with the addition of some carbonatitic components for the nephelinites.
The origin of the late Eocene-early Miocene magmatic activity along the Periadriatic Lineament and along the Sardinia Trough can be interpreted as the result of melting due the presence of H_2O and K-rich fluids during unloading processes. Such processes are commonly associated with the Ligurian–Provençal back-arc, extensional process in Sardinia and with the Alpine slab break-up in the case of the Periadriatic Lineament (Macera et al., 2008 and references therein). It is interesting to note that in the late Eocene-early Oligocene (that is before the opening of the Ligure-Provençal basin), the Periadriatic Lineament and the Sardinian Trough were nearly continuous along a common ENE-WNW direction. Also, considering the magmatic affinity between the two regions, we wonder if both might be related to the onset of extensional processes in the Western Mediterranean basin, with consequent lithospheric unloading. In such a case, the Periadriatic Lineament, corresponding to a pre-existing high-angle, northward-dipping, crustal discontinuity between the north-verging and the south-verging Alpine system, might have acted as a transfer fault to allow the eastward shift of the Adria foreland and the progressive opening of the Provençal basin.

6. Radiogenic isotopic compositions

The variation of the Italian igneous rocks in terms of Sr, Nd and Pb radiogenic isotopic is extreme, reflecting both depleted and very enriched sources, not only on a regional, but also on a local scale (Bell et al., 2005; Lustrino and Wilson, 2007; Lustrino et al., 2011; Bell et al., submitted). The compositional variations can be defined by a limited number of end-members that form well-defined binary mixing array (Fig. 6), that possibly reflect the magmatotectonic domains (Fig. 2). Two end-members are of widespread distribution and are considered to be plume-related. They are FOZO (Focus Zone) with low $^{87}Sr/^{86}Sr$ (0.7025), high $^{143}Nd/^{144}Nd$ (0.51315), moderate $^{206}Pb/^{204}Pb$ (19.40) and HIMU (High μ = high $^{238}U/^{204}Pb$) with low $^{87}Sr/^{86}Sr$ (0.70285), high $^{143}Nd/^{144}Nd$ (0.51285), high $^{206}Pb/^{204}Pb$ (22.00) (Hart et al., 1992, and reference therein). FOZO is considered to be a ubiquitous component of all of the analysed rocks from Italy, whereas an HIMU-like component is mainly restricted to occurrences lying within the Padan-Adriatic-Pelagian foreland and the Sicily Channel. Two other end-members range from very radiogenic (ITEM) to moderately radiogenic (possibly EM1) pointing to metasomatised mantle sources. ITEM (ITalian Enriched Mantle) with very high $^{87}Sr/^{86}Sr$ (0.7200), very low $^{143}Nd/^{144}Nd$ (0.51185), low $^{206}Pb/^{204}Pb$ (18.70) is widespread in the Italian mantle (Bell et al., 2005), while the EM1 component, with moderate $^{87}Sr/^{86}Sr$ (0.70530), low $^{143}Nd/^{144}Nd$ (0.51236) and very low $^{206}Pb/^{204}Pb$ (17.50), is mainly observed in northern and central Sardinia.

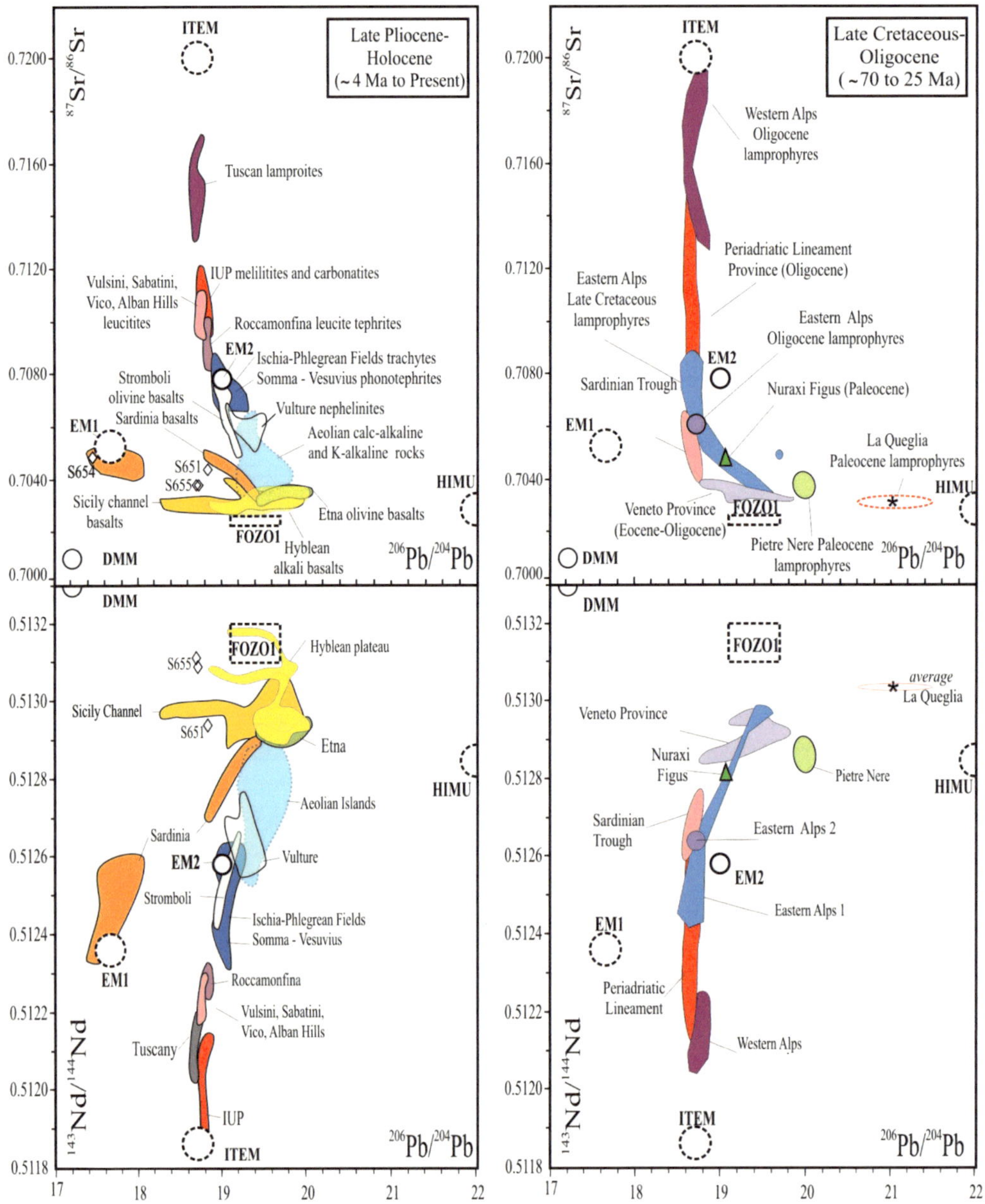

Fig. 6. Plots of $^{87}Sr/^{86}Sr$ and $^{143}Nd/^{144}Nd$ against $^{206}Pb/^{204}Pb$ for the Italian young (left side) and old (right side) volcanic occurrences. The diagram on the left side is from Bell et al., 2005; the diagram on the right side is from Bell et al., 2005 integrated with data from Owen, 2008; Beccaluva et al., 2007; Lustrino et al., 2011 and Bell et al., submitted. The identified isotopic distribution is compared with known world-wide, mantle end-members (EM1, EM2, HIMU, FOZO, (Hart et al., 1992, and references therein) and with the mantle Italian component ITEM (Bell et al., 2005 and 2006). We use the term FOZO1 in the diagram to refer to Hart et al's 1992 values since there are others in the literature.

The main mixing array is defined by FOZO-ITEM and includes data from the Tyrrhenian, the Apennine and the Alpine magmatotectonic domains (from Etna in Sicily to the Alps). A relevant increase in $^{87}Sr/^{86}Sr$ (from 0.703 to 0.720), with a corresponding relevant decrease in $^{143}Nd/^{144}Nd$ and a slight increase in $^{206}Pb/^{204}Pb$ (from 19 to 20) is observed moving along the binary mixing array from ITEM to FOZO (Fig. 6). This mixing array essentially traces the isotopic compositions of a wide range of rocks, outcropping from north to south along the length of Italy and ranging in age from Late Cretaceous to Quaternary (Fig. 6 a and b). The most radiogenic Sr compositions are shown by the early Oligocene western and eastern Alpine lamprophyres ($^{87}Sr/^{86}Sr$ ~0.72); progressively decreasing values are observed moving towards the Plio-Quaternary Tuscany lamproites, the Oligo-Miocene Periadriatic calc-alkaline rocks, the Quaternary IUP carbonatites (~0.712), the Roman Province leucitites, the Campanian Province phonotephrites, the Vulture nephelinites and the Stromboli alkali basalts, down to the Aeolian calc-alkaline products and the alkali basalts from Etna (Fig. 6).
The FOZO-HIMU-like mixing line almost exclusively consists of data from outcrops belonging to the Padan-Apulia-Pelagian foreland domain. Moving from HIMU-like to FOZO, we observe a relevant decrease in $^{206}Pb/^{204}Pb$, from an average value of ~21.5 at la Queglia to a minimum of ~18.30 at the Sicily Channel and rather constant values in $^{87}Sr/^{86}Sr$ ratios (0.703) and high radiogenic $^{143}Nd/^{144}Nd$ (0.512-0.513). Along this array, we move from the Paleocene lamprohyres of La Queglia and Punta delle Pietre Nere in the Apulia foreland, to the Paleocene-Oligocene Na-alkali basalts of the Veneto Province in the Padan foreland, to the late Miocene-Pliocene Na-basalts in the Hyblean mountains within the Pelagian foreland, to the Na-alkali basalts of the Sicily Channel.
A possible FOZO-EM1 mixing line constitutes a third subordinate array. It almost exclusively consists of data from Pio-Quaternary rocks from northern and central Sardinia.

7. The southern Tyrrhenian Benioff plane

The Tyrrhenian extensional basin is commonly considered a back arc-basin, developed at the rear of the NW-subducted Ionian lithosphere, with the Aeolian islands offshore of Calabria being considered the associated insular arc (Malinverno and Ryan, 1986; Doglioni et al., 1997). Such an interpretation is largely based on the presence of deep-focus earthquakes (up to depths of ~500 km) supposedly associated with a westerly-dipping subduction plane (D'Agostino and Selvaggi, 2004; Chiarabba et al., 2005). As a matter of fact, the Southern Tyrrhenian Benioff Plane (STBP), analysed in terms of size, strain deformation pattern and spatial relationships with the overlying Aeolian volcanic arc appears to be in conflict with the classic geometric and kinematic configurations predicted by both active and passive subduction-related models (Isacks and Molnar, 1971).
The STBP depth contour lines in Fig. 7a were obtained by analysing the depth distribution of the sub-crustal (>35 km) seismicity, in the time interval 1978 to 2008. The data were extracted from the International Seismological Centre database and projected along a set of radial sections across the plane, with an average semi-amplitude of 25 km. The hypocenters projected in Fig. 7b were extracted from the same dataset, assuming a semi-width of 60 km along the trace of the section.
In the map of Fig. 7a, the STBP depth contour lines show an average NE-SW direction in the southern sector, which then turns to NNE-SSW. The lack of a geometric correspondence with the shape of the Aeolian volcanic arc, which is tightly curved, is evident. In the section view of Fig. 7b, the Benioff plane is sub-vertical to deeply SE-dipping down to nearly 200

km and dips to the northwest, at an average dip angle of about 50°, at higher depths. Large portions of the slab are characterized by the presence of aseismic domains which prevail at depths greater than 350 km. The Aeolian volcanoes, projected along the trace of the section, are positioned both above the seismic and aseismic plane segments. The various volcanoes lie at different depths above the plane, from a minimum of ~150-200 km for Vulcano-Lipari, to depths of ~250-300 km for Filicudi-Alicudi and still more for Enarete-Sisifo (~400 km). It is evident that the Aeolian magma sources do not form a more or less continuous linear zone at a constant depth along the top of the subduction zone, as should be if they were connected to a down-going slab (Moores and Twiss, 1995).

Available focal mechanisms associated with the STBP show a predominant normal-faulting kinematics (Chiarabba et al., 2008). T-axes turn in azimuth perpendicularly to the slab direction and P-axes are usually parallel to the average plane dip direction (blue and red arrows in Fig. 7 b). Such evidence of slab down-dip compression enables us to exclude any model of retreating slab, which would be characterized by down-dip extension, due to the negative buoyancy of the subducting lithosphere. On the other hand, a process of active subduction, consistent with the down-dip compression, would imply fault plane solutions showing thrust faulting in the shallower depth range (0-100) km in the vicinity of the convergent boundary which are missing in the study area (Chiarabba et al., 2005).

The Southern Tyrrhenian Benioff zone is also unusual in terms of its size. Its lateral extension (nearly 250 km) is one of the smallest on Earth and its along strike length/along dip-length ratio is very low (about 0.5) unlike the circum-Pacific subduction planes (about 20). Furthermore, an along-strike length of a few hundred kilometres cannot help explain the length of the Apennine-Maghrebian belt, that would represent the associated accretionary prism extending for nearly 3500 km from northern Italy to the Gibraltar Arc, unless 90% percent of the subduction plane is considered aseismic.

Another interesting point concerns the balance between the along-dip length of the STBP (maximum length of 500 km) and the amount of shortening of the Apennine fold-and-thrust belt system. The volume of the entire Apennine crust in Calabria is smaller than the volume of the upper crust that would be involved in the formation of an accretionary prism, assuming that the upper crust had been scraped off during subduction (Doglioni *et al.*, 1999). In general, the compressional structures of both the central and the southern Apennines do not show the thin-skinned geometries typical of subduction-related complexes, but rather they are characterized by a thick-skinned style, typical of ensialic deformations, with basement largely involved in the deformation and with only limited amounts of horizontal shortening (van Dijk et al., 2000; Barchi et al., 2001; Noguera and Rea, 2000; Lavecchia et al., 2003).

Tomographic models of the mantle beneath the Apennines and the Tyrrhenian show the presence of a highly discontinuous, intra-asthenosphere, high-velocity body, usually assumed to be the Ionian lithosphere subducted in the course of the Apennine compressional phase, but images are very different in length, position and continuity (Spakman *et al.*, 1993; Cimini and De Gori, 2001; Piromallo and Morelli, 2003; Piromallo and Faccenna, 2004). Some alternative, very speculative, hypotheses have been advanced. Some authors interpret the structure of the Calabro-Sicilian Arc to krikogenesis rather than subduction (Wezel, 1981); others associate the STBP to seismic shearing along intra-asthenospheric remnants of a pre-existing Alpine subduction plane (Lavecchia and Creati, 2006), to asthenosphere dragging (Locardi and Nicolich, 1988), to deep-seated reverse faulting (Choi, 2004), and to dense mantle material rising towards the surface from a large body of lower mantle material trapped in the transition zone (Scalera, 2006).

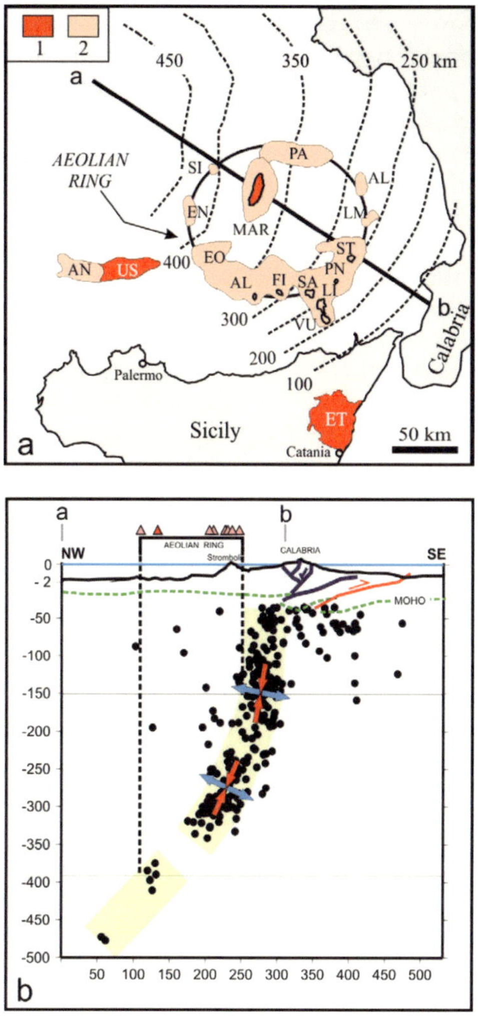

Fig. 7. a) Location of Quaternary igneous rocks in the southern Tyrrhenian Sea compared with the geometry of the Benioff plane off Calabria (from Bell et al., 2005). Key: 1) alkali basalts to trachybasalts; 2) olivine basalts, trachybasalts with shoshonites and calc-alkaline rocks; PA) Palinuro, AL) Alcione, LM) Lamentini, ST) Stromboli, PN) Panarea; SA) Salina, LI) Lipari, VU) Vulcano, FI) Filicudi, AL) Alicudi, EO) Eolo, EN) Enarete, SI) Sisifo, MAR) Marsili, AN) Anchise, US) Ustica, ET) Etna; dashed lines = reconstructed average depth contours of the Benioff plane labelled in kilometres.
b) Section view of the Southern Tyrrhenian Benioff plane along a SW-NE striking, transect assuming a semi-width of 60 km. The trace of the a-b segment of the section is given in the map. The hypocentral data set is from the International Seismological Centre (ISC); it consists of sub-crustal events registered at depth ≥35 km in the time interval 1978-2008 (M_b 2.0 to 5.7). The red and blue arrows refer to average P- and T- axes, respectively, calculated from focal mechanisms extracted from the CMT catalogue and the RCMT catalogues, available on line. The pink triangles above the section represent the islands of the Aeolian volcanic ring projected along the trace of the section.

No matter which of the above alternative models is favoured, the ring-like configuration of the Aeolian islands at the surface is not governed by the geometry of the slab at depth, which is much more linear (Fig. 7). The Aeolian magmatism might simply be due to lithospheric stretching, unloading and associated partial melting in the thinned and stretched areas surrounding the Marsili mantle core complex, independently from any subduction.

8. Proposed model

8.1 The origin of the mantle radiogenic components

Among the isotopic mantle end-members identified in Italy, FOZO is ubiquitous and has been involved in the generation of all the igneous products since Late Cretaceous times (Fig. 6). It is very close in composition to FOZO of Hart et al. (1992) and is similar to other European and Mediterranean end-members, such as EAR (European Asthenospheric Reservoir, Granet et al, 1995; Wilson and Patterson, 2001), LVC (Low Velocity Component, Hoernle et al., 1995) and CMR (Common Mantle Reservoir, Lustrino and Wilson, 2007). The world-wide FOZO component was first identified on the basis of isotopic data from oceanic island basalts; it is anorogenic, independent of subduction and normally associated with intra-plate magmatism. The Italian ubiquitous FOZO end-member, first introduced by Bell et al (2003), has been considered a pure deep mantle component entrained within the Mediterranean lithosphere and asthenosphere *via* upwelling plumes (Bell et al., 2006; Cadoux et al, 2007). It might also represent an ancient phase of regional rifting which pre-dated the Late Triassic-Early Jurassic continental break-up of Europe and Africa (Bell et al., submitted).

In the Alps, the Apennines and the Tyrrhenian Sea, the ITEM component is just as ubiquitous as FOZO, and is involved to different degrees in the generation of almost all of the igneous rocks. Conversely, there is no trace of ITEM within the Alpine and Apennine foreland domain (Figs. 2 and 6). The ITEM involvement is lowest in Sicily and highest in the Alps. Given its high $^{87}Sr/^{86}Sr$ ratio it has commonly been interpreted as a crustal component released within the mantle during an inferred Apennine subduction process (Peccerillo, 1999). ITEM, however is most prevalent in the most primitive and extreme ultra-alkaline Italian rocks (carbonatites, melilitites, lamprophyres) that are not related to subduction. Its presence not only in the early Oligocene Alpine lamprophyres, but also in the Late Cretaceous-Paleocene lamprophyres from the western Alps and possibly in the Early-Late Cretaceous lamprophyres of southern Tuscany means that whatever process generated the mantle enrichment, it had to date back at least to Cretaceous times. This rules out any relationship with the perceived subduction of continental crust during the Mediterranean extensional phase.

In general, ITEM increases with an increase in Mg number, with an increase in K content and with an increase in depth of the magmatogenetic source. This might reflect depth-controlled variation in partial melting and in the mantle source composition, with ITEM being more common at deeper levels. The ITEM signature always co-exists with FOZO, requiring two distinct mantle reservoirs that are contiguous and hence able to communicate with one another. A vertical compositional variation, with a prevailingly depleted lithosphere and an underlying metasomatised, ITEM-rich, asthenosphere enriched by plume-derived fluids is one possible solution, or a heterogeneous mantle plume another. According to Lavecchia and Creati (2006), the ITEM end-member could originate within the D" layer, at the lower mantle/core transition, where it could have evolved to its extreme isotopic compositions and could have been transported within the asthenosphere *via* plumes, in recent times.

The HIMU-like composition is not as widespread as ITEM and FOZO, but is still regionally, fairly important since it is present in all of the occurrences within the Adriatic foreland. Beccaluva et al. (2007) explains the isotopic characteristics of the mantle sources within the Adriatic foreland regions, as due a variable mixing of HIMU and, to a lesser extent, EM2 metasomatic components with a pristine depleted-mantle lithosphere. Given its extreme isotopic compositions in radiogenic lead, HIMU has been interpreted as a lower mantle component which has experienced considerable ageing during long-term residence (Collerson et al., 2010). In the case of Italy, the HIMU-like end-member is found, together with FOZO, within the sodic-alkaline and ultra-alkaline suites of the Foreland domain and of the Sicily Channel domain. We could hypothesise a present residence of the HIMU-like component within the asthenosphere, in areas that have not been affected by the plume-generated ITEM metasomatic components.

The EM1-like component is rare in Italy and is found in the Plio-Pleistocen basalts of northern and central Sardinia. It is similar to that of the eastern Atlantic, northern Africa, and central Europe (Macera et al., 2003) and might be ascribed to melting of metasomatised veins enclosed within the European continental lithosphere.

Summarising, the FOZO component might reside within the lithosphere of both the Tyrrhenian-Tuscan stretched domain and the surrounding unstretched areas (Foreland domain and Sardinia-Corsica domain in Fig. 2) where perhaps it might have developed during plume-driven, rifting processes pre-dating the Pangea break-up. Conversely, both the ITEM and HIMU-like components could reside within the asthenosphere, in geographically and tectonically distinct magmatotectonic domains. ITEM could occur beneath the Tuscan-Tyrrhenian domain (Fig. 2) and, in general, beneath the widely extended Western and Central Mediterranean domain. The HIMU-like component might lie beneath the stable foreland domain, and beneath the "narrow-rift" domains, such the Sicily Channel and the Central Europe Rift System.

8.2 The Mediterranean trapped plume model

We do not consider the Western and Central Mediterranean regions as a back-arc basin developed on the rear of westward-subducting lithosphere, but instead propose an alternative hypothesis that involves lithospheric stretching driven by mantle asthenosphere expansion due to the growth of a plume head (Fig. 8) (Lavecchia and Creati, 2006; Bell et al., submitted). Because most plumes involve thermal highs, elevated topography and flood basalts, missing in the Mediterranean region, the plume head would have to be trapped within the transition zone between the asthenosphere and the mesosphere (~410 to 670 km depth). Such a mantle plume would carry radiogenic fluids/melts from the deep mantle in order to generate metasomatic agents with an ITEM signature that affected the host Mediterranean asthenosphere. Influxes in the upper mantle of fluids/melts associated with a plume enriched in CaO-CO_2-K_2O would generate a source capable of producing ultra-alkaline magmatism, and the deep-seated CO_2 emissions found in peninsular Italy. The progressive eastward growth of a large plume head trapped within the transition zone and the consequent asthenospheric volume increase would also cause stretching and large scale extension of the overlying lithosphere. The consequent lithospheric unloading would, in turn, control the Tyrrhenian and peri-Tyrrhenian magmatogenetic activity.

Given the Late Cretaceous age of the oldest Italian occurrences carrying the ITEM signature, the birth of such a plume should be older than the onset of the Mediterranean extensional phase, and perhaps related to the Alpine Tethys Jurassic extensional phase. We hypothesise a pulsating plume activity (Bell et al., submitted). We argue that during the Alpine compressional phase, the plume was relatively quiescent and only produced minor upwellings that allowed the hot, low viscosity material to escape to upper levels, thus generating a number of isolated lamprophyric occurrences on both the European and the African sides of the Alpine orogenic zone, implying that their emplacement was not controlled by slab migration. During the Mediterranean phase, the plume was active again and controlled the magmatotectonic evolution of the Ligure-Provençal and Tyrrhenian basins.

The interpretation of the Mediterranean region in the framework of plume-driven mantle expansion also implies an unusual interpretation of the large-scale high-velocity anomaly, shown by tomographic data within the transition zone (410 to 670 km) (Piromallo and Morelli, 2003). The high-velocity body might not represent accumulated Alpine and Apennine subduction material, but rather the trapped plume head (Lavecchia and Creati, 2006). In such a case, the velocity anomaly would not result from thermal variations (cold subducted lithosphere), but rather from chemical/compositional variation (upwelled deeper mantle material) (Fig. 8). The low-velocity tomographic anomaly that characterizes the Mediterranean upper mantle down to 400 km, could in turn represent asthenospheric material that has been enriched, metasomatised and softened by the fluids released by the plume head.

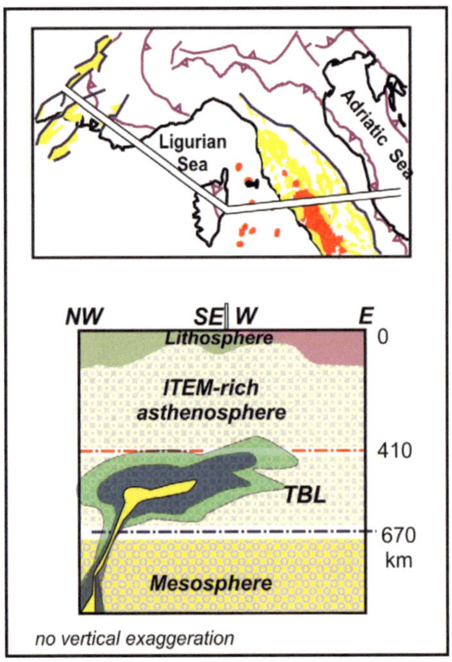

Fig. 8. Proposed interpretation of the mantle structure beneath the lithospheric section of Fig. 1b, assuming the trapped plume scenario discussed in the paper. The shape of plume head although speculative (from Lavecchia and Creati, 2006) is largely derived from Fig. 6 in Brunet and Yuen (2000). The plume head substantially coincides with a high-velocity anomaly highlighted by tomographic data beneath the Western and Central Mediterranean region (Piromallo and Morelli, 2003). In our sketch, the colours of plume head mark an hypothesised progressive outward increase of the Vp velocities, due to the loss of volatiles and light elements released within the overlying asthenosphere.

9. Final remarks

The geology of the Western and Central Mediterranean region, with all of its paradoxes and contradictions, presents an exceptional natural laboratory for assessing the cause-effect relationships between the tectonic/magmatic processes associated with crust, the mantle lithosphere and the underlying mantle. Rather than fitting all of the evidence that has been acquired during the last decades into a standard subduction model, we have attempted to find alternative solutions that might more easily explain the available data. The simple models of orogenic *versus* anorogenic, and active rifting *versus* passive rifting, although useful, may be more complex than we imagine, and we have sought alternative, more encompassing solutions that might explain the diverse tectonic and magmatic activity in Italy, during Tertiary and Quaternary times.

In the case of the progressive opening since Oligocene times of the Central and Western Mediterranean "wide-rift" basins, a model assuming the eastward growth of a plume head fed from the deep mantle and pinched within the transition zone (410 to 670 km depth) may be a reasonable alternative to one involving back-arc opening at the rear of a retreating subduction slab. In fact, closer evaluation of the evidence for subduction is not so convincing when looked at in detail. The unusual tectonic, petrographic and geochemical characters, such as extension dominating over compression, high potassic to ultra-alkaline suites predominating over calc-alkaline occurrences, and plume-related isotopic compositions, have been casually dismissed by subductionists as "anomalous" and "exotic" (Lustrino et al., 2011).

A trapped plume model may well explain (i) the fast lithosphere extensional rate (up to 5-6 cm/y, Faccenna et al., 2004), (ii) the cold continental geotherm (~1300°C at the lithosphere-asthenosphere boundary, Cella and Rapolla, 19987), (iii) the regional subsidence (bottom sea level reaching depths of less than 3000 km), and (iv) the lack of extensive basaltic magmatism. Because the plume head in our model does not reach the base of the lithosphere, it does not directly control the lithospheric tectonics and magmatism. Instead, the plume growth within the transition layer exerts an indirect control on the lithosphere tectono-magmatic activity. The highly radiogenic character (ITEM signature) of the fluids injected within the Mediterranean astenosphere, and carried up to the surface during igneous activity, implies a long residence time in an isolated source that may coincide with the transition zone between the asthenosphere and the mesosphere (670 km depth) and/or with the D" boundary layer at the core mantle/transition (2900 km depth).

10. References

Appleton, J.D. (1972). Petrogenesis of potassium-rich lavas from the Roccamonfina volcano, Roman region, Italy. *Journal of Petrology*, Vol. 13, pp. 425-456.

Bailey, D.K. and Collier, J.D. (2000). Carbonatite-melilitite association in the Italian collision zone and the Ugandan rifted craton: significant common factors. *Mineralogical Magazine*, Vol. 64, pp. 675-682.

Barchi, M., Landuzzi, A., Minelli, G. and Pialli, G. (2001). Outer northern Apennines. In: *Anatomy of an orogen: the Apennines and Adjacent Mediterranean Basins*, Vai, G. and Martini, I. (Eds.), Kluwer Academic Publishers, pp. 215-254, Great Britain.

Beccaluva, L., Bianchini, G. and Siena, F. (2004). Tertiary-Quaternary volcanism and tectono-magmatic evolution in Italy. In: *Geology of Italy*, Crescenti, U., D'Offizi, S., Merlino,

S., and Sacchi, L. (Eds.), Special volume of the Italian Geological Society for the IGC 32 Florence, Societa Geologica Italiana, Roma, pp. 153-160.

Beccaluva, L., Bianchini, G., Bonadiman, C., Coltorti, M., Milani, L., Salvini L., Siena F. and Tassinari, R. (2007). Intraplate lithospheric and sub-lithospheric components in the Adriatic domain: Nephelinite to tholeiitic magma generation in the Paleogene Veneto volcanic province, southern Alps. *Geol. Soc. Am. Spec. Paper*, Vol. 418, pp. 131-152.

Beccaluva, L., Siena, F., Coltorti, M., Di Grande, A., Lo Giudice, A., Macciotta, G., Tassinari, R., and Vaccaro, C. (1998). Nephelinitic to tholeiitic magma generation in a transtensional tectonic settino: an integrated model for the Iblean Volcanism, Sicily, *Journal of Petrology*, Vol.39, pp. 1547-1576.

Bell, K., Castorina, F., Lavecchia, G., Rosatelli, G. and Stoppa, F. (2003). Large scale, mantle plume activity below Italy: Isotopic evidence and volcanic consequences. *Geophysical Research Abstract*, EAE03-A-14217.

Bell, K., Castorina, F., Lavecchia, G., Rosatelli, G. and Stoppa, F. (2004). Is there a mantle plume below Italy?. *Eos*, Vol. 85, pp. 541, 546-547.

Bell, K., Castorina, F., Rosatelli, G. and Stoppa, F. (2006). Plume activity, magmatism and the geodynamic evolution of the central Mediterranean. *Annals of Geophys*ics, Vol. 49, pp. 357-371.

Bell, K., Lavecchia, G. and Stoppa, F. (2005). Reasoning and beliefs about Italian geodynamics. *Bollettino della Società Geologica Italiana,* Vol. 5, pp. 119-127.

Bell, K., Lavecchia, G., Rosatelli, G. and Stoppa, F. (submitted). Cenozoic Italian magmatism - isotope constraints for plume-related activity. *Journal of South America Earth Sciences.*

Brunet, D. and Yuen, D. (2000). Mantle plumes pinched in the transition zone. *Earth and Planetary Science Letters*, Vol. 178, pp. 13-27.

Cadoux, A., Blichert-Toft, J., Pinti, D.I. and Albarède, F. (2007). A unique lower mantle source for the Southern Italy volcanics. *Earth and Planetary Science Letters*, Vol. 259, pp. 227-238.

Carminati, E., Wortel, M.J.R., Meijer, P.T. and Sabadini, R. (1998). The two-stage opening of the western-central Mediterranean basins; a forward modelling test to a new evolutionary model. *Earth and Planetary Science Letters*, Vol. 160, pp. 667-679.

Cella, F., M. Fedi, G. Florio and A. Rapolla (1998). Gravity modelling of the litho-asthenosphere system in the central Mediterranean. *Tectonophysics*, 287, 117-138.

Cherchi, A., Mancin, N., Montadert, L., Murru, M., Putzu, M.T., Schiavinotto, F. and Verrubbi, V. (2008). The stratigraphic response to the Oligo-Miocene extension in the western Mediterranean from observations on the Sardinia graben system (Italy). *Bullettin de la Societe Geologique de France*, Vol.179, pp. 267–287.

Chiarabba, C., de Gori, P. and Speranza, F. (2008). The southern Tyrrhenian subduction zone: deep geometry, magmatism and Plio-Pleistocene evolution. *Earth and Planetary Science Letters*, Vol. 268, pp. 408-423.

Chiarabba, C., Jovane, L. and Di Stefano, R. (2005). A new view of Italian seismicity using 20 years of instrumental recordings. *Tectonophysics*, Vol. 395, pp. 251-268.

Choi, D.R. (2004). Deep tectonic zones and structure of the Earth's interior revealed by seismic tomography. *Newsletter - New Concepts in Global Tectonics*, Vol. 30, pp. 7-14.

Cimini, G.B. and De Gori, P. (2001). Nonlinear P-wave tomography of subducted lithosphere beneath central-southern Apennines (Italy). *Geophysical Research Letters*, Vol. 28, pp. 4387-4390.

Cohen, K.M. and Gibbard, P.L. (2009). Major subdivisions of the Global Statotype Section and Point (GSSP) -http://www-qpg.geog.cam.ac.uk/

Collerson, K.D., Williams, Q., Ewart, A.E. and Murphy, D.T. (2010). Origin of HIMU and EM-1 domains sampled by ocean island basalts, kimberlites and carbonatites: The role of CO_2-fluxed lower mantle melting in thermochemical upwellings. *Physics of the Earth and Planetary Interiors*, Vol. 181, pp. 112-131.

Conticelli, S. and Peccerillo, A. (1992). Petrology and geochemistry of potassic and ultrapotassic volcanism in central Italy: petrogenesis and inferences on the evolution of the mantle source. *Lithos*, Vol. 28, pp. 221-240.

Cundari, A. and Ferguson, A.J. (1994). Appraisal of the new occurrence of götzenitess, khibinskite and apophyllite in kalsilite-bearing lavas from San Venanzo and Cupaello (Umbria), Italy. *Lithos*, Vol. 31, pp. 155-161.

D'Agostino, N. and Selvaggi, G. (2004). Crustal motion along the Eurasia-Nubia plate boundary in the Calabrian arc and Sicily and active extension in the Messina Strait from GPS measurements. *Journal of Geophysical Research*, Vol. 109, pp.1-16.

de Ignacio, C., Munoz, M., Sagredo, J., Fernandez-Santin, S. and Johansson, A. (2006). Isotope geochemistry and FOZO mantle component of the alkaline-carbonatitic association of Fuerteventura, Canary Islands, Spain. *Chemical Geology*, Vol. 232, pp. 99-113.

Doglioni, C., Gueguen, E., Sabat, F. and Fernandez, M. (1997). The western Mediterranean extensional basins and the Alpine orogen. *Terra Nova*, Vol. 9, pp. 109-112.

Doglioni, C., Harabaglia, P., Merlini, S., Mongelli, F., Peccerillo, A. and Piromallo, C. (1999). Orogens and slabs vs. their direction of subduction. *Earth-Science Reviews*, Vol. 45, pp. 167-208.

Faccenna, C., Piromallo, C., Crespo-Blanc, A., Jolivet, L. and Rossetti, F. (2004). Lateral slab deformation and the origin of the western Mediterranean arcs. *Tectonics*, Vol. 23, TC1012, doi: 10.1029/2002TC001488.

Faraone, S. and Stoppa, F. (1990). Petrology and regional implications of Early Cretaceous alkaline lamprophyres in the Ligure-Maremmano Group (southern Tuscany, Italy): an outline. *Ofioliti*, Vol. 15, pp. 45-59.

Federico, C. and Pauselli, C. (1998). Thermal evolution of the Northern Apennines (Italy). *Memorie della Società Geologica Italiana*, Vol. 52, pp. 267-274.

Finetti, I.R., Boccaletti, M., Bonini, M., Del Ben, A., Geletti, R., Pipani, M. and Sani, F. (2001). Crustal section based on CROP seismic data across the North Tyrrhenian-Northern Apennines-Adriatic Sea. *Tectonophysics*, Vol. 343, pp. 135-163.

Francalanci, L., Tommasini, S. and Conticelli, S. (2004). The volcanic activity of Stromboli in the 1906-1998 AD period: mineralogical, geochemical and isotope data relevant to the understanding of the plumbing system. *Journal of Volcanology and Geothermal Research*, Vol. 131, pp. 179-211.

Granet, M., Wilson, M. and Achauer, U. (1995). Imagining a plume benearth the French Massif Central. *Earth and Planetary Science Letters*, Vol. 136, pp. 281-296.

Hart, S.R., Hauri, E.H., Oschmann, L.A. and Whitehead, J.A. (1992). Mantle plumes and entrainment: isotopic evidence. *Science*, Vol. 256, pp. 517-520.

Hoernle, K., Zhang, Y.S. and Graham, D. (1995). Seismic and geochemical evidence for large-scale mantle upwelling beneath the eastern Atlantic and western and central Europe. *Nature*, Vol. 374, pp. 34-39, doi:10.1038/374034a0.

Jones, A.P., Kostoula, T., Stoppa, F. and Woolley, A.R. (2000). A geotherm estimated from mantle xenoliths in Mt. Vulture volcano, Southern Italy. *Mineralogical Magazine*, Vol. 64, pp. 593-613.

Lavecchia, G. (1988). The Tyrrhenian-Apennines system: structural setting and seismotectogenesis. *Tectonophysics*, Vol. 147, pp. 263-296.

Lavecchia, G. and Creati, N. (2006). A mantle plume pinched in the transition zone beneath the Mediterranean: a preliminary idea. *Annals of Geophysics*, Vol. 49, pp. 373-387.

Lavecchia, G. and Stoppa, F. (1989). Il "rifting" tirrenico: delaminazione della litosfera continentale e magmatogenesi. *Bollettino della Società Geologica Italiana*, Vol.108, pp. 219-235.

Lavecchia, G. and Stoppa, F. (1996). The tectonic significance of Italian magmatism; an alternative view to the popular interpretation. *Terra Nova*, Vol. 8, pp. 435-446,

Lavecchia, G. Creati, N. and Boncio, P. (2002). The Intramontane Ultra-alkaline Province (IUP) of Italy: a brief review with considerations of the thickness of the underlying lithosphere. *Bollettino della Società Geologica Italiana* Volume Speciale 1, pp. 87-98.

Lavecchia, G., Boncio, P. and Creati, N. (2003). A lithospheric-scale seismogenic thrust in Central Italy. *Journal of Geodynamics*, Vol. 36, pp. 79-94.

Lavecchia, G., Federico, C., Karner, G.D. and Stoppa, F. (1995). La distensione tosco-tirrenica come possibile motore della compressione appenninica. *Studi Geologici Camerti*, Vol. Spec. 1995/1, pp. 489-497.

Lavecchia, G., Ferrarini, F., de Nardis, R., Visini, F. and Barbano, M. S. (2007). Active thrusting as a possible seismogenic source in Sicily (sourthern Italy): Some insights from integrated structural-kinematic and seismological data. *Tectonophysics*, Vol. 445, pp. 145-167.

Lavecchia, G., Stoppa, F. and Creati, N. (2006). Carbonatites and kamafugites in Italy: mantle-derived rocks that challenge subduction. *Annals of Geophysics*, Vol. 49, pp. 389-402.

Locardi, E. and Nicolich, R. (1988). Geodinamica del Tirreno e dell'Appennino centro-meridionale: la nuova carta della Moho. *Memorie della Società Geologica Italiana*, Vol. 41, pp. 121-140.

Lustrino, M. and Wilson, M. (2007). The circum-Mediterranean anorogenic Cenozoic igneous province. *Earth Science Reviews*, Vol. 81, pp. 1-65.

Lustrino, M., Duggen, S. and Rosenberg, C.L. (2011). The Central-Western Mediterranean: Anomalous igneous activity in an anomalous collisional tectonic setting. *Earth-Science Reviews*, Vol. 104, pp. 1-40.

Macera, P., Gasperini, D., Piromallo, C., Blichert-Toft, J., Bosch, D., Del Moro, A. and Martin, S. (2003). Geodynamic implications of deep mantle upwelling in the source of Tertiary volcanics from the Veneto region (South-Eastern Alps). *Journal of Geodynamics*, Vol. 36, pp. 563-590.

Macera, P., Gasperini, D., Ranalli, G. and Mahatsente, R. (2008). Slab detachment and mantle plume upwelling in subduction zones: An example from the Italian South-Eastern Alps. *Journal of Geodynamics*, Vol. 45, pp. 32-48.

Malinverno, A. and Ryan, W.B.F. (1986). Extension in the Tyrrhenian Sea and shortening in the Apennines as result of arc migration driven by sinking of the lithosphere Source. *Tectonics*, Vol. 5, pp. 227-245.

Mascle, G., Lemoine, M., Mascle, J., Rehault, J.P. and Tricart, P. (1991). Ophiolites and the oceanic crust: New evidence from the Tyrrhenian sea and the Western Alps. *Journal of Geodynamics*, Vol. 13, pp. 141-161.

Mongelli, F., Zito, G., Ciaranfi, N. and Picri, P. (1989). Interpretation of heat flow dcnsity of the Apennine Chain, Italy. *Tectonophysics*, Vol. 164, pp. 267-280.

Moores, E.M. and Twiss, R.J. (1995). Tectonics. Freeman and Company, New York, 415 pp.

Noguera, A.M. and Rea, G. (2000). Deep structure of the Campanian-Lucanian Arc (Southern Apennine, Italy). *Tectonophysics*, Vol. 324, pp. 239-265.

Olafsson, M. and Eggler, D.H. (1983). Phase relations of amphibole, amphibole-carbonate, and phlogopite-carbonate peridotite: petrologic constraints on the asthenosphere. *Earth and Planetary Science Letters,* Vol. 64, pp. 305-315.

Owen, J.P. (2008). Geochemistry of lamprophyres from the Western Alps, Italy: Implications for the origin of an enriched isotopic component in the Italian mantle. *Contribution to Mineralogy and Petrology*, Vol. 155, pp. 341-362.

Panza, G.F. and Suhadolc, P. (1990). Properties of the lithosphere in collisional belts in the Mediterranean-a review. *Tectonophysics*, Vol. 182, pp. 39-46.

Pasquale, V., Verdoya, M., Chiozzi, P. and Ranalli, G. (1997). Rheology and seismotectonic regime in the northern central Mediterranean. *Tectonophysics*, Vol. 270, pp. 239-257.

Peccerillo, A. (1999). Multiple mantle metasomatism in central-southern Italy: geochemical effects, timing, and geodynamic implications. *Geology*, Vol. 27, pp. 315-318.

Peccerillo, A. and Manetti, P. (1985). The potassium-alkaline volcanism of Central-Southern Italy: a review of the data relevant to petrogenesis and geodynamic significance. *Transactions of the Geological Society of South Africa*, Vol. 88, pp. 379-384.

Piromallo, C. and Faccenna, C. (2004). How deep can we find the traces of Alpine subduction?. *Geophysical Research Letters*, Vol. 31, L06605, doi:10.1029/2003GL019288

Piromallo, D., and Morelli, A. (2003). P-wave tomography of the mantle under the Alpine-Mediterranean area. *Journal of Geophysical Research,* Vol. 108, 2065, doi: 10.1029/2002JB001757

Rehault, J.P., Moussat, E. and Fabbri, A. (1987). Structural evolution of the Tyrrhenian back-arc basin. *Marine Geology*, Vol. 74, pp.123-150.

Rosatelli, G., Wall, F. and Stoppa, F. (2007). Calcio-carbonatite melts and metasomatism in the mantle beneath Mt. Vulture (Southern Italy). *Lithos*, Vol. 99, pp. 229-248.

Rotolo, S.G., Castorina, F., Cellura, D. and Pompilio, M. (2006). Petrology and geochemistry of submarine volcanism in the Sicily Channcl Rift. *Journal of Geology*, Vol. 114, pp. 355-365.

Savelli, C. (2002).Time-space distribution of magmatic activity in the western Mediterranean and peripheral orogens during the past 30 Ma (a stimulus to geodynamic considerations). *Journal of Geodynamics*, Vol. 34, pp. 99-126.

Scalera, G. (2006). The Mediterranean as a slowly nascent ocean. *Annals of Geophysics*, Vol 49, pp. 451-482.

Serri, G., Innocenti, F. and Manetti, P. (1993). Geochemical and petrological evidence of the subduction of delaminated Adriatic continental lithosphere in the genesis of the Neogene-Quaternary magmatism of central Italy. *Tectonophysics*, Vol. 223, pp. 117-147.

Spakman, W., van der Lee, S. and van der Hilst, R. (1993). Travel-time tomography of the European-Mediterranean mantle down to 1400 km. *Physics of the Earth and Planetary Interiors*, Vol. 79, pp. 3-74.

Stampfli, G.M., and Borel, G.D. (2004). The TRANSMED transects in space and time; constraints on the paleotectonic evolution of the Mediterranean domain. In: *The TRANSMED Atlas*, Cavazza et al. (Eds.), Springer, pp. 53-80.

Stoppa, F. (2008). Alkaline and ultramafic lamprophyres in Italy: Distribution, mineral phases, and bulk rock data. In: Deep-seated magmatism, its sources and plumes, Vladykin, N.V. (Ed.), pp. 209-238, ISBN/ISSN: 978-5-94797-130-9, Irkustk.

Stoppa, F. and Lupini, L. (1993) Mineralogy and petrology of the Polino monticellite calciocarbonatite (Central Italy). *Mineralogy and Petrology*, Vol. 49, pp. 213-231.

Stoppa, F. and Principe, C. (1998). Eruption style and petrology of a new carbonatitic suite from the Mt. Vulture (Southern Italy): The Monticchio Lakes Formation. J. Volcanol. Geoth. Res., Vol. 80, pp. 137-153.

Stoppa, F., and Woolley, A.R. (1997). The Italian carbonatites: field occurrence, petrology and regional significance. *Mineralogy and Petrology* ,Vol. 59, pp. 43-67.

Stoppa, F., Jones, A.P. and Sharygin, V.V. (2009). Nyerereite fromcarbonatite rocks at Vulture volcano: implications for mantle metasomatism and petrogenesis of alkali carbonate melts. *Central European Journal of Geosciences*, Vol. 1, pp. 131-151, doi: 10. 2478/v10085-009-0012-9

Suhadolc, P., Panza, G.F. and Mueller, S. (1990). Physical properties of the lithosphere-asthenosphere system in Europe. *Tectonophysics*, Vol. 176, pp. 123-135.

Trua, T., Serri, G., Marani, M.P., Rossi, P.L., Gamberi, F. and Renzulli, A. (2004). Mantle domains beneath the southern Tyrrhenian: constraints from recent seafloor sampling and dynamic implications. In: *A showcase of the Italian research in petrology: magmatism in Italy*, Conticelli, S. and Melluso, L. (Eds.), *Periodico di Mineralogia*, Vol. 73, pp.53-73.

Van Dijk, J.P., Bello, M., Brancaleoni, G.P, Cantarella, G., Costa, V., Frixa, A., Golfetto, F., Merlini, S., Riva, M., Torricelli, S., Toscano, C. and Zerilli, A. (2000). A regional structural model for the northern sector of the Calabrian Arc (southern Italy). *Tectonophysics*, Vol. 324, pp. 267-320.

Vichi, G., Stoppa, F. and Walls, F. (2005). The carbonate fraction in carbonatitic Italian lamprophyres. *Tectonophysics*, 85, 154-170.

Wezel, F.C. (1981). The structure of the Calabro-Sicilian Arc: krikogenesis rather than subduction. In: *Sedimentary Basin of Mediterranean Margins*, Wezel, F.C. (Ed.), 485-487, Bologna.

Wilson, M., and Patterson, R. (2001). Intraplate magmatism relate to short-wavelength convective instabilities in the upper mantle: evidence from the Tertiary-Quaternary volcanic province of western and central Europe. In: *Mantle plumes: their identification through time*, Ernst, R.E. and Buchan, K.L. (Eds), *Geol. Soc. America Special Paper* 352, pp. 37-58.

Zhou, S. (1996). A revised estimation of the steady-state geotherm for the continental lithosphere and its implication for mantle melting. *Terra Nova*, Vol. 8, pp.514-524.

Ziegler, P.A. (1992). European Cenozoic rift system. *Tectonophysics*, Vol. 208, pp. 91-111.

Part 3

Applied Volcanology

8

Identification of Paleo-Volcanic Rocks on Seismic Data

Sabine Klarner and Olaf Klarner
PGS Reservoir & Klarenco
Germany

1. Introduction

While exploring for hydrocarbons in rift related basins, volcanics, volcaniclastic deposits or their erosional products are common lithologies. The presence of rock types derived from volcanism and/or affected by post-volcanic re-deposition may lead to complex lithologies with complex diagenetic overprints at the reservoir level. Partial or complete reservoir substitution, alteration by circulating hot fluids and addition of mineral components have led to a number of unsuccessful wells, both in exploration and field development projects. Furthermore, volcanic rocks may form lateral seals or migration barriers, providing both positive and negative impact on the petroleum system. Non-permeable volcanic layers can seal the top of the reservoir, preventing it from breaching, or they can build a migration barrier for the fluid on its way from the source rock into the trap. In most cases, highly varied lithologies with wide ranges of inherent rock properties occur. It is therefore essential to understand the distribution of volcanics in the vicinity of the reservoir.

Published examples of volcanic reflectors identifiable from seismic data are still very sparse, and there is no systematic compilation of information and knowledge available. Free air gravity maps, as well as other potential field methods, may show anomalies caused by volcanic and sub volcanic bodies if they are really massive geobodies. Unfortunately, the majority of volcanic features are rather thin (some tens of meters thick). At a reservoir depth of more than 3 km the resolution of standard potential field data is not good enough to delineate the individual units.

A close cooperation of specialists from different disciplines is required in order to resolve specific problems related to seismic interpretation and reservoir prediction within such a complex environment. Sedimentology and fieldwork studies of recent analogues provide the basis for recognition of the depositional environment used for seismic interpretation. Special seismic processing must be utilized to calculate attributes from prestack data, showing amplitude variations with increasing offset from the energy source (commonly referred to as AVO analysis). These data are interpreted using results from rock physics analysis - elastic parameters of different lithology types and seismic forward modeling. The AVO analysis is used to support the lithology identification from seismic geometries. The input for the rock physics is delivered by the petrophysicist – log derived properties and interpretations. Petrography is used for provenance analysis, as well as analysis of diagenesis and fabric. In many cases, this turns out to be the key to understanding the anomalous AVO behavior of sandstones. Finally, sequence stratigraphy relates the different levels of sandstones to periods with and without volcanic activity.

Our experience from working in different basins around the world has shown that there are certain distinct features that can assist in the identification of volcanics and volcanic related lithologies from seismic data. These features can be divided in two main groups: (1) geometry, specifically in comparison with observed morphology of recent volcanics and (2) amplitude expression, particularly the AVO behavior. Modern seismic interpretation technologies, such as 3D imaging, spectral decomposition, analysis of elastic rock properties and AVO analysis in combination with sound geological understanding of recent volcanism reveal completely new insights into the geological past.
In the current chapter both geometry and amplitude behavior of volcanics, as seen in seismic and well data will be discussed using a number of case studies from several hydrocarbon bearing basins around the world. The chapter summarizes previously presented ideas with a special emphasis on the multidisciplinary approach required to succeed with seismic interpretation.

2. Seismic geometries of paleo-volcanics

2.1 Overview

In this section, we will discuss typical shapes and images seen on seismic data resembling geometries observed in recent areas with volcanic activity. The best images stem from seismic data offshore Brazil, from the Upper Cretaceous reservoir section in the Santos Basin. The Santos Basin, offshore Brazil, is an ideal place to study the seismic signature of volcanic rocks embedded in a clastic section. The volcanism of the Santos Basin is a product of the late phase of rifting, which led to the opening of the Atlantic. The Santos basin itself is underlain by stretched continental crust with deep-reaching faults supporting the ascent of mafic melts from the upper mantle (Mohriak et al, 2002).

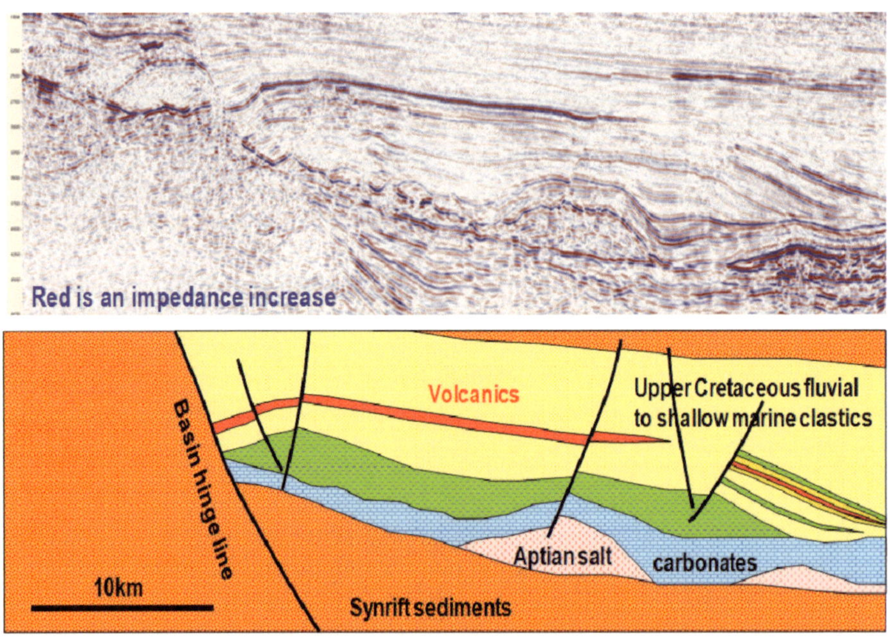

Fig. 1. Schematic interpretation of a NW-SE striking seismic section from the Santos Basin, crossing the basin hinge line

The basalts encountered in exploration wells are located along the basin hinge line (Fig. 1) that controlled basin geometry and sedimentation during the Upper Cretaceous (e. g. the Aptian salt distribution). This gave rise to the assumption that this fault trend was very active and formed an ideal conduit for the magma. Interestingly, the occurrences of magma not only follow the main fault, but also smaller fractures associated with it. Within the area of interest, at least four stages of Upper Cretaceous volcanic activity have been recognized. The age of the volcanics ranges between 80 and 65 Mio years and they occur at a depth range of 3200 - 4600 m.

2.2 Volcanic geometries

The geometries of the mapped individual volcanic bodies display a wide range of features, well known from recent surface volcanoes. Within the area of investigation, four different types of volcanic features have been mapped:

1. Massive basalt flows, indicative of sub aqueous cones and small plateaus with a thickness of several tens to hundreds of meters, covering an area of up to 400km^2.

Most of the sub aqueous volcanoes consist, to a large extent, of pillow lava flows. Lava tubes occur at low to medium eruption rates presumably from central chimneys and short conveyor cracks. The size of the lava pipes decreases with increasing height of the volcano. Pillows tend to solidify completely, later compaction will not substantially change their internal structure and the amplitude response will be unambiguous. As a pillow forms, a new eruption phase will cover the predecessor over the top and then flow outward. This process builds steep sided mounds in contrast to sheet flows that occur at higher eruption rates and lead to flatter shield volcano features. It is assumed that an alternation of sheet flows and pillow lava commonly occurs with a ratio of 30 to 70% (Schminke, 2000). Extension of these sheets can reach hundreds of square kilometers with considerable thicknesses. Core data and Formation Micro Images (FMI) data from an exploration well in the target area show pillow lava sections, but also layers of hyaloclastic rocks within basalt lava flows. These were deposited as debris from the cooling pillow lava flow. Due to their fragility and re-deposition processes, only remnants may have been preserved.
The geomorphology of volcanic lava or pyroclastic flows can in many cases display similar features to siliciclastic facies like turbidities or fluvial delta deposits (Fig. 2). This can lead to serious pitfalls in hydrocarbon exploration.

2. Massive aerial basalt flows with single lava flow geometries

Depending upon viscosity, flow velocity, and eruption rate, for monogenetic basalt flows a number of different surface features can be observed. In general the surface is glassy, and tends to peel off due to the brittleness of the rock. If viscosity and shear strain rise, the more liquid Pahoehoe lava will convert into the blocky Aa lava (Schminke, 2000). During the cooling phase, a regular fracture system is developed. The more rapid cooling of the surface of a lava flow results in the lower third of the flow developing regular pillars which are overlain by a layer with vaulted and less regular offsets. The upper layer is more brittle than the lower and can be sheared off during later tectonic movements. This is important for hydrocarbon exploration. If fractures remain open, hydrocarbons can escape through these pathways leaving the potential target empty. In thick lava flow deposits, the pillar part can be subdivided into a lower part with broad pillars, a middle "Entablature" zone with thin pillars, and the unstructured upper part. Due to the low viscosity of the Pahoehoe Lava, we can observe relatively thin bedded flow units, which can be stacked on top of each other (Fig. 3,4), jointly cooled down leading to complex flow units (Schminke, 2000). These complex flow units develop different internal facies distribution with different properties such as brittleness and shear strain (Nelson 2009).

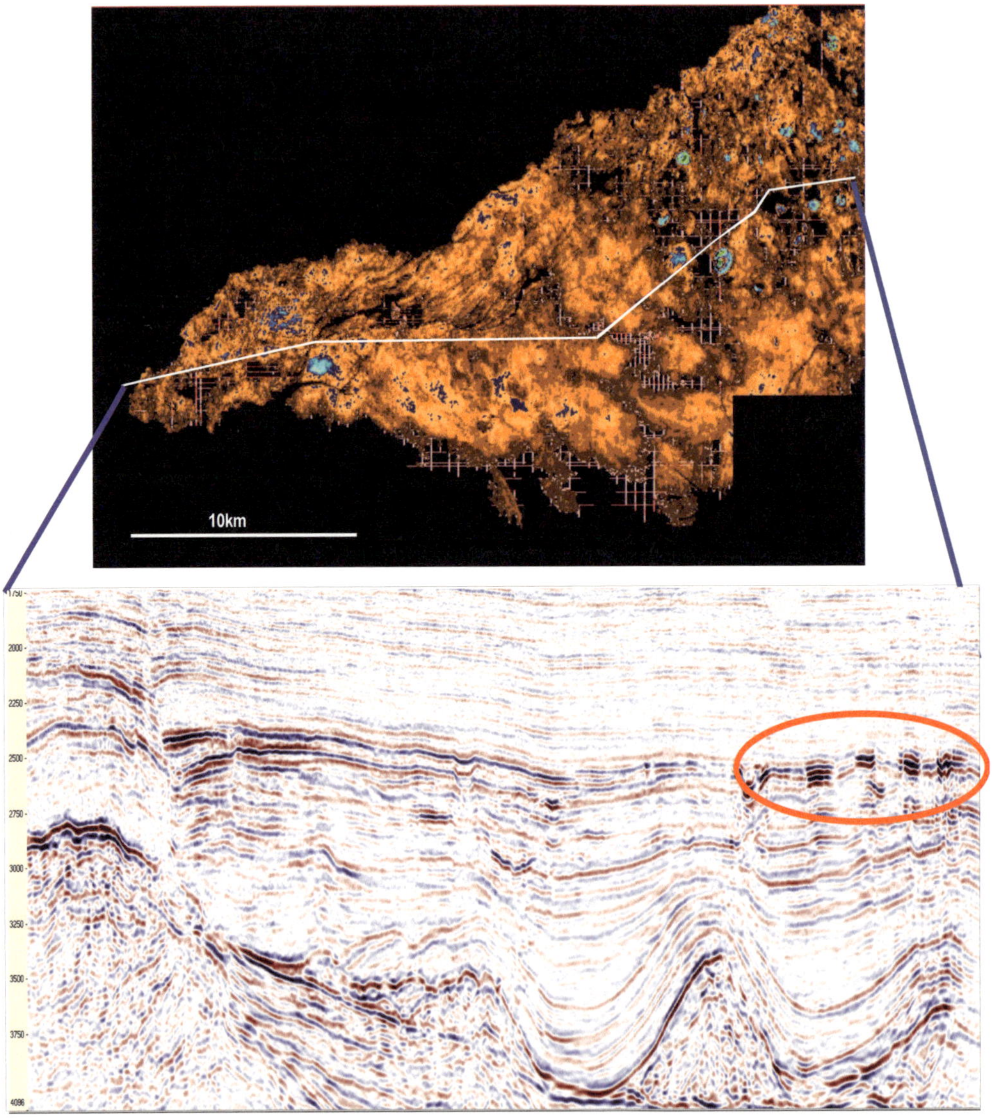

Fig. 2. Amplitude map of a sharply defined massive mainly sub aerial lava flow consisting of several separate units and exhibiting numerous vents (circular features with strong amplitudes)

Fig. 3. Outcrop example of multiple stacked lava flows, North East coast Tenerife, 2006, Klarner

The outcrop sample from Tenerife serves as an analogue for the observed stacked lava flows in deep water exploration wells in the Santos Basin, Brazil (Fig. 4).

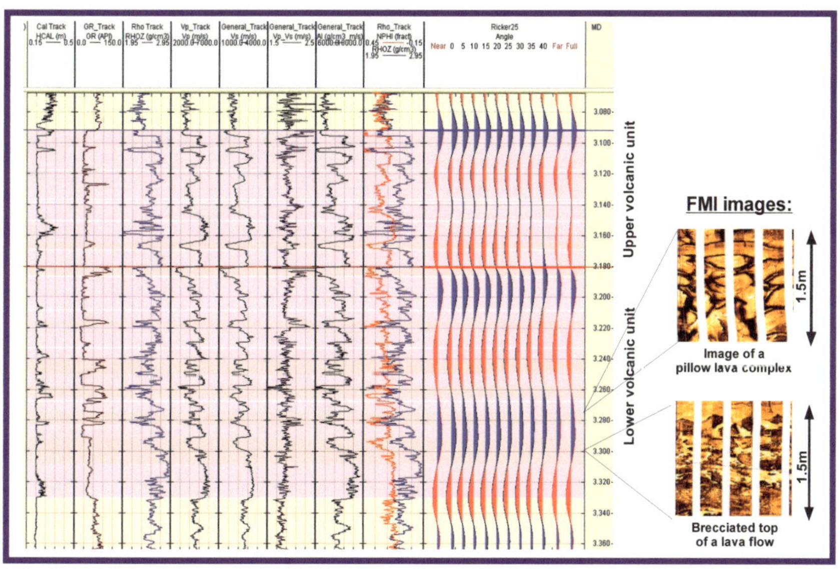

Fig. 4. Upper Cretaceous multiple stacked lava flows, drilled in the Santos basin. The saw-tooth log pattern reflects the individual flow units, each beginning with massive basalt and ending with a brecciated (weathered) top. The lower unit contains pillow lavas, indicating subaqueous deposition

In addition, we need to account for the limited resolution capability of seismic data. Individual effusive events have thicknesses from a single meter up to 10 or maximum 15 meters. We see this in the outcrop (Fig.3), but also within the well log data (Fig.4). The seismic resolution depends on the velocity of the seismic waves and the frequency of the signal, and for a p-wave velocity of 3500 m/s and a lead frequency of 35 Hz we can assume a vertical resolution power of about 25 m (equals a quarter of the wavelength), for 5000 m/s and 25 Hz we can see cycles of about 50 m thickness. This means, that in reality many effusive events are always compiled into one phase of the seismic data - as seen on the right column of Fig. 4. The resolution can be slightly enhanced by procedures like spectral decomposition or inversion, revealing details of the internal structure, but there is still a significant resolution limit inherent in the data.

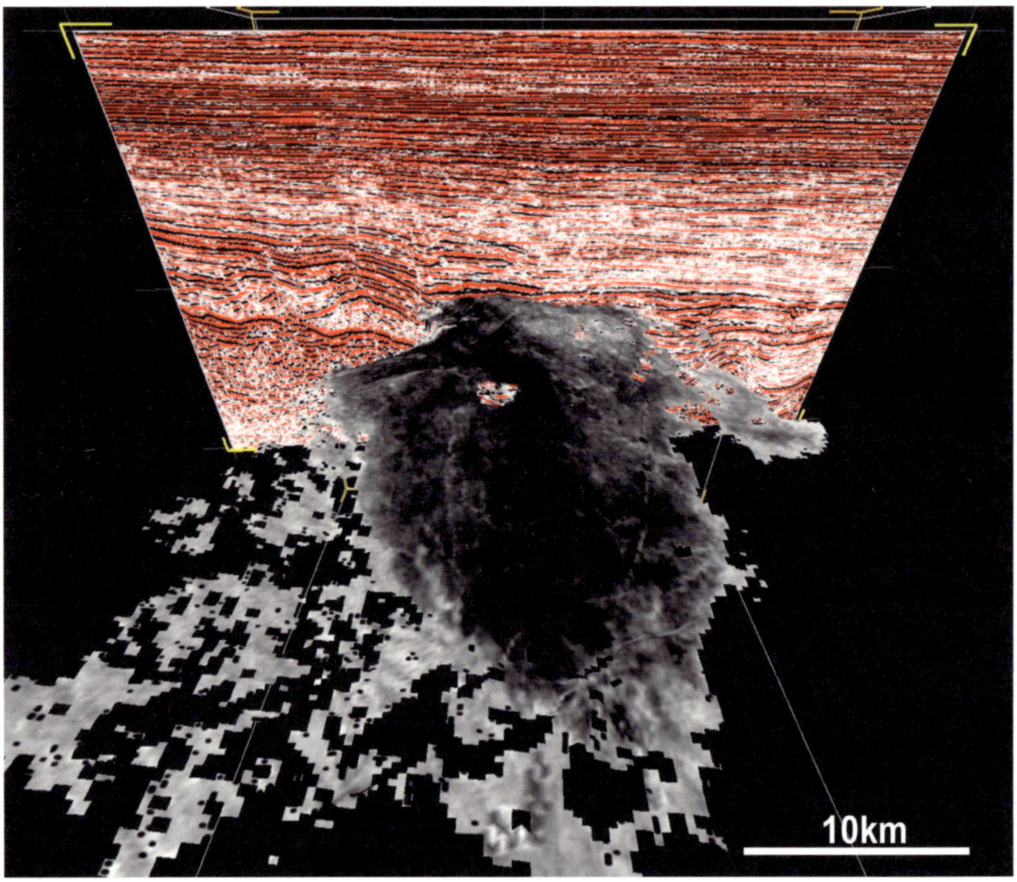

Fig. 5. 3D seismic image of a small shield volcano, covered by over 3000 m sediments (Santos Basin)

Typical features of low viscosity basalt flows are pipe vesicles and vesicle cylinders. We believe that these surface features do not play a role after compaction and subsidence of several thousand meters (Fig. 5). The analogs known from surface are important for a good understanding of the subsurface features, but a one to one comparison might not be adequate. Hence, our model of paleo-volcanoes requires some adjustment in terms of subsidence, lithostatic pressure and the respective diagenetic alterations. Under surface or subaqueous conditions, the lateral extent of low viscous lava flows is normally short due to the increasing viscosity caused by decreasing temperature. However, the reach can be substantial in the case of transportation through lava pipes. Hence, volcanic rock features can occur miles away from identified sources like a volcano or main fault system.

In many areas with predominantly mafic volcanism, the Aa-lava type is prevalent (Schminke, 2000). These lava flows are characterized by cindery, sharp-edged boulders, which can be strongly welded. Such flow units have a lower part with boulders and scoria, whereas the upper part consists of successive flow units. Aa- flow units do not have a large aerial extent and will be found closer to the volcanic source (Fig. 6).

Fig. 6. Caldera filled by repeated ejecta of viscous lava, Pico de Teide 2006, Klarner

As demonstrated by Peterson and Tilling 1980, shear stress plays an important role in the lava type, which is to a large extent driven by the value of the shear strength, and not only by the viscosity. Hence, it is important to understand the paleo-morphology during the volcanic phases in order to predict the possible extent of paleo-volcanics in seismic.

Fig. 7 nicely shows that local lava tongues may reach only a few hundred of meters extent from origin. They are exposed to surface erosion processes and might not have been preserved as geomorphologic relevant features. However, the erosion products will be deposited together with potential reservoir sandstones, leading to a deterioration of their quality mainly in terms of the porosity-permeability properties due to immature components and their products of diagenesis.

3. Smaller individual cones and lava flows (local events with small vents)

These individual volcanic features could be seen as remnants of hyaloclastic deposits, explaining the small thicknesses encountered in the exploration wells. They were quickly eroded and typical concentric features are left, which can be recognized in the NE of Fig. 8 b. Phreatic and Surtseyan eruptions typically include steam and rock fragments; the extrusion of massive lava is unusual, ashes are distributed by currents. The eruptions can create broad, low relief craters appearing as concentric features. Typical surface features such as spatter cones, ash or tuff cones possess a short life time due to quick erosion. Their remnants can be sedimented with siliciclastics or carbonates. Preservation of ash cones in paleo-volcanics seems to be unlikely. The observed concentric features might represent secondary volcanic cones on the flank of bigger complexes, filled with resistant rocks and being preserved over time. This could also explain the number of different smaller lava flows observed in Fig. 8 and 11.

Fig. 7. Local lava tongue, Pico de Teide 2006, Klarner

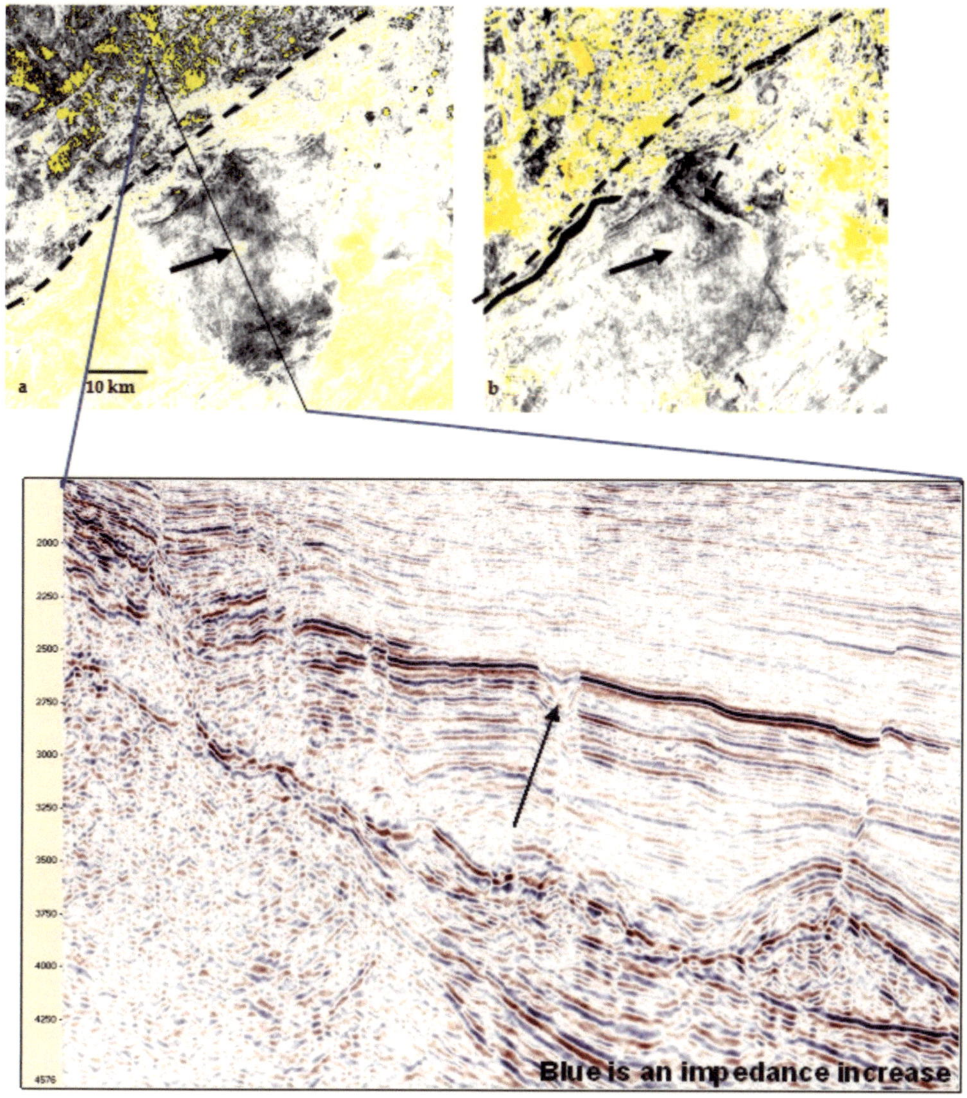

Fig. 8. Amplitude map (a) of the top of a massive 400 km² basalt flow, indicative of a small mainly subaqueous shield volcano. The dashed line marks the basin hinge line. The arrow indicates the main crater, terminating a deep reaching chimney. In between the seismic layers, representing the volcanic unit, on one of the horizon slices (b), a single channel-like feature can be observed (dashed arrow). As there were no intercalated sediments penetrated, this must be interpreted as an aerial low viscosity lava channel, as they can be observed on dipping flanks of recent volcanoes. Both images show a number of circular features, interpreted as vents and craters

4. Single dykes as well as dyke and sill complexes.

Fig. 9 demonstrates the possible density of sills and dykes along the main fault pattern. Sills can develop several levels of occurrence. They can occur as radially symmetrical sill complexes consisting of a saucer-like inner sill at the base with an arcuate inclined sheet or as bilaterally symmetrical sill complexes. Both types are sourced by magma diverted from a magma conduit feeding an overlying volcano and can reach considerable distances. Both sill complex types can appear as isolated bodies but commonly occur in close proximity and consequently merge, producing hybrid sill complexes (Thomson and Hutton, 2003).

The spatial density of rocks with higher densities and Vp velocities can influence the amplitude behavior to a large extent. The occurrence of a sill - dyke complex can substantially deteriorate reservoir properties or lead to complete reservoir extinction in hydrocarbon production. Recognition of sills and dykes on seismic is not an easy task due to their thinness and geological position (Fig. 10).

Fig. 9. Sills, laccolites and dykes, NE coast Tenerife, 2006, Klarner

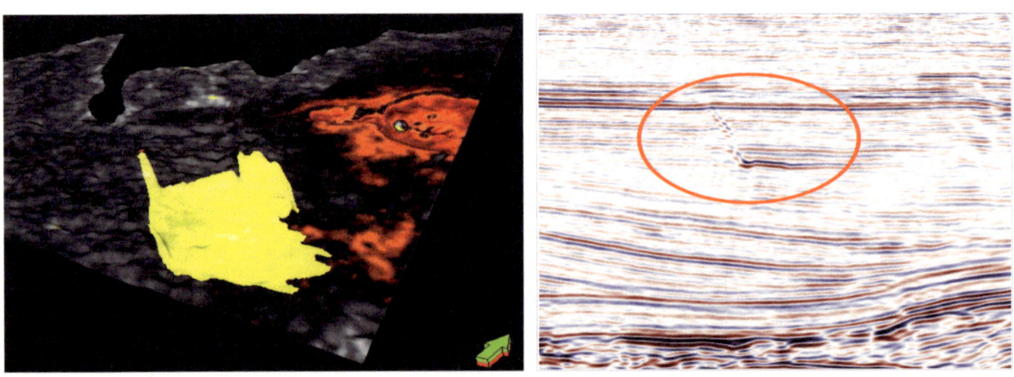

Fig. 10. Image of a dyke, discharging at a depth where a well defined double crater is observed; this implies that both features are fed by the same source

2.3 Delineation of structural details by spectral decomposition

Many of the volcanic bodies show typical crater geometries. A few vents can be traced as chimneys through the section, but many of them could be delineated only by the use of spectral decomposition. Fig. 11 shows an example of several lava complexes being stacked to form a small basalt shield. The older vents are covered by younger lava flows, but can be imaged by spectral decomposition. Spectral decomposition also revealed excellent examples of vent chains, sitting along conduit fractures. This is well demonstrated in Fig. 11 showing two separated centers with lava flows in different directions sourced by the volcanoes, but also from fractures. Size of each complex is about 100 km².

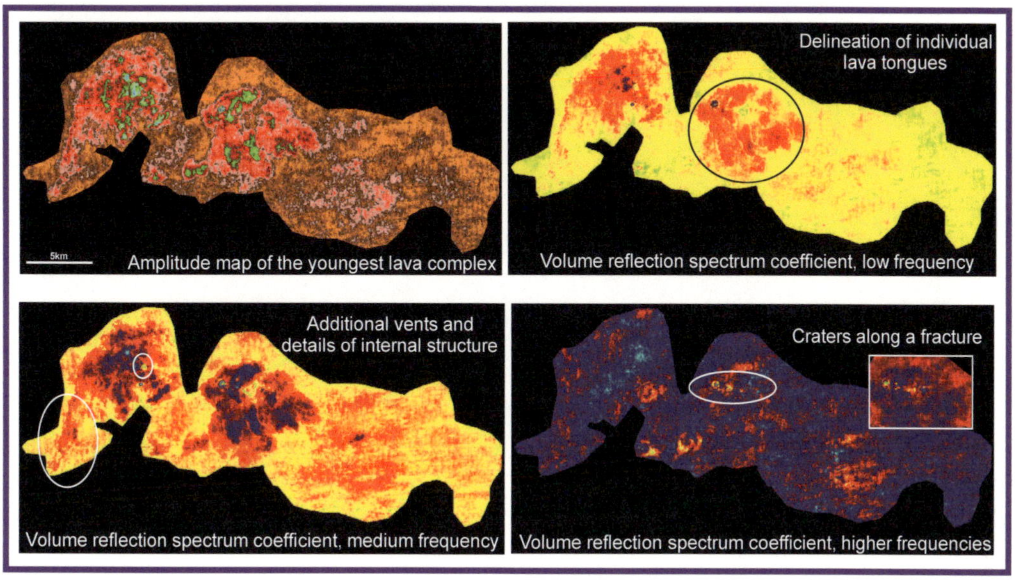

Fig. 11. Examples of spectral decomposition delineating shield volcanoes, lava flows and conduits

The same feature can be visualized by an amplitude image (Fig. 12), displaying a very comparable picture. The volcanic complex is recognizable on the corresponding seismic line. However, an understanding of the true nature of this event is not possible without the applied spectral decomposition and amplitude mapping.

3. Identification of volcanics using amplitude variations

3.1 Elastic rock properties, introduction

To interpret the seismic amplitude response of different rock types correctly, a sound understanding of their elastic rock properties is required. Seismically effective parameters are the p-wave velocity Vp, the shear wave velocity Vs and the density ρ. Physically, these parameters influence the rock's (In)Compressibility $\lambda = f(Vp, \rho)$, responding to lithology and pore fluid, the (Un)Elasticity or Rigidity $\mu = f(Vs, \rho)$ responding to lithology and the Bulk Density RHOB (ρ), responding to porosity and pore fluid. Commonly used rock properties are the Acoustic Impedance ($AI = Vp*\rho$) and the Shear Impedance ($SI = Vs*\rho$), and, derived from them, the Vp/Vs ratio (= AI/SI). Quartz sandstones usually are far more

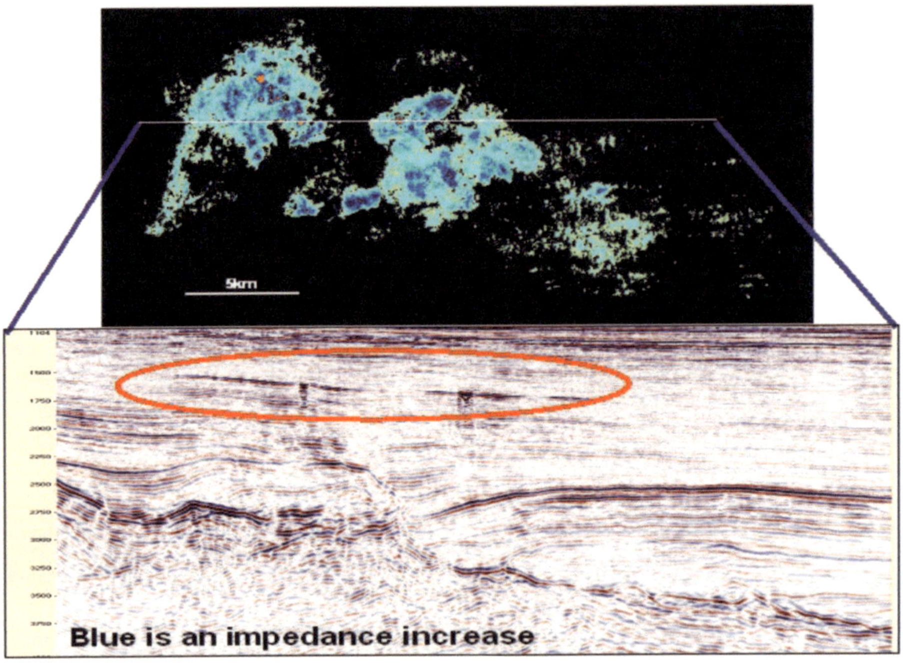

Fig. 12. Amplitude image of the youngest volcanic activity in the area: two sources fed several lava flows through numerous vents and stacked into two small shield volcanoes

compressible then shales, but much less elastic, and in the AI versus Vp/Vs ratio cross plot domain the explorationist looks for lithologies with a local minimum indicating hydrocarbon bearing clastic reservoirs. A more detailed description of applied rock physics can be found in Avseth et al (2005). Different types of seismic inversion are used to extract rock properties like AI, Vp/Vs, λ, μ, ρ from seismic data. In this work, we will use AI and Vp/Vs, as they help to describe the specifics of volcanics and volcaniclastics versus their embedding lithologies.

3.2 Elastic rock properties, application

Volcanics display a wide range of physical properties. This is dependent on the chemistry and differentiation stage of the feeding magma as well as on the depositional environment (subaqueous, aerial). In addition, volcanics are mineralogically quite unstable and undergo rapid alteration under the influence of weathering processes and diagenesis. Consequently, they vary in mineral composition and texture. This causes significant differences in the acoustic and elastic behavior. Our understanding of the properties of recently deposited volcanic rocks is becoming better and better but there is a distinct lack of published information dealing with paleo-volcanics, now buried under several kilometers of sediment. However, having worked in different rift related basins around the world, we observe certain trends for the elastic properties of mafic volcanics and related rock types. We are going to show three independent examples to demonstrate this.

3.2.1 Volcanics from the Sirte Basin, onshore Libya

In the Hameimat trough (Eastern part of the Sirte Basin), a number of hydrocarbon targeting exploration and production wells at a depth range around 3600 - 3800 m have penetrated volcanics forming part of a Pre-Upper-Cretaceous (PUC) intracratonic rift system. The actual reservoir consists of quartz sandstones that are syn-rift sediments of the Late Jurassic-Early Cretaceous Sarir Group, deposited in a fluvio-lacustrine paleo-environment (Ottesen et al. , 2005). To compound the problem, fluvial sandstones have in parts been invaded by a variety of extrusive basalts and intrusive dolerites, but also volcaniclastic facies types, related to lahar flows confined by the fluvial depositional system. The goal of several reservoir characterization studies was to quantify critical reservoir properties and to identify the occurrence of non reservoirs, such as the volcaniclastics. In 2002, De Vincenzi et al. reported that for some of the basalt varieties their high acoustic impedance was a physical differentiator that could be used to identify them from seismic data (Fig. 13).

By cross plotting all available petrophysical data at well locations they established subtle relationships between porosity and acoustic impedance. The volcanics are characterized by extremely low porosities (between 0. 5 and 5%). The pay zone exhibits average porosities between 9. 1 and 11. 6 percent; single units can reach 14%. A significant number of basaltic layers cross plot with clearly higher acoustic impedances than all other lithologies. Seismic response modeling was carried out to verify whether or not the thickness of the individual basaltic layers would be large enough to allow a seismic resolution of this lithotype. Time-migrated seismic data were transformed into their principal components within the target time window and a number of seismic attributes were extracted. Using neural network algorithms in commercially available software, a number of independent workflows for seismic waveform clustering were tested both for unsupervised and supervised classifications, as well as for the supervised property (porosity) prediction. The synthetic impedance traces showed that especially the basaltic layers could be identified at least qualitatively at the reservoir level. This in turn allowed searching the seismic for basaltic rock types.

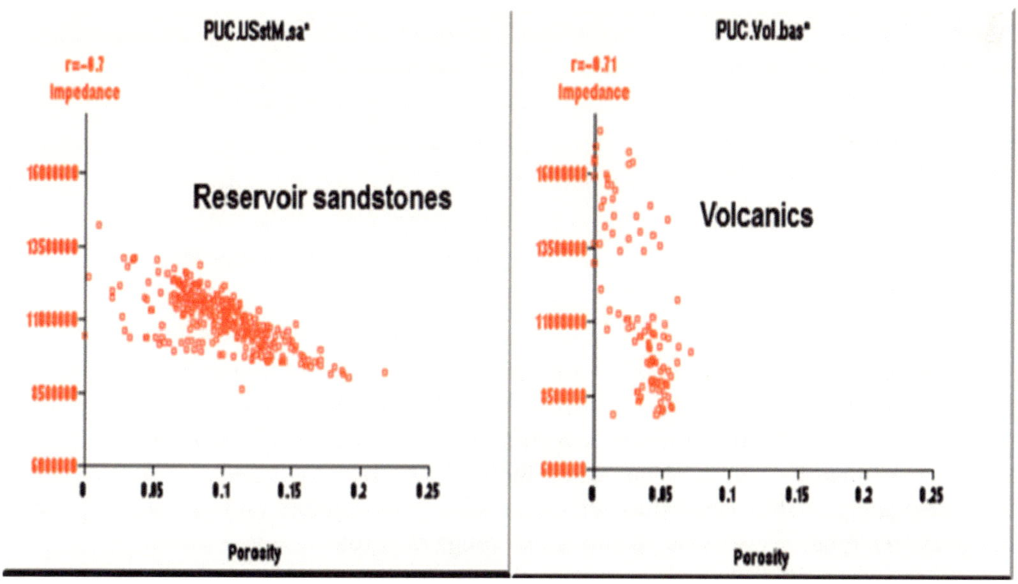

Fig. 13. Cross plots Porosity versus Acoustic Impedance; right - reservoir sandstones; left - volcanic (high impedance basalts and low impedance lahar type volcaniclastics)

Seismic data were inverted into a cube of acoustic impedance, and by the use of a neural network algorithm transformed into average porosity. Using the correlation of acoustic impedance and porosity, within the reservoir section we identified areas with very low average porosity (Fig. 14), interpreted as indicating presence of volcanic lithofacies types (basalt, dolerite). The data also demonstrate that the amplitudes interpreted as volcanics seem to follow a trend parallel to the main fault direction. Therefore, a deep seated fault is interpreted to have served as a conduit for the magma.

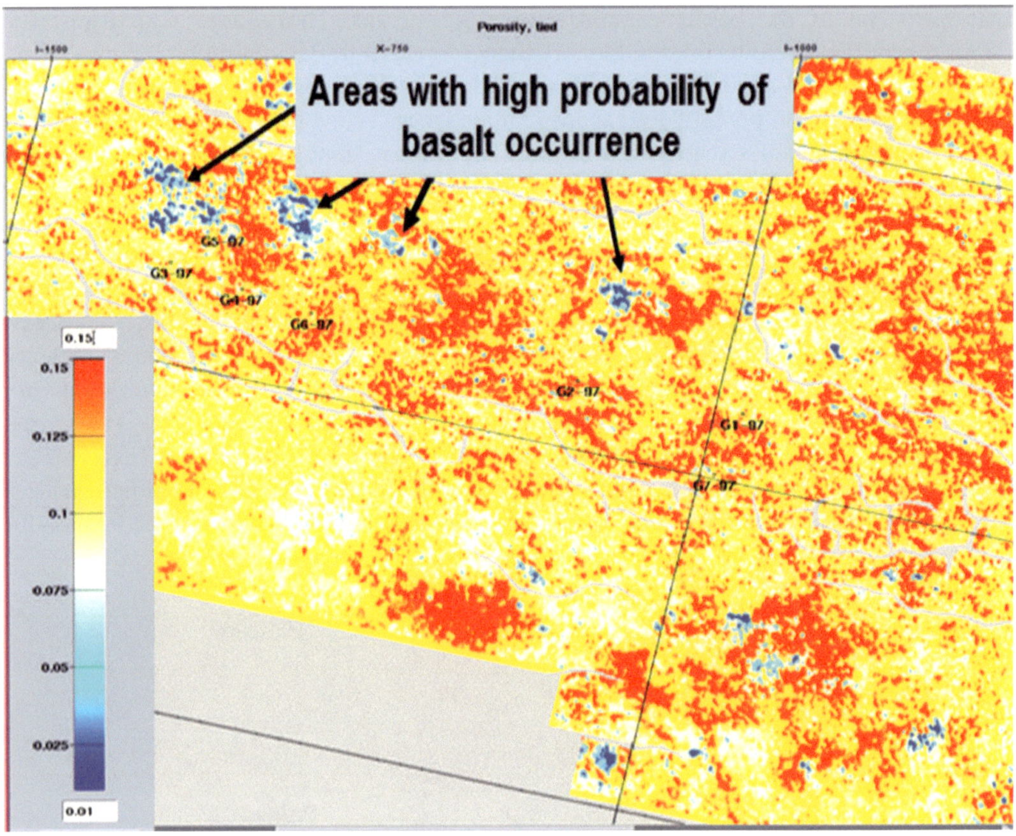

Fig. 14. Supervised porosity prediction (in volume fractions) from acoustic impedance data; in blue: supposed volcanics

However, the low impedance lahar flows encountered in one of the wells could not be separated from the porous reservoir sandstones using impedance data alone. We had to look for an additional parameter, distinguishing the reservoir from non-reservoir units. Some of the elastic parameters are presented in Fig. 15. We can clearly see that the volcanic facies types (pink and purple dots) exhibit high density, but also a broad scatter for the P-wave velocity, partially overlapping with the reservoir sandstones (orange dots). Especially the lahar rock type covers the same impedance range as porous reservoir sandstone.

Nevertheless, we see a clear separation in the Vp/Vs domain: both volcanic facies types exhibit very low shear wave velocities (compare Hanitzsch et al. 2007, who then used this parameter for the application of dual inversion applied to 2D multi-component seismic data onshore Libya). Interestingly, the volcanics seem to follow the empirical trend for Vp/Vs established by Greenberg and Castagna (1992) for limestones.

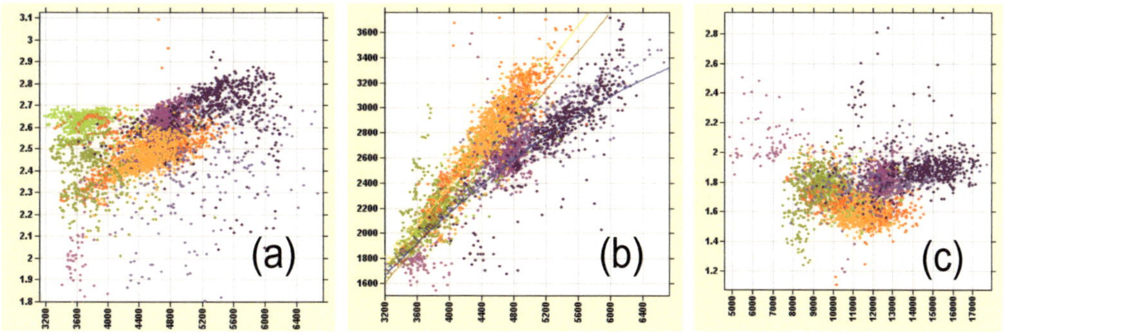

Fig. 15. Cross plots of various physical parameters for different lithotypes from the Sirte basin: orange - quartz sandstones; green - shales; pink - volcaniclastics/lahar flows; purple - basalt. (a) p-wave velocity (m/s, horizontal axis) versus density (g/cm^3, vertical axis); (b) p-wave velocity (m/s, horizontal axis) versus shear wave velocity (m/s, vertical axis) with Greenberg-Castagna's empirical trends for sandstones (orange), shale (brown) and limestone (blue); (c) acoustic impedance (horizontal axis) versus Vp/Vs ratio (vertical axis)

3.2.2 Volcanic example from the Central North Sea

Another example of elastic parameters is demonstrated for Jurassic volcanics from the North Sea. Graverson (2006) gave a comprehensive overview for the evolution of the Jurassic-Cretaceous Rift Dome in the Central Graben in the North Sea. A few exploration wells, targeting Jurassic sandstones for hydrocarbon exploration, encountered both, Mid-Jurassic basalts and Mid-Jurassic tuffites. We can look at their elastic properties using data from block 15/22, representing a depth range from 3000 - 3200 m (Fig. 16).

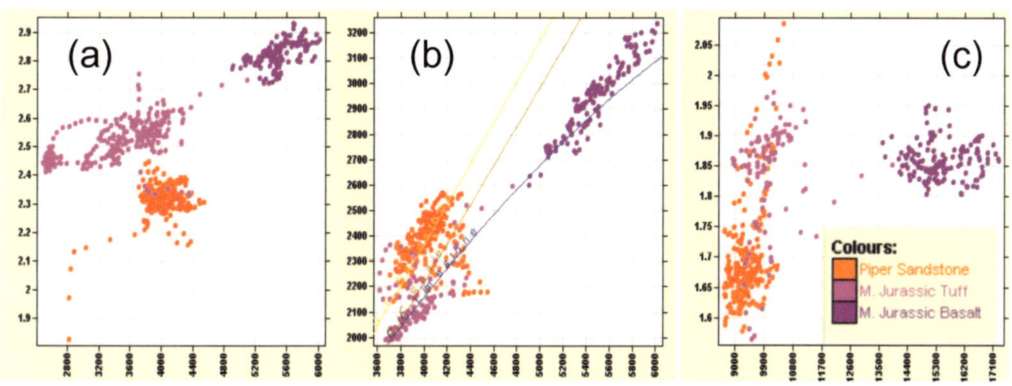

Fig. 16. Cross plots of various physical parameters for different lithotypes from the Central North Sea: orange - quartz sandstones; pink - volcaniclastics/tuffites; purple - basalt. (a) p-wave velocity (m/s, horizontal axis) versus density (g/cm^3, vertical axis); (b) p-wave velocity (m/s, horizontal axis) versus shear wave velocity (m/s, vertical axis) with Greenberg-Castagna's empirical trends for sandstones (orange), shale (brown) and limestone (blue); (c) acoustic impedance (horizontal axis) versus Vp/Vs ratio (vertical axis)

The pore fluid of the Jurassic sandstone has been substituted to water using the Gassmann fluid substitution algorithm to ensure that we only look at rock properties (comp. Smith et al. 2003). We can see again, that the volcanic rock types have a higher density then the porous reservoir sandstone, but at least the volcaniclastics overlap again in the p-wave domain. The basalts can be identified by their high acoustic impedance. For the tuffites we need the Vp/Vs ratio as the discriminating physical property. The relatively slow shear wave velocity is also typical for the encountered basalts, and as in the North African example the shear wave can be rather well approximated using the empirical limestone trend from Greenberg-Castagna.

3.2.3 Volcanics from the Santos Basin offshore Brazil

The Santos Basin provides a unique setting for the investigation of paleo-volcanics and their interaction with the embedding sediments. Although a number of wells along the basin hinge line have penetrated Upper Cretaceous basalts, only few of them have had a shear sonic log acquired. Examples from those wells have been included in the current study, showing volcanics from a depth range of 3200 - 4600 m.

The heterogeneity of the volcanics produces a wide scatter of petrophysical properties. Looking at the cross plots in Fig. 17, we can observe several trends in the acoustic and elastic properties: Different basalt units are compact, non-weathered volcanics and show higher impedances than the clastic lithologies. Weathered and/or re-deposited basalts (shown in purple) may have lower impedances, mainly due to far lower velocity values. But both types show high Vp/Vs ratios, distinguishing them from most of the clastic sediments. There is a group of sandstones (orange colors) that also exhibit high Vp/Vs ratios. It has been shown that these sandstones contain a high amount of volcanic rock fragments that are erosional products from the basalts nearby (Klarner et al 2005, 2008). Fortunately, this sandstone type shows significantly lower impedances than all other types of lithology and in this case will produce softer reflections than the basalts.

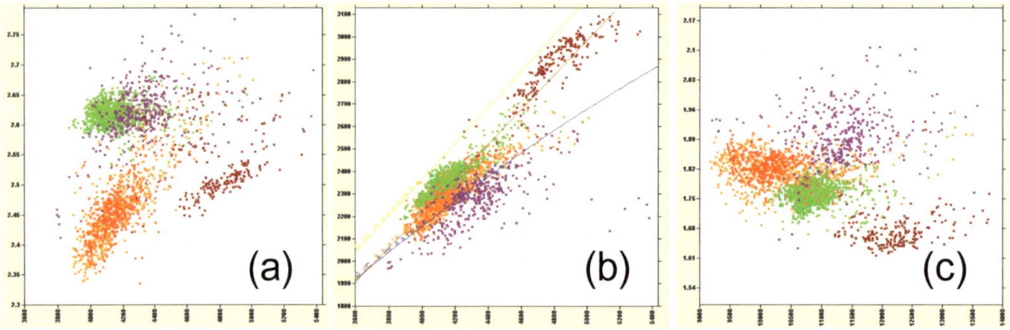

Fig. 17. Cross plots of various physical parameters for different lithotypes from the Santos basin: brown - reservoir sandstones; orange - sandstones with a high percentage of re-deposited volcaniclasts; green - shales; purple - basalt. (a) p-wave velocity (m/s, horizontal axis) versus density (g/cm3, vertical axis); (b) p-wave velocity(m/s, horizontal axis) versus shear wave velocity (m/s, vertical axis) with Greenberg-Castagna's empirical trends for sandstones (orange), shale (brown) and limestone (blue); (c) acoustic impedance (horizontal axis) versus Vp/Vs ratio (vertical axis)

The cross plot of Vp versus Vs in one of the wells (Fig. 17) shows that the volcanics (in this case diabase/dolorite and volcanic breccia) exhibit Vp/Vs ratio similar to a typical limestone. The thin section analysis and FMI data show that the wells encountered basalts and their weathering products the chemical composition comprising (Fig. 18): silicified vitreous alkali olivine basaltic rocks; olivine alkali dolerite (diabase); pebbly volcanic conglomerate; volcaniclastic litharenite.

Fig. 18. Thin section images of different Upper Cretaceous volcanic rock types from the Santos Basin: (a) conglomerate with fragments of volcanic and non-volcanic rocks, cemented by displacive calcite, (b) dolerite (diabase) with trachytoidal texture, (c) olivine basalt with intergranular texture of reddish altered olivine crystals

3.3 Elastic rock properties - conclusions

All the investigated volcanic rock types show a broad scatter of acoustic properties. Whereas tight basalts and dolerites may exhibit high acoustic impedances and, thus, may stand out on the seismic amplitude section, there are numerous examples of volcanics (weathered basalts, hyaloclastic rocks, tuffites, lahar deposits etc), which overlap with reservoir sandstones in the acoustic impedance domain. Measured velocity ranges from over 3000 m/s to over 6000 m/s (Fig. 19 left). As to be expected for magmatic rock types with their heterogeneous internal structure and texture, we do not see a depth dependent (compaction) trend for velocity for any of the lithotypes (Fig. 19 right).

However, all investigated volcanic lithotypes show a distinct shear weakening compared to average clastic rocks, and the ratio of p-wave velocity to shear-wave velocity is abnormally high - in many cases comparable to limestones. This is valid for all investigated volcanic lithotypes and can be used for prediction of shear wave velocity from p-wave velocity in cases where no shear log information is available. Only for very high velocities, the volcanics diverge from the empiric Greenberg-Castagna trend for carbonates, and can be described by the linear regression

$$Vs=0.51Vp+148 \ (R=0.86) \text{ in m/s} \tag{1}$$

Looking at the thin section images, we currently believe that the commonly observed extremely low shear velocities for all investigated volcanic related rock types are mainly a texture effect: the microcrystalline fabric of the volcanics as well as the chlorite-calcite matrix of the volcaniclastics react in a similar way to the microcrystalline matrix in carbonates.

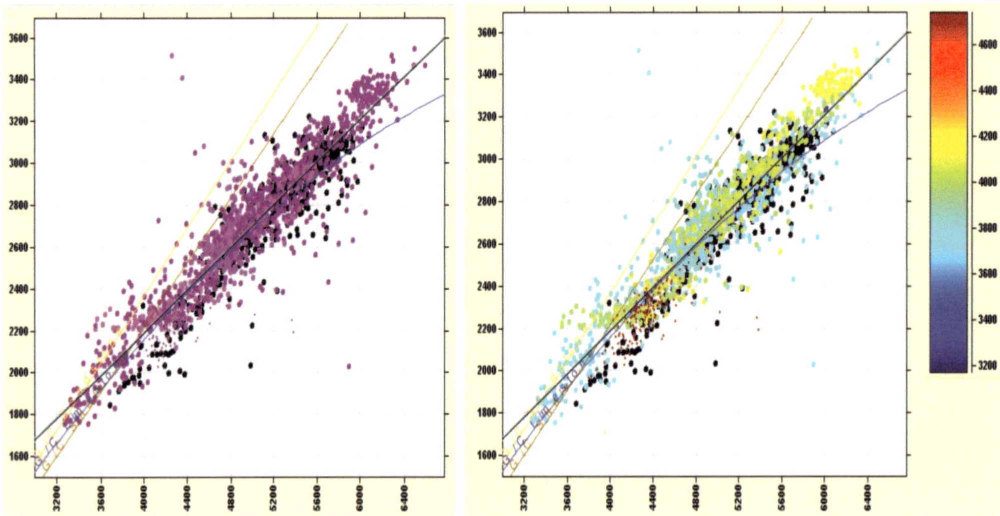

Fig. 19. P-wave velocity (m/s, horizontal axis) versus shear wave velocity (m/s, vertical axis) for volcanic from well data from different regions (Central North Sea, Sirte basin onshore Libya, and Santos Basin offshore Brazil). Left: different lithotypes (pink - volcaniclastics, purple - basalt); right: depth of penetration in m. The linear correlation is Vs=0. 51Vp+148 (R=0. 86)

3.4 Amplitude behavior

Although the impedances cover a wide range of values, the individual volcanic bodies identified in the seismic mostly appear as strong bright hard reflectors. If carbonates are not involved, a basalt body can be traced by its significantly higher Vp/Vs ratio compared to other lithology types. This is especially important when we deal with weathered basalts which show a lower density and therefore lower impedance. Our interpretation is also confirmed by the occurrence of typical volcanic geometries which are found in the specific tectonic context of a rift basin. Some of the individual bodies could be misinterpreted as the top of a carbonate cemented delta or terminal lobe or are reminiscent of fluvial channels. But the sum of the observed features within the given tectonic setting, including chimneys, craters, stacked lava flows and dykes gives us confidence in classifying these geometries as being volcanic in nature.

3.4.1 AVO analysis

In hydrocarbon exploration, AVO analysis is a common tool used for better understanding the lithology and the pore fluid of potential oil and gas bearing reservoirs. AVO means analysis of amplitude changes with offset, e. g. with the distance from the seismic energy source. Very often a high porosity gas or oil charged sandstone reservoir, due to a very low Vp/Vs ratio, shows a soft reflection, increasing with offset, whereas the embedding lithology, such as shales or brine sands, may dim out due to the dispersion of the wave energy. Therefore, the seismic interpreter compares the amplitudes at different offsets from the source to identify potential anomalies. To enhance the data, selected offset ranges are usually stacked into so called angle stacks. Detailed explanations of the method can be found in Castagna et al. (1993) and many other publications.

Due to their anomalously high Vp/Vs ratio when compared with other lithologies, acoustically hard volcanic layers also usually show amplitude strengthening with offset. At shallower levels (~2,000 m), we can observe an absolute increase of the positive amplitude related to the top of the basalt. At greater depths (>3,000 m) the amplitude with offset dims much less than the background and we still observe a relative amplitude anomaly. This effect, being observed from the interface sediment-basalt and basalt-sediment, can be stronger than any variation (lithology or fluid) within the clastic section. Fig. 20 shows an example of an AVO model from a location in the Santos basin demonstrating this effect: both reflections top and bottom of the basalt layer - show an absolute increase of amplitude with increasing distance from the energy source.

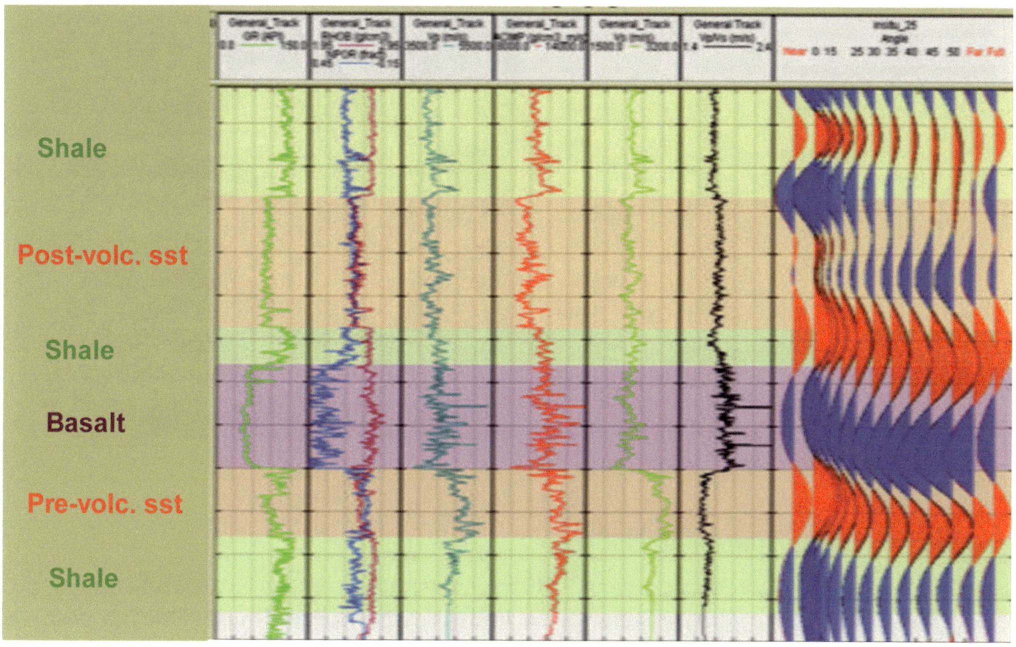

Fig. 20. Demonstration of the AVO behavior (strengthening of amplitude with increasing incidence angle) at top and bottom of a basalt reflection

A real seismic example is seen in Fig. 21. On the Near angle stack we see a prominent reflection (red is the positive polarity top of the geobody, meaning an increase in impedance, black the negative polarity bottom), which becomes stronger on the Mid and on the Far angle stack, whereas the embedding clastic lithology on the section dims out. The 3D image of this reflection reveals a geobody, interpreted as a volcanic event: we see a vent (circular feature), from which a number of stacked lava tongues discharged. The whole geobody extends over an area of about 5X5 km.

Based on the AVO response of lithologies with anomalous values of Vp/Vs ratio, Gidlow & Smith (1992) developed the so called fluid factor approach. The Fluid Factor attribute is based on the deviation of analyzed points from a background mudrock line that should be locally derived from the given data set. Negative values mean that due to the low Vp/Vs ratio the deviation is into the negative intercept - negative gradient direction, which is typical for hydrocarbon bearing sandstones. An example for a gas bearing sandstone is represented in Fig. 22. For our volcanics, the high Vp/Vs ratio means that we see a deviation from the background towards positive intercept and positive gradient, hence a positive fluid factor. Similar observations have been reported by Dillon et al (2004), where they compared the strong positive feature on the Fluid Factor Stack at the top of a volcanic layer to carbonate response. Again, we have to use the amplitude geometry to distinguish between carbonate and volcanic response.

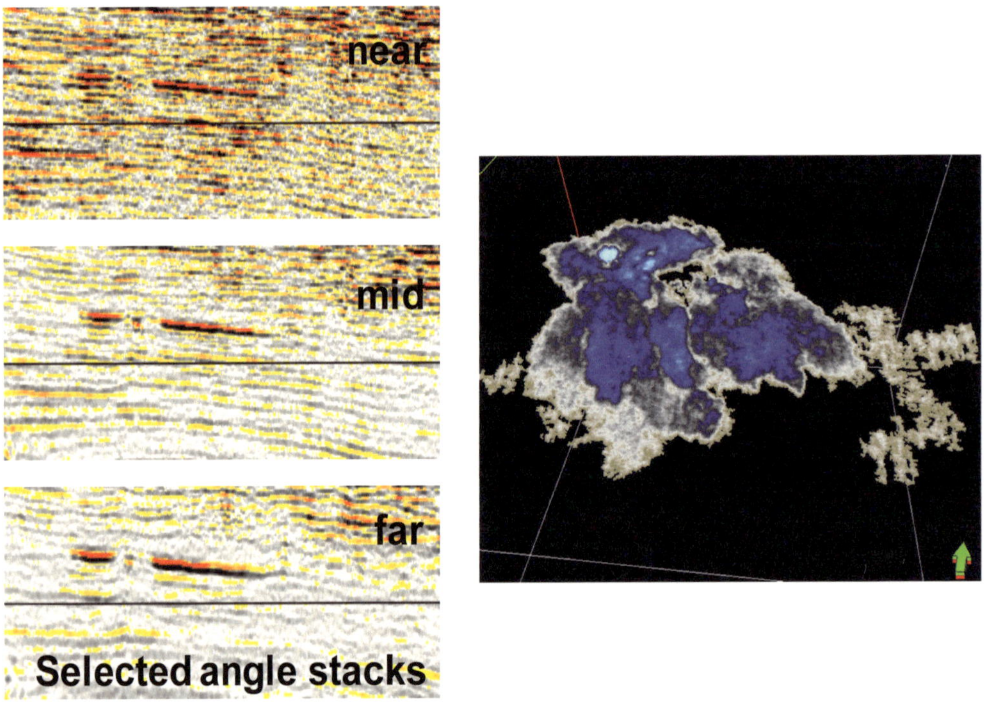

Fig. 21. The images of a volcanic layer on different seismic attributes: Near, Mid and Far Angle Stack (left); 3D amplitude image of the Full Angle Stack Amplitude (blue is a positive reflector)

Another useful tool to discover high Vp/Vs ratios relative to a regional background trend is the simultaneous inversion of different angle stacks. The output is usually cubes of acoustic impedance, shear impedance and, in the case of good data quality, density, which can be converted to additional products like the Vp/Vs ratio. We do not show an example here, but the method is a proven industry tool for lithology classification and shows similar results to those demonstrated in Fig. 22.

3.5 Implication for hydrocarbon exploration

The economic impact of volcanic rock types for the geology of fields and prospects in many exploration and production areas is tremendous: in the past, partial or complete reservoir substitution, complex diagenetic alteration by circulating hot fluids and addition of mineral components have led to alteration of rock properties and misinterpretation of seismic data.

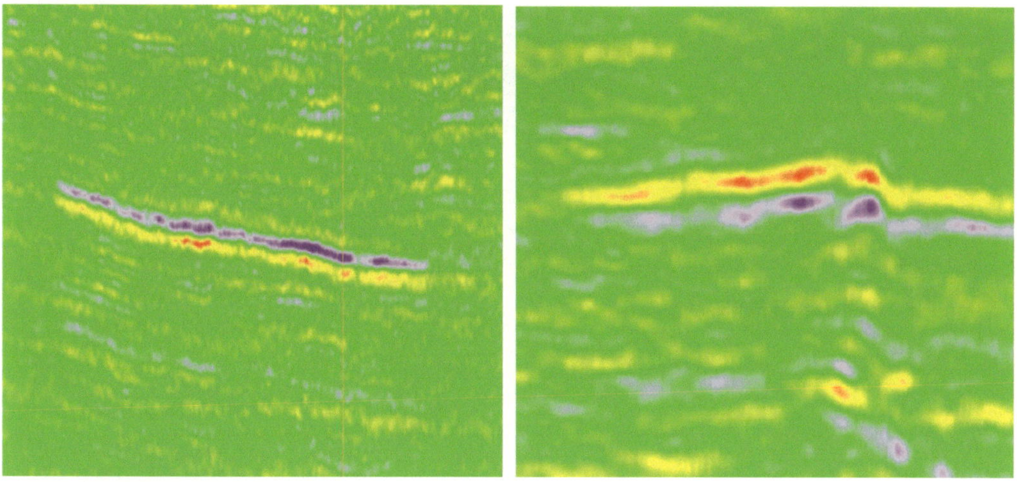

Fig. 22. Fluid factor stacks of the basalt layer shown in Fig. 21 (left) and a gas bearing sandstone layer (right). Purple is a positive deviation from the greenish background trend, orange is a negative deviation

On the other hand, fractured mafic intrusions are either acting as a reservoir, or have thermally cracked the embedding lithology, making it a producing reservoir, as in the Filon example in Argentina (compare Ortin et al. 2005).

In addition, volcanic rocks may form migration barriers, working both in our favor and against us: non-permeable volcanic layers can seal the top of the reservoir, preventing it from breaching, or they can build a migration barrier for the fluid on its way from the source rock into the trap.

We encountered Rotliegendes, Jurassic, Cretaceous and partially Tertiary volcanics in areas of interest in Argentina, Brazil (Fig. 23), Libya, Northern Germany and the Dutch offshore, UK North Sea. Colleagues report presence of volcanics in E&P projects offshore Australia (O'Halloran et al. 2001), onshore China (Zhao et al. 2002; Wu et al. 2006, Yang et al. 2008), and in Mexico (Pena et al. 2009), to name only a few.

The majority of rock physics publications on clastic reservoirs deal with the properties of subfacies in the sand(stone) - shaly sand(stone) - sandy shale - shale environment, where grain-supported quartz sandstones, representing one of the end members of the succession, are well understood. The elastic properties of shales are the current focus of a number of research groups. Based on generalized velocity-porosity models calibrated to local well data, theoretical models and empirical relationships help us to understand the seismic facies, to calculate missing logs and to predict the bulk properties under reservoir conditions.

However, in many basins around the world, much more complex lithologies are encountered. Synsedimentary volcanic activity and its erosional products may lead to a complex mineralogy, including volcanic rock fragments, and consequently to a very complex diagenetic history. As a result, the texture and thus the elastic properties of the rocks are affected. Known relationships between P-wave velocity and S-wave velocity are not applicable. Instead, anomalously low shear velocities may lead to Vp/Vs ratios for sandstones that are higher than in the bounding shales, and the seismic response may be very specific.

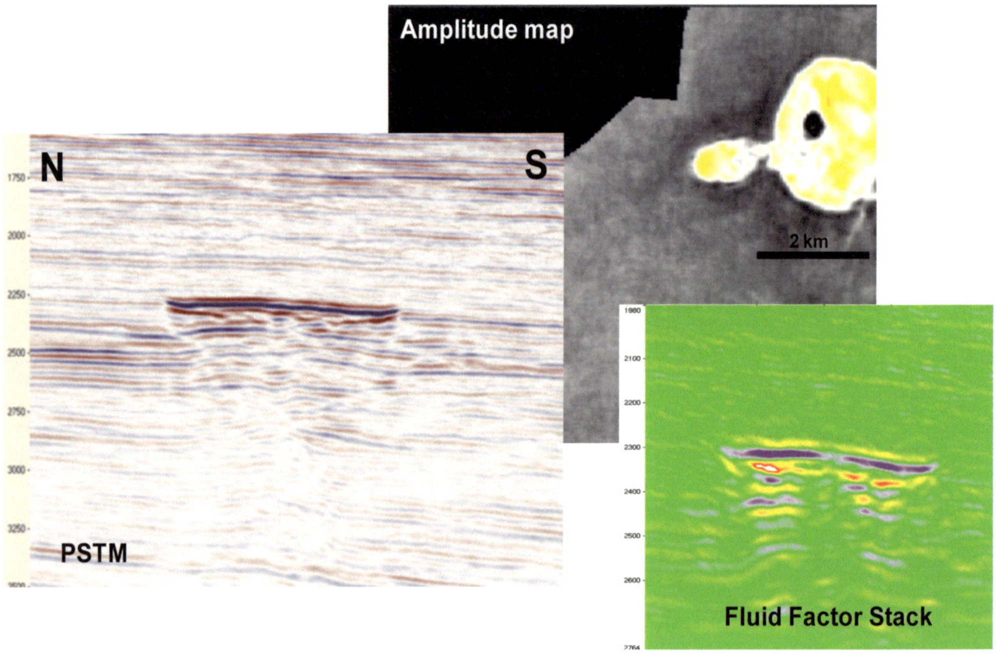

Fig. 23. Another example of amplitude expression, geometry and AVO response of a volcanic feature

One of the most important implications for seismic amplitude interpretation of the presence of rift volcanics in the sedimentary section is the fact that the erosional products from unstable volcanic rocks tend to contaminate the reservoir sandstones in the nearer and wider vicinity. Wind and water spread the particles around, mixing quartz and acid feldspar grains with mafic elements, thus completely changing the mineralogy of the sediment and subsequently causing different diagenetic alterations. A perfect example has been described from Cretaceous reservoirs offshore Brazil in Klarner et al. (2008). The pre-volcanic sandstones - an arkose with about 50% quartz content, 14% detrital feldspar grains and 4% volcanic rock fragments, classification after Folk (1968) - in the thin section images showed up as a grain supported stiff rock, showing the typical low Vp/Vs ratio compared to the embedding shales. It is represented by brown dots in Fig. 17. The post-volcanic sandstone had much less quartz (30%), 13% volcanic rock fragments and the grains were embedded into a pseudo matrix of secondary chlorite and calcite. The rock (represented in Fig. 17 by orange dots) exhibits relatively low velocities and an extremely high Vp/Vs ratio, with the inherent risk of pitfalls for AVO interpretation.

Similar phenomena have been observed and reported for some Tertiary turbidities offshore Sakhalin (Klarner et al. 2009) and in the Triassic section of the Barents Sea (Polyaeva et al. , 2011).

It is therefore crucial to understand the distribution of these volcanics in and around the reservoir. 3D seismic can be extremely helpful, if well understood.

4. Conclusions

Examples in the literature, dealing with volcanics and volcanic related rock types on seismic data used for exploration and production of hydrocarbons are still rare. There has been a lot of activity centering upon sub-basalt imaging, but this does not really help to overcome the problems we have in identifying and predicting volcanics from 3D seismic data at reservoir level. AVO investigations have typically concentrated on sandstone-shale intercalations and carbonates. The pure visual identification of volcanics within the seismic is very difficult. Either they appear to be similar to carbonates (basalts) or to siliciclastic rocks (volcaniclastics). However, due to the different acoustic and elastic properties of clastic and carbonate reservoir rocks versus volcanics, there is a realistic chance to develop an approach for the identification of volcanics in seismic data:

1. Basalts, even relatively thin layers, show typical geometries known from surface observations that can be identified on seismic data.
2. Basalts may display extremely high acoustic impedances and high vertical heterogeneity. They can be responsible for attenuation effects, which would be discovered by spectral decomposition/ spectral ratio analysis.
3. Volcanic layers and small edifices could be identified by differential compaction analysis, using curvature and high frequency structure images.
4. Volcaniclastics are multimineral, proximal, immature, poorly sorted sediments with a matrix-supported fabric. They exhibit significantly higher Vp/Vs ratios than cleaner and better sorted grain-supported siliciclastics.
5. Due to the high Vp/Vs ratio, basalts and volcaniclastics are likely to produce increasing AVO effects on hard kicks (high acoustic impedance layers) - porous sandstones usually show these effects on soft kicks.
6. The images and seismic response modeling of intrusive bodies, varying in dip and thickness, will help the seismic interpreter to identify them on his data set. In frontier areas, they could give indications as to whether we are dealing with volcanics.

5. Acknowledgment

The authors would like to thank all our colleagues (see references) for their friendly cooperation and fruitful discussions of the results.

6. References

Anjos, S. M. C., De Ros, L. F. & Silva, C. M. A. (2003). Chlorite authigenesis and porosity preservation in the Upper Cretaceous marine sandstones of the Santos Basin, offshore eastern Brazil, In: Worden, R. H. & Morad, S. (eds.) Clay Cements in Sandstones: IAS Special Publication, International Association of Sedimentologists - Blackwell Scientific Publications, Oxford, UK, 34, 291-316.

Avseth, P., Mukerij, T., Mavko, G. & Tysskvam, J. A. (2001). Rock physics and AVO analysis for lithofacics and pore fluid prediction in a North Sea oil field. The Leading Edge, April 2001.

Avseth P., Mukerji T., & Mavko G. (2005). Quantitative Seismic Interpretation: applying rock physics tools to reduce interpretation risk; Cambridge University Press.

Castagna, J. P. (1993). Amplitude-versus-offset analysis; tutorial and review. In: J. P. Castagna and M. M. Backus, Eds., Offset-Dependent Reflectivity - Theory and Practice of AVO Analysis, Society of Exploration Geophysicists.

De Ros, L. F., Mizusaki, A. M. P., Silva, C. M. A. & Anjos, S. M. C. (2003). Volcanic rock fragments of Paraná Basin provenance in the Upper Cretaceous sandstones of

Santos Basin, Eastern Brazilian Margin (abstract). 2° Congresso Brasileiro de P&G em Petróleo e Gás, Rio de Janeiro, Extended abstracts.

De Vincenzi, L. & Klarner, S. (2002). Seismic reservoir characterisation of a fluvial reservoir, Nakhla oil field (Libya, onshore). EAGE 64th Conference & Exhibition, Florence, Extended abstracts.

Dillon, L. D., Vasquez, G. F., Nunes, C. M., Neto, G. S., & Velloso, R. Q. (2004). Atributos DHI (Indicadores Diretos de Hidrocarboneto) obtidos a partir do dado sísmicopré-empilhamento: uma análise integrada da rocha à sísmica. Boletim de Geociencias da Petrobras, Rio de Janeiro, 12(1), 149-173.

Folk, R. L. (1968). Petrology of Sedimentary Rocks. Hemphill's, Austin, University of Texas Publication.

Gassmann, F. 1951. Über die Elastizität poroeser Medien, Vierteljahresschrift der Naturforschenden Gesellschaft in Zürich, 96, 1-23.

Garten, P., Houbiers, M., Planke, S., & Svensen, H. (2008). Vent Complex at Heidrun. SEG Las Vegas 2008 Annual Meeting, Expanded abstracts.

Gidlow, P. M., Smith, G. C., & Vail, P. J. (1992). Hydrocarbon detection using fluid factor traces: A case history, Extended abstracts of the Joint SEG/EAEG Summer Research Workshop on "How useful is Amplitude-Versus-Offset (AVO) Analysis?", 78-89.

Graversen, O. (2006). The Jurassic-Cretaceous North Sea Rift Dome and Associated Basin Evolution. Search and Discovery Article #30040, Posted February 19, 2006.

Greenberg, M. L. & Castagna, J. P. (1992). Shear-wave velocity estimation in porous rocks; theoretical formulation, preliminary verification and applications. Geophysical Prospecting, 40, 195-209.

Hanitzsch, C., deVincenzi, L., Heerde,W., Michel, J. M. & Semond, D. (2007). Dual inversion applied to 2D multi-component seismic data onshore Libya; *first break* volume 25, April 2007, pp 49-54.

Klarner, S., Culpan, R. & Smith, T. (2005). An enhanced AVO model for lithologically complex sandstones in the Santos Basin, Brazil. 75th Meeting, Society of Exploration Geophysicists, Houston, Expanded abstracts.

Klarner, S., Klarner, O. & Ujetz, B. (2006). Rock properties of complex lithologies - similarities between Offshore Brazil and West Siberia. EAGE Conference Saint Petersburg, Extended abstracts.

Klarner, S., Ujetz, B., Fontana, R. & Altenkirch, J. (2006). Seismic Signature of Upper Cretaceous Volcanics; Santos Basin, Brazil. EAGE 68th Conference & Exhibition, Vienna, Extended abstracts.

Klarner, S., Culpan, R., Fontana, R. & Bankhead, B. (2006). Identification of Upper Cretaceous Volcanics using AVO Attributes; Santos Basin, Brazil. EAGE 68th Conference & Exhibition, Vienna, Extended abstracts.

Klarner, S., Ujetz, B. & Fontana, R. (2008). Enhanced depositional and AVO models for lithologically complex sandstones in the Santos Basin, offshore Brazil. *Petroleum Geoscience*, Vol. 14 /3. 2008, pp. 235-243.

Klarner, S. & Zabrodotskaya, O. (2009): Rock physics in complex reservoirs - the key to understanding their seismic responses. EAGE/SEG Conference Tyumen, Extended abstracts.

Klarner, S., Kruglyak, V. F. & Lewis A. J. (2009): Anomalous elastic properties - an example from offshore Sakhalin. EAGE Conference and Exhibition Amsterdam, Extended abstracts.

Li, L., Wu, Q., Zhao, H. & Wu, X. (2009). Laboratory analysis of seismic parameters of volcanic rock samples from Daqing Oilfield. SEG Houston 2009 International Exposition and Annual Meeting. Expanded abstracts.

Mello, M. R., Mohriak, W. U., Koutsoukos, E. A. M. & Bacoccoli, G. (1994). Selected Petroleum Systems in Brazil: AAPG memoir, 60, 499-512.

Mohriak, W. U., Rosendahl, B. R., Turner, J. T. & Valente, S. C. (2002). Crustal architecture of the South Atlantic volcanic margins. In: Menzies, M. A., Klemperer, S. L., Ebinger, C. J., & Baker, J. (eds.) Volcanic Rifted Margins. Geological Society of America Special Paper, Boulder, Colorado, 362, 159-202.

Mukerji,T., Gonzalez, E. , Cabos, C., Hung, E., Mavko, G. (2002). Understanding amplitude anomalies and pitfalls in offshore Venezuela; quantifying the effects of geologic heterogeneities using statistical rock physics. Society of Exploration Geophysicists, International exposition and 72nd annual meeting, Salt Lake City, UT, United States. Oct. 6-11, 2002; Technical program, Expanded abstracts with authors' biographies. 72; Pages 2439-2442.

Nelson, C. E., Jerram, D. A. & Hobbs, R. W., (2009). Flood basalt facies from borehole data: implications for prospectivity and volcanology in volcanic rifted margins. Petroleum Geoscience, Vol. 15 2009, pp. 313–324.

O'Halloran, G. J. & Johnstone, E. M. (2001). Late Cretaceous rift volcanics of the Gippsland Basin, SE Australia - new insights from 3D seismic. In: Hill, K. C. & Bernecker, T. (eds) 2001. Eastern Australasian Basins Symposium, A Refocussed Energy Perspective for the Future, PESA, Special Publication, 353-361.

Ortin, A., Sanchez, E. & Bustos, U. (2005). Reservoir characterization in volcaniclastics; EAGE 67th Conference & Exhibition — Madrid, Spain, 13 - 16 June 2005, extended abstracts.

Ottesen, C., Heerde, W. & Weihe, T. (2005). Tectono-Stratigraphic Evolution and Controls on Volcanics in Early Cretaceous in Hameimat Area, Libya EAGE 67th Conference & Exhibition — Madrid, Spain, 13 - 16 June 2005. Extended abstracts.

Pena, V., Sarkar, S., Marfurt, K. J. & Chávez-Pérez, S. (2009). Mapping Igneous Intrusive and Extrusive from 3D Seismic in Chicontepec Basin, Mexico. SEG Houston 2009 International Exposition and Annual Meeting, Expanded abstracts.

Peterson, D. W., & Tilling, R. I., (1980). Transition of basaltic lava from pahoehoe to aa, Kilauea Volcano, Hawaii: field observations and key factors: Journal of Volcanology and Geothermal Research, v. 7, p. 271-293.

Planke, S., Alvestad, E., & Eldholm, O. (1999). Seismic characteristics of basaltic extrusive and intrusive rock, The Leading Edge, March 1999, pp 342-348.

Polyaeva, E., Lowrey, C. J., Klarner, S. & Zabrodotskaya, O. (2011). Depth dependent rock physics trends for Triassic reservoirs in the Norwegian Barents Sea EAGE 73th Conference & Exhibition incorporating SPE EUROPEC — Vienna, 2005. Extended abstracts.

Porębski, S. J. & Steel, R. J. (2003). Shelf-margin deltas: their stratigraphic significance and relation to deepwater sands. Earth-Science Reviews, 62, 283-326.

Porębski, S. J. & Steel, R. J. (2006). Deltas and Sea-Level Change. Journal of Sedimentary Research, 76 (3), 390-403.

Ross, C. P. 2002. Comparison of popular AVO attributes, AVO inversion, and calibrated AVO predictions. The Leading Edge, March 2002, 244-252.

Rutherford, S. R. & Williams, R. H. (1989). Amplitude-versus-offset variations in gas sands. Geophysics, 54, 680-688.

Schminke, H. -U. (2000). Vulkanismus. Wissenschaftliche Buchgesellschaft Darmstadt.

Scotchman, I. C., Carr, A. D. & Parnell, J. (2002). Hydrocarbon Generation Modelling along the UK North eastern Atlantic margin. AAPG Hedberg Conference, Hydrocarbon habitat of volcanic rifted passive margins, Stavanger, Abstracts.

Smith, T. M., Sondergeld, C. H. & Rai, C. S. (2003). Gassmann fluid substitutions: a tutorial. *Geophysics*, 68, 430-440.

Thomson, K. & Hutton, D. 2004. Geometry and growth of sill complexes: insights using 3D seismic from the North Rockall Trough. Bull Volcanol 66:364–375.

Thomson, K. (2005). Volcanic features of the North Rockall Trough: application of visualisation techniques on 3D seismic reflection data Bull Volcanol 67:116–128.

Vernik, L. (1994). Predicting lithology and transport properties from acoustic velocities based on petrophysical classification of siliciclastics. Geophysics, 53(3), 420-427.

Wu, C., Gu, L., Zhang, Z., Ren, Z., Chen, Z., & Li, W. (2006). Formation mechanisms of hydrocarbon reservoirs associated with volcanic and subvolcanic intrusive rocks: Examples in Mesozoic-Cenozoic basins of eastern China. AAPG Bulletin, v. 90, no. 1 (January 2006), pp. 137–147.

Yang J., Mao, H., Chang, X., Zhu, M., Wang, X. & Zou, Y. (2008). Mu-rho direct inversion for volcanic rock reservoir prediction: a case study of the Dinan Field, Junggar Basin. SEG Las Vegas Annual Meeting. Expanded abstracts.

Zhao G. L., & Y. Q. Zhang, (2002). Seismic reflection character of volcanic reservoir of Daqing and the comprehensive prediction technology: Petroleum Exploration and Development, 29, 44–46.

9

Multiscale Seismic Tomography Imaging of Volcanic Complexes

Ivan Koulakov
Institute of Petroleum Geology and Geophysics,
Siberian Branch of Russian Academy of Sciences,
Russia

1. Introduction

Immense scales of volcanic eruptions attracted vital interest of many people since ancient times. It was a long dream of volcano explorers to take a glance inside volcanoes to see the mechanisms responsible for their activity. However, only since the last decades, the development of technologies in geophysics has begun to provide images of the earth structure beneath volcanoes. One of the most powerful methods of geophysical investigations of volcanoes is seismic tomography, a relatively young method, which made a revolutionary breakthrough in the deep earth imaging during the last decades. To study the depths of tens and hundreds kilometers, the geophysicists mostly use the natural sources of seismic signal, earthquakes. These events generate seismic waves which propagate through the earth and accumulate the information on the inner structure of the Earth. The purpose of the seismic tomography method is to decipher this information and to provide the 3D distributions of seismic parameters. In the case of using earthquakes as natural sources an additional problem consists in determining their location coordinates and origin times. Seismic tomography usually provides the distributions of two types of seismic parameters: velocities of compressional and shear (P and S) waves. These parameters have a different behavior depending on several petrophysical conditions; so their joint interpretation may give important informations about the temperature, rheology and composition of deep rocks. Besides two seismic velocities, 3D distribution of seismic wave attenuation (for P and/or S waves) is also reported in some cases. Some recent tomographic studies provided the distributions of seismic wave anisotropy which may give a hint about dynamical state of rocks.

In this chapter I will make an overview of different-scale tomographic studies oriented to better understanding the nature of volcanic processes. I do not claim to present in details all variety of different tomographic studies in volcano regions. I will start with a description of various tomographic schemes which can be useful for studying volcano processes on different scales. Then I will provide some real examples of tomographic studies related to different volcanic areas of the world, mostly based on results of projects in which I was personally involved.

2. Tomographic schemes for studying deep structures beneath volcanoes

The tomographic investigations of deep structure beneath volcanic areas can be performed on different scales. The regional structures in the mantle, which may achieve dimensions of hundreds and thousands kilometers, can be studied using worldwide catalogues of seismic data. In the earth, hundreds of earthquakes are recorded daily by seismic stations located all over the Globe and reported by global agencies, such as International Seismological Center (ISC, 2001). In total, their catalogues contain several millions of travel time picks collected for more than 50 years. Based on these data, a number of global and regional models were constructed. If in a selected area, there are stations and/or sufficiently high seismicity, this enables the presence of many seismic rays travelling in different directions in the target volume (Figure 1). Inversion of these data using the tomographic approaches provides the 3D distributions of P and S heterogeneities in the mantle beneath the study area. For example, recent seismic models beneath Europe (Koulakov et al., 2009a) and Asia (Koulakov, 2011) were constructed using this approach.

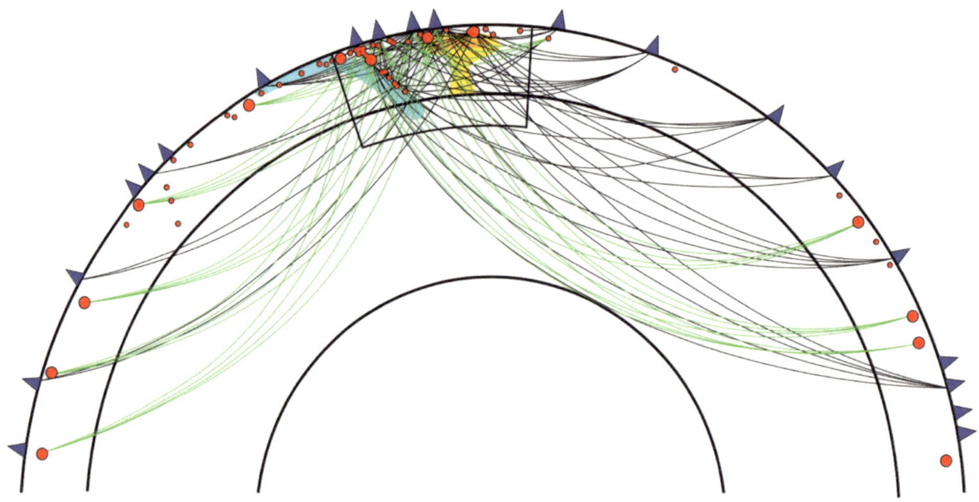

Fig. 1. Schematic distribution of data in the regional tomographic studies. Red circles are the earthquakes; blue triangles are seismic stations of the worldwide network; shaded zone is the target area; green rays correspond to out-of-network events recorded by stations in the target area; black lines depict all available rays from events in the target area recorded by the worldwide stations.

Global and regional tomographic studies based on worldwide seismic catalogues are most appropriate for retrieving the shapes of the slab in different subduction zones where most intensive seismic activity is concentrated. Other contrasted features in the earth having relations with volcanism, mantle plumes, can also be studied by regional tomography, but they are much harder objects to investigate by seismic tomography than slabs. First of all, in plume areas the seismicity is usually much weaker and is located only at shallow depths. Second, due to Fermat principle, seismic wave tends to travel mostly through high velocity patterns and avoid low-velocity plume-related anomalies. It is believed that plumes in the mantle have the shape of thin vertical channels of about 100 km thickness with contrasted low-velocity properties inside. In this case, seismic rays would overturn the plume and would accumulate almost no information about it. Even in cases of ideal ray coverage, regional tomography would only be able to retrieve the indirect evidences of plume existence, such as large warmer areas around plumes, but not the shape of the plume itself.

In some extent, this problem can be solved if instead of ray theory, one use finite frequency approach (Montelli et al., 2004). In this approximation, the seismic waves propagate along «fat» rays (areas with variable sensitivity) which may accumulate more rich information about plumes than line-representation of seismic rays.

Talking about regional studies I have to shortly mention surface wave tomography studies. Surface waves propagate along the Earth surface; the amplitudes of particle motions due to surface wave are maximal at the surface and decrease with depth. Two distinct types of surface waves are observed: Love waves, whose associated displacement is parallel to the free surface and perpendicular to the direction of propagation (toroidal), and Rayleigh waves, with displacement on a plane perpendicular to Love-wave displacement (spheroidal). The main feature of surface waves that allows using them for tomography is that their velocity strongly depends on the frequency. The velocity of surface wave is an integral value of the shear velocity distribution in the layer where the wave effectively propagates; lower frequency of the signal corresponds to the thicker layer (a kind of skin effect which is observed in different branches of wave physics). Thus, if the velocity increases with depth, low-frequency modes of surface wave have higher speed, as embracing thicker layers of the Earth. As a result, multi-frequency surface wave packets are clearly observed to disperse as they propagate away from their source. From the point of view of tomography, this means that surface-wave signal filtered at different frequencies will correspondingly provide information about different depth ranges of the Earth. These properties of surface waves are widely used in global and regional tomographic studies.

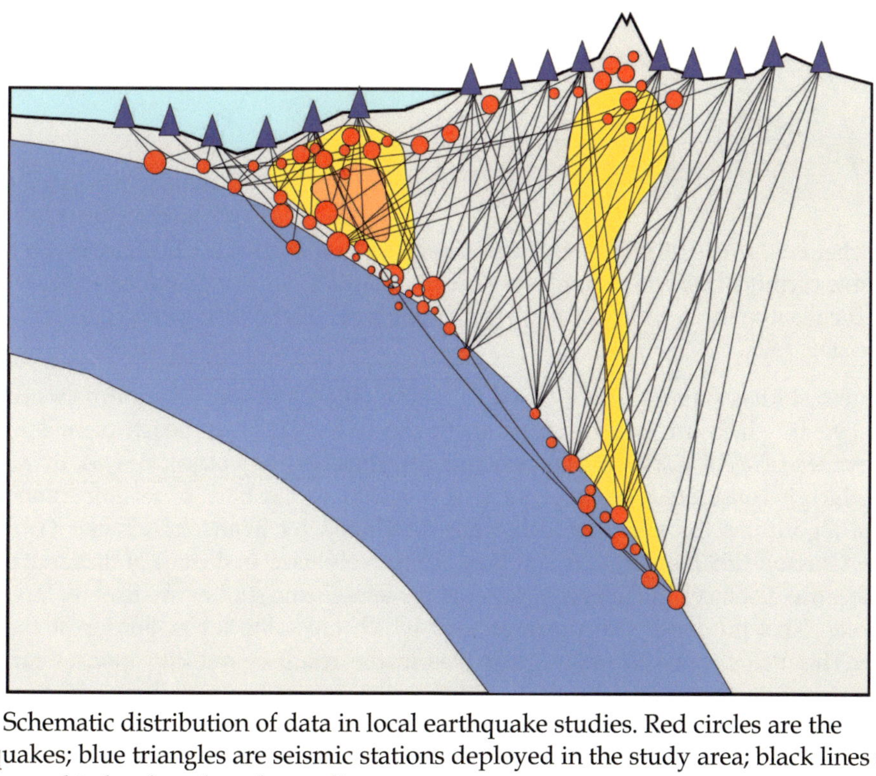

Fig. 2. Schematic distribution of data in local earthquake studies. Red circles are the earthquakes; blue triangles are seismic stations deployed in the study area; black lines depict all rays used in local earthquake studies.

For smaller scale studies, temporary networks are installed in target areas and operate for a limited time period continuously recording seismic signal. Among variety of approaches used to process data of local seismic networks, in our studies we mostly use two tomographic schemes: local earthquake and teleseismic tomography. Local earthquake tomography (Figure 2) uses the travel times of P and S seismic waves from sources occurred in the study area. In this case, the inversion should be performed simultaneously for seismic velocities and source parameters (coordinates and origin times). The location of sources and configuration of the ray paths strongly depend on the unknown velocity distribution. Thus, this problem is strongly non-linear and its solution is usually performed by iterative repetition of several linearized steps. This scheme is especially productive for areas with intermediate depth seismicity, such as in subduction zones (see an example in Figure 2). This gives us a possibility to investigate the 3D seismic structure above seismic clusters.

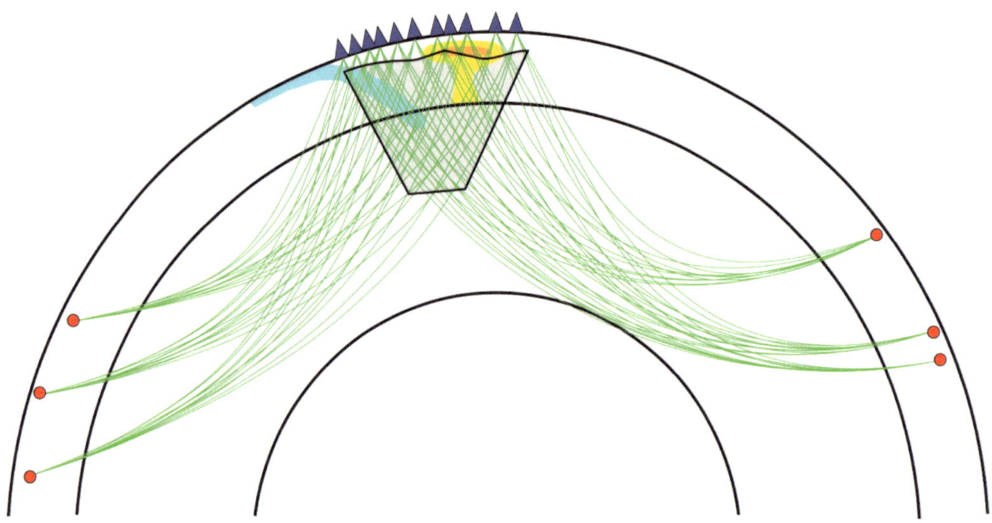

Fig. 3. Schematic data distribution in teleseismic tomography studies. Red circles are the teleseismic earthquakes; blue triangles are seismic stations of the regional network; shaded zone is the resolved area where rays intersect in different directions; green rays correspond to teleseismic rays.

Teleseismic studies (Figure 3) are based on using records of local station networks from remote events which are mostly based on the so called ACH approach named after the author names (K. Aki, A. Christoffersson and E.S. Husebye) of a classical work by Aki et al, (1977) which is considered as one of the first practical applications of seismic tomography. Practical algorithms for this approach were developed by Evans & Achauer (1993), van Decar & Crosson (1990) and others and they are actively used in dozens of different studies. The epicentral distances to the events selected for teleseismic studies are usually larger than 30 degrees. This produces steep seismic rays which cross the lower border of the study volume. This helps to avoid any regular bias in the resulting velocity models caused by heterogeneities in the upper mantle in neighboring areas. The waveforms of teleseismic signal recorded simultaneously by the stations of the network usually have similar shapes, but arrive with some delays which can be identified by cross correlation. These relative delays are the input data for the tomographic inversion which provides relative seismic anomalies, but incapable to give any constraint on absolute velocities. The major problem of teleseismic scheme is considerably lower vertical resolution in respect to horizontal, which is caused by steep ray orientation. Taking into account the fact that tomographic inversion can work only in areas where rays cross in different direction, the upper limit of the resolved area in teleseismic studies is roughly at a depth where "ray tubes" from neighboring stations start to intersect. This depth approximately corresponds to the average distance between stations in the network. With depth, the size of the resolved area decreases like an inverted cone; the maximal depth is roughly equal to the diameter of the network.

According to the above paragraphs, teleseismic and local earthquake schemes use completely different data. In most local earthquake studies all the events located outside the network perimeter are not taken into consideration. In teleseismic studies, the selection of the events starts only from the epicentral distances of 25-30 degrees (~3000 km). So when processing data from temporary networks, no data corresponding to regional distances between the network perimeter and about 3000 km are used. In my opinion, this is obviously shame, because regional data contain very important information about structure in the target area which can be deciphered by tomography. Furthermore, these data enable strong variety of ray orientations in the target area that is very important to increase the resolution capacity of tomography. I believe that all these data can be inverted together that may provide the results of better quality than in cases of separate local earthquake and teleseismic tomography. The advantages of using data from regional events in local earthquake tomography scheme are illustrated by Koulakov (2009b).

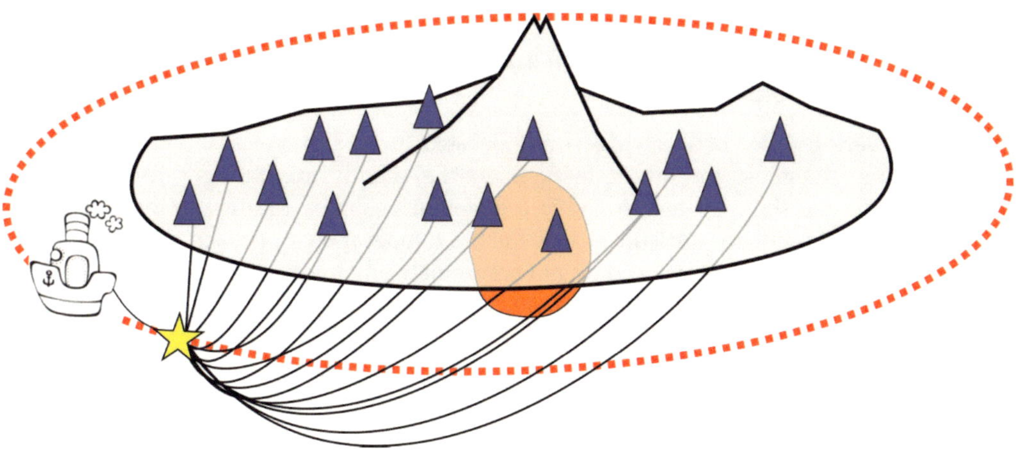

Fig. 4. Example of 3D active source studies based on marine airgun shots recorded by onshore stations. Blue triangles depict stations; yellow star is one of the shots along the path of the vessel (red dotted line). Black lines represent seismic rays.

In last years some alternatives of travel time tomography appeared. One of the tomography schemes which do not use seismic rays is ambient noise tomography which cross-correlate the microseismicity signal in different combinations of stations that finally gives the distribution of S-velocities. This method has been applied in some volcanic areas. For example, Stankiewicz et al., (2010) used ambient noise tomography and data from 40 temporary seismic stations to clearly image the magma chamber beneath Lake Toba caldera, one of the largest Quaternary calderas on Earth. Another method is based on surface waves recorded in local stations. Based on different group and phase velocities, one can get the 3D distribution of S velocity beneath the target area.

Active source studies with the use of artificial seismic sources are also used to investigate some volcanic systems. There were some examples of onshore studies in volcanic areas with the use of several explosions, mostly for profile observations (e.g. Aoki et al., 2009). However, the signal produced in such experiments was fairly weak to penetrate deeply. Seismic waves are strongly scattered in highly heterogeneous structures beneath volcanoes and, in most cases, seismic processing of these data hardly provides any reasonable structures. In addition, strong land explosions are forbidden in most volcanic areas. A most commonly used way in active source observations is generating seismic signal by airguns installed on research vessels (Figure 4). This scheme can be used for studying underwater and island volcanoes, or those located close to the Sea shore. A great amount of airgun shots enables special data processing which allows retrieving the signal of relatively high quality. In recent years these kinds of experiments have increased and provided high resolution images of volcanic regions, for example, Vesuvius (Italy) (di Stefano & Chiarabba, 2002, Zollo et al., 2002), Campi Flegrei (Italy) (Zollo et al., 2003), Deception Island (Zandomeneghi et al. 2009), and Montserrat Island (Paulatto et al., 2010).

Below I consider several examples of real studies that give information about deep sources of volcanism in different regions. In the worldwide scientific literature, there is a huge amount of papers on this topic, and obviously I do not pretend to cover all of them. In the following sections I will make an overview of some studies in which I was personally involved.

3. Regional-scale studies (1000s km size): Link between hot mantle and intracontinental volcanism

When considering links between deep mantle structures and volcanic complexes on a regional scale, the most important features in the mantle are thought to be plumes, convective cells and slabs. As mentioned in the previous chapter, plumes, as thin contrasted channels in the mantle, are difficult objects for the tomographic approach, because seismic rays tend to overpass the plumes and do not accumulate any information about them. At the same time, indirect evidences of plumes, such as heating of the mantle, can be detected robustly as large low-velocity anomalies. Most regional and global tomographic models of the upper mantle reveal close correlation between the distributions of intra-plate Cenozoic volcanic fields and low-velocity anomalies which can be interpreted as areas of higher temperature in the upper mantle.

3.1 Mantle structure beneath volcanic fields in Europe

An example of the distribution of Pn velocity anomalies just below the crust in Europe and Mediterranean is shown in Figure 5. This model was computed by Koulakov et al., (2009a) from inversion of several millions of travel times from the ISC catalogue based on the regional tomographic scheme (Figure 1). The calculations were separately performed in 12 overlapping circular windows and then the results were composed by averaging in one model. The reliability of this model was thoroughly verified by a number of different tests presented in Koulakov et al., (2009a). Note that this model seems to be consistent with regional and global models constructed by other authors (e.g. Piromallo & Morelli, 2003, Bijwaard et al., 1998).

Comparison of the tomographic model with Cenozoic tectonic units in Europe presented in Figure 5 shows that most of the volcanic fields appear to be associated with low velocity anomalies in the uppermost mantle. For example, the volcanic complex of the French Massif Central (MC) perfectly fits with an isometric low-velocity pattern. Similar features, although not as sharp, are located beneath the Cenozoic volcanic fields in Eastern Pyrenees, Eifel area, Bohemian and Pannonian massifs. These volcanic fields seem to be related to overheated parts of the upper mantle which can be attributed to ascending flows in active mantle convection. We propose that this convection can be activated by complex subduction features in the collision zones between European and African plates. At the same time, we do not see any evidences that the volcanism in these areas of Europe can be caused by plumes, thin highly contrasted channels in the upper mantle.

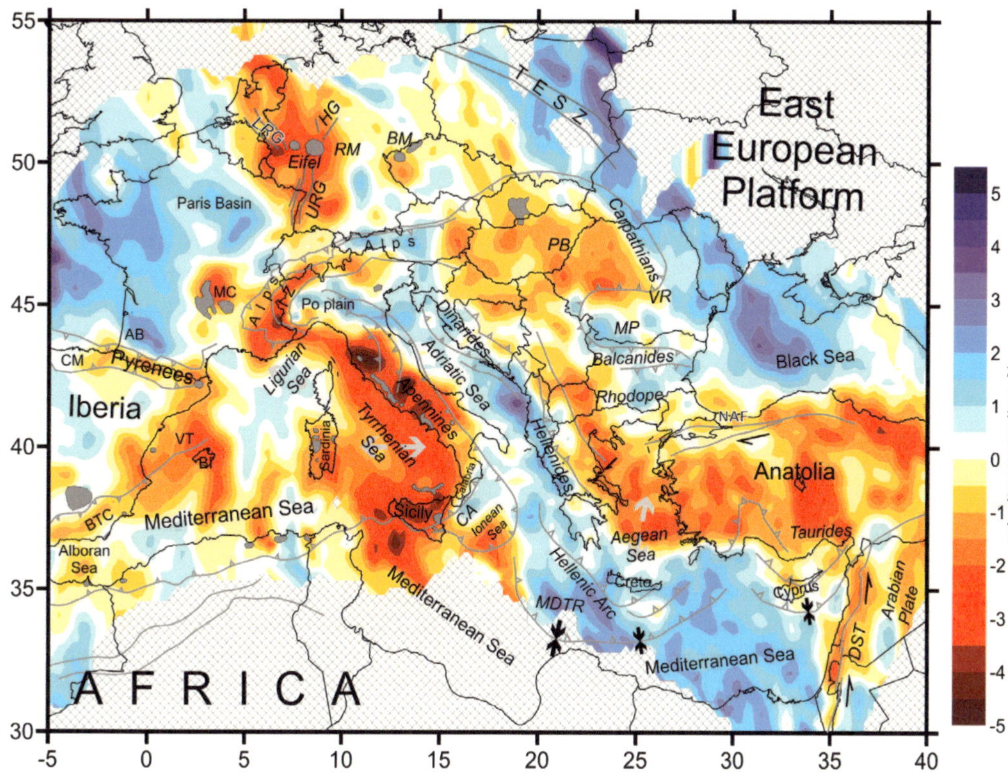

Fig. 5. The distribution of P velocity anomalies in the uppermost mantle beneath Europe and Mediterranean, just below Moho, from regional tomographic study by Koulakov et al., (2009a). Main geographic and tectonic elements in the study area modified after Faccenna et al (2003). Grey spots indicate the main Cenozoic volcanic fields. Abbreviations: AB - Aquitaine Basin, BI - Balearic Islands, BM - Bohemian Massif, BTC - Betic Cordillera, CA - Calabrian Arc, CM - Cantabrian Mountains, DST - Dead Sea Transform, HG - Hessian Graben, IZ - Ivrea Zone, LRG - Lower Rhine Graben, MP - Moesian Platform, MC - Massif Central, MDTR - Mediterranean Ridge, NAP - North Anatolian Fault, PB - Pannonian Basin, PRM - Rhenish Massif, TESZ - Trans European Suture Zone; URG - Upper Rhine Graben, VR - Vrancea, VT - Valencia Trough, WC - West Carpathians.

Some of the mentioned volcanic fields were previously studied in more details using temporary seismic networks and based on the teleseismic approach. For example, French Massif Central was studied by Granet et al., (1995). This model was used to estimate the temperature beneath MC (Sobolev et al., 1997). Another example of "searching for a plume" by tomographic approaches is teleseismic inversion for the area of Eifel volcanic field by Ritter et al., (2001). These and other studies in areas of Cenozoic volcanism in Europe found low velocity anomalies beneath the volcanic fields; however they look rather as large continuous features than sharp plume-shaped channels. It can be due to limited resolution enabled by ray-based tomography which is especially serious when long wavelength teleseismic rays were used. In this context, frequency-dependent tomography would be especially important; however it was not implemented for these areas. All these observations make us doubt in existence of plumes in Europe.

Many authors attribute the volcanism in the Central Mediterranean area to subduction-related magmatism (e.g. Dercourt et al., 1986; Lustrino, 2000) caused by closing of the Tethys Ocean. However, the composition variability of volcanic products around the Tyrrhenian Sea (e.g., Bell et al., 2004, 2006) may indicate to both mantle and subduction-related sources of magmatism. For example, carbonatites and kamafugites in Intramontane Province of Italy can only be explained in the frame of a plume model (Lavecchia et al., 2006). Based on geochemical analyses, Laveccia and Stoppa (1996) claim that a mechanism of intra-continental passive rifting in the Tyrrhenian and Adriatic Seas, which drives mantle upwelling, is sufficient to satisfy the petrological and geochemical constraints and the observed tectonic environment without requiring a subduction plane. At the same time, the facts of mantle genesis of some volcanic rocks in Mediterranean cannot automatically distinguish whether they are related to plumes or to large overheated areas in zones of ascending convective mantle flows.

Zones of active volcanism around the Tyrrhenian Sea, which includes volcanic fields in Sicily, volcanoes along the SW coast of Appennines and Cenozoic volcanic manifestations in Sardinia, are lying along the perimeter of a large low-velocity anomaly which covers the entire Tyrrhenian basin. We propose that these volcanic manifestations are related with the subduction processes. It is thought that beneath the Tyrrhenian Sea two subductions occur in different directions. The volcanoes along the SW coast of Apennines can be a kind of volcanic arc above the Adriatic plate subducting to SW direction. Volcanoes of Sicily probably form the arc above the Calabrian slab which subducts to NW direction. The other centers of volcanism in the Tyrrhenian and Adriatic Seas can be due to mantle upwelling and microplumes (e.g., Lavecchia et al., 2006, Bell et al., 2006). The latter are too small to be resolved by tomography.

3.2 East African Rift system: Plumes or convective flows?

In this section I will present an example of regional tomography study in the area of the East African Rift System (EARS) and will discuss its relationship with the volcanic processes. EARS is the largest continental rift system on the Earth where active volcanic manifestations occur in Cenozoic. In the Afar Triple Junction (ATJ), fast extension of the lithosphere causes the origin of two oceanic basins, Red Sea and Golf of Aden, and the Ethiopian segment, which is the unique place on the earth where spreading of the lithosphere takes place on-shore.

The origin and mechanisms of processes in EARS is a hot topic which is actively debated in the literature (see an overview in Chang and Van der Lee 2011). In particular, one of the main questions is whether the entire system is driven by one plume beneath ATJ, or by several flows along the EARS. To answer this question, a lot of different geological and geophysical studies were performed. In particular, deep seismic structure beneath different segments of EARS was investigated in various seismic studies. Regional travel-time tomography by Benoit et al. (2006) shows presence of a single plume beneath the Ethiopia. Results of SKS-splitting analysis (Hansen et al., 2006) and tomography Park et al. (2007, 2008) suggested presence of the channeled mantle flow from the Afar hot spot. A number of studies are dedicated to investigations of relatively small segments of the EARS using different tomographic methods (e.g. Daly et al., 2008; Jakovlev et al., 2011; Park and Nyblade, 2006; Tongue, 1992).

Here I present a regional tomographic model of the upper mantle beneath the EARS constructed by Koulakov (2007) and based on global catalogues using the regional tomographic scheme shown in Figure 1. The data amount and quality in this study is much lower than in Europe, and the spatial resolution in this case is much poorer. This study presented several tests, including synthetic modeling with realistic shapes of patterns, which directly demonstrate the limitations of the resolution with this dataset. These tests show that we cannot detect a plume, although very large; we are only able to see traces of indirect effect of the plume upon the surrounding areas.

In Figure 6, two horizontal sections, which correspond to the uppermost mantle and bottom of the upper mantle, are shown. In the shallower section, the tomographic image clearly highlights general geological features of the region. The Tanzania Craton is represented by high-velocity anomaly; volcanic fields in Western and Eastern Branches of the EARS surrounding the Craton coincide with low-velocity belts. Beneath the Ethiopian segment of the EARS, where most active rift processes and related volcanism are observed, the tomographic model provides intensive low-velocity anomalies. In the Afar Triple Junction (ATJ), the situation seems to be more complicated. Just in the middle of ATJ, we observe higher P-velocity anomaly, whereas the S anomaly is negative in the same location. One of possible explanations for higher P-velocity in the uppermost mantle could be the presence of very thin crust in the Red Sea and Gulf of Aden which is responsible for negative delays of seismic rays. In S-velocities this pattern is not seen due to significantly lower resolution. Another explanation can be related to petrophysical properties of rocks inside the plume. Higher P-velocity, which is more sensitive to the composition of rocks, might represent the existence of deeper rocks brought by the plume from the mantle. S-velocity is more sensitive to the presence of fluids and melts; thus low S-anomaly may reflect the rocks with high melting degree located in the top of the plume.

In deeper sections we observe two separate isometric low-velocity patterns of similar size (about 700 km width, highlighted with red dotted lines) which are located beneath the Tanzania Craton and ATJ. These anomalies, having the width compatible with height, cannot be directly called plumes. One explanation is that these are ascending branches of convective flows in the upper mantle. We cannot exclude that inside these flows there are thin plumes which affect chemically or thermally the surrounding rocks. However, this tomographic model does not have enough resolution to confirm of disprove this statement.

In the lower plot of Figure 6, a qualitative interpretation of the result along the profile passing through Western Rift, Tanzania Craton, Eastern Rift, Ethiopian Rift and ATJ. We claim that there are two approximately equal ascending flows beneath Tanzania Craton and ATJ which behave differently in shallower part of the mantle. Beneath ATJ and the

Ethiopian segment, this hot flow causes strong extension of the lithosphere which is manifested by spreading, rifting and volcanic activity. The flow beneath Tanzania Craton meets the bottom of thick lithosphere, but it is not capable to destroy it. As a result, this flow follows the bottom of the lithosphere and reaches the surface at the perimeter of the craton. The volcanic and rifting processes in the Western and Eastern branches are considerably weaker than those around the ATJ and the Ethiopian segment.

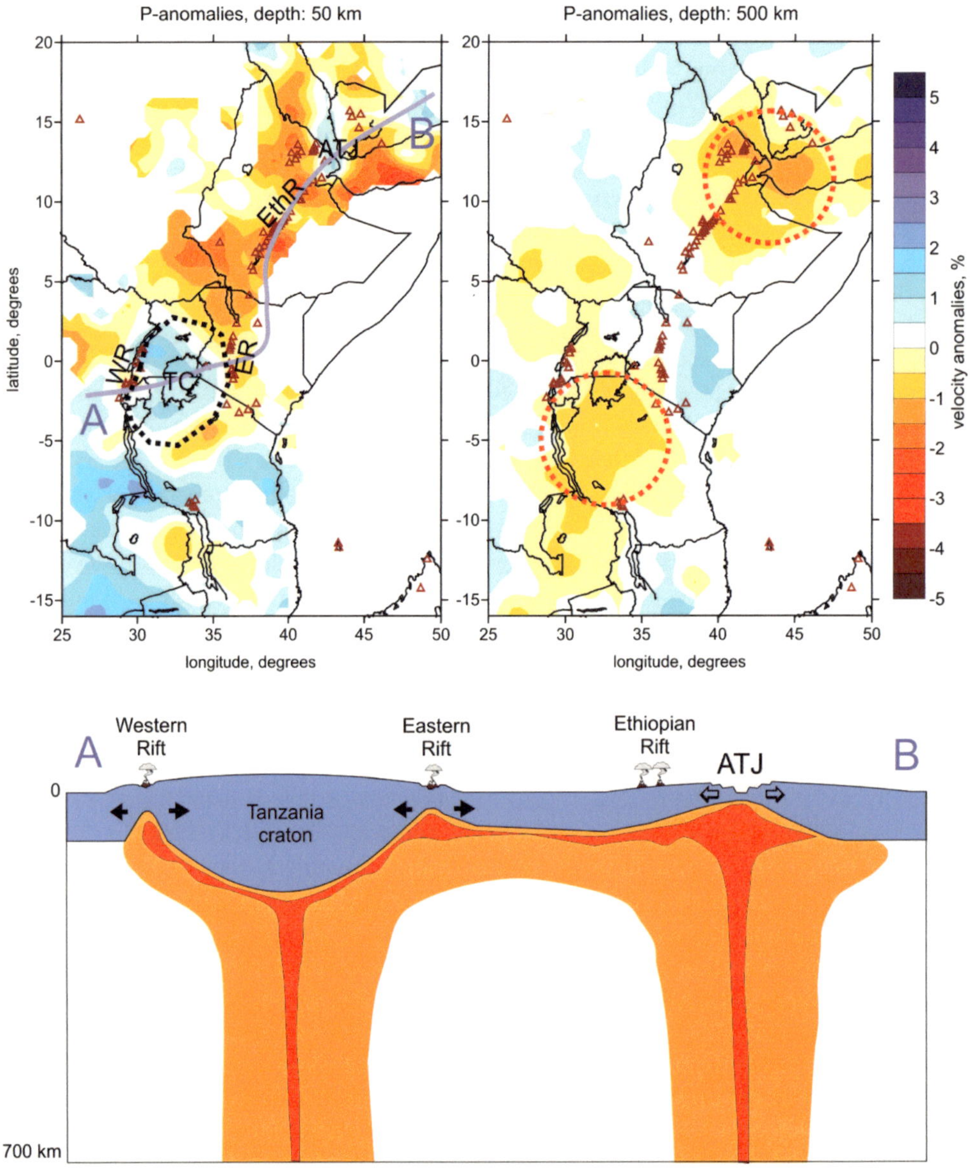

Fig. 6. Upper plots: results of regional tomographic study by Koulakov (2007) in the area of the East African Rift System at 50 km and 500 km depth; lower plot: qualitative interpretation along the profile A-B (violet line) which pass through Western Rift (WR), Tanzania Craton (TC), Eastern Rift (ER), Ethyopian Rift (ER) and Afar Triple Junction (ATJ).

3.3 Siberia and Mongolia: Overheated mantle beneath basaltic fields

The last example of a regional tomographic study related to zones of intraplate Cenozoic volcanism corresponds to the area of Siberia and Mongolia. Logachev et al. (1996) and Logachev (2005) evaluated the total amount of Cenozoic erupted material in Southern Siberia and Mongolia as 6000 km^3 (yellow fields in Figure 7). Roughly half of this amount corresponds to the Sayan area along the southern border of the Siberian craton. It is interesting that in Lake Baikal, which is the main rift depression of Siberia, no signature of Cenozoic volcanism has been observed (e.g., Eskin et al., 1978). Another volcanic field is related to the Hangay dome in Mongolia. Analysis of magma compositions in both areas has shown that they have generally similar properties (e.g., Yarmoluk et al., 1990), which probably indicates their common source. Rasskazov et al. (1993) estimated the depth of their origin as 50-150 km, while the geochemical analysis by Ashepkov et al. (1996) provided an argument for a greater depth of the magma sources. The youngest manifestations of volcanism are observed in Eastern Sayan.

Here I present the results of tomographic inversion by Koulakov and Bushenkova (2010) which was based on travel time data from global catalogues. Most of the area of Siberia is seismically passive, and a very few stations are installed there. The regional schemes presented in Figure 1 can be used for studying the seismically active southern part of Siberia and Mongolia (Koulakov, 2008). However it cannot provide any resolution in most parts of the Northern Siberia where neither stations, nor seismicity are presented. To fill the gaps in ray coverage, we used the PP phases which corresponded to seismic rays with one reflection from the Earth surface. When considering both PP and P phases, their difference gives information about seismic anomalies around the reflection point. For this study, we selected all PP rays having reflection points inside the study area. Note that these data are quite noisy, and their distribution does not allow computing high-resolution models. For areas with active seismicity and/or dense seismic networks, PP data had smaller weight compared to regular data. However, in case of absence of events and/or stations, these data play the major role for constructing velocity models.

In tomographic images, Koulakov and Bushenkova (2010) observed a clear correlation between the main seismic patterns and Cenozoic volcanic fields (Figure 7). This correlation was especially clear at shallower depths (50-100 km), where the amplitudes of anomalies beneath the volcanic areas reached 2%. For deeper sections, the anomalies were much weaker (approximately 0.5%) and displaced with respect to the volcanic fields.

The presence of several plumes beneath Mongolia and southern Siberia, as well as their link with Cenozoic volcanism, was proposed by Zorin et al. (2002, 2003) based on long wavelength gravity anomalies. However, these individual plumes were not visible in the tomographic inversion results reported here and elsewhere. Taking into account the deep nature of the magmas revealed by geochemistry studies (e.g., Ashepkov et al., 1996), we believe that the plumes may exist, but their sizes are too small to be resolved by seismic tomography. These plumes might provide the general anomalous thermal background beneath Sayan and Mongolia that is observed in our tomograms as one regional low velocity anomaly.

One of these plumes is probably located beneath the southern part of the Siberian craton. It was visible as a low velocity anomaly in our tomograms at depths of 400 km and deeper. This plume may accumulate hot material on the bottom of the cratonic lithosphere. While reaching a critical mass, this material starts moving up toward the border of the Siberian craton, where it produces fields of volcanic activity.

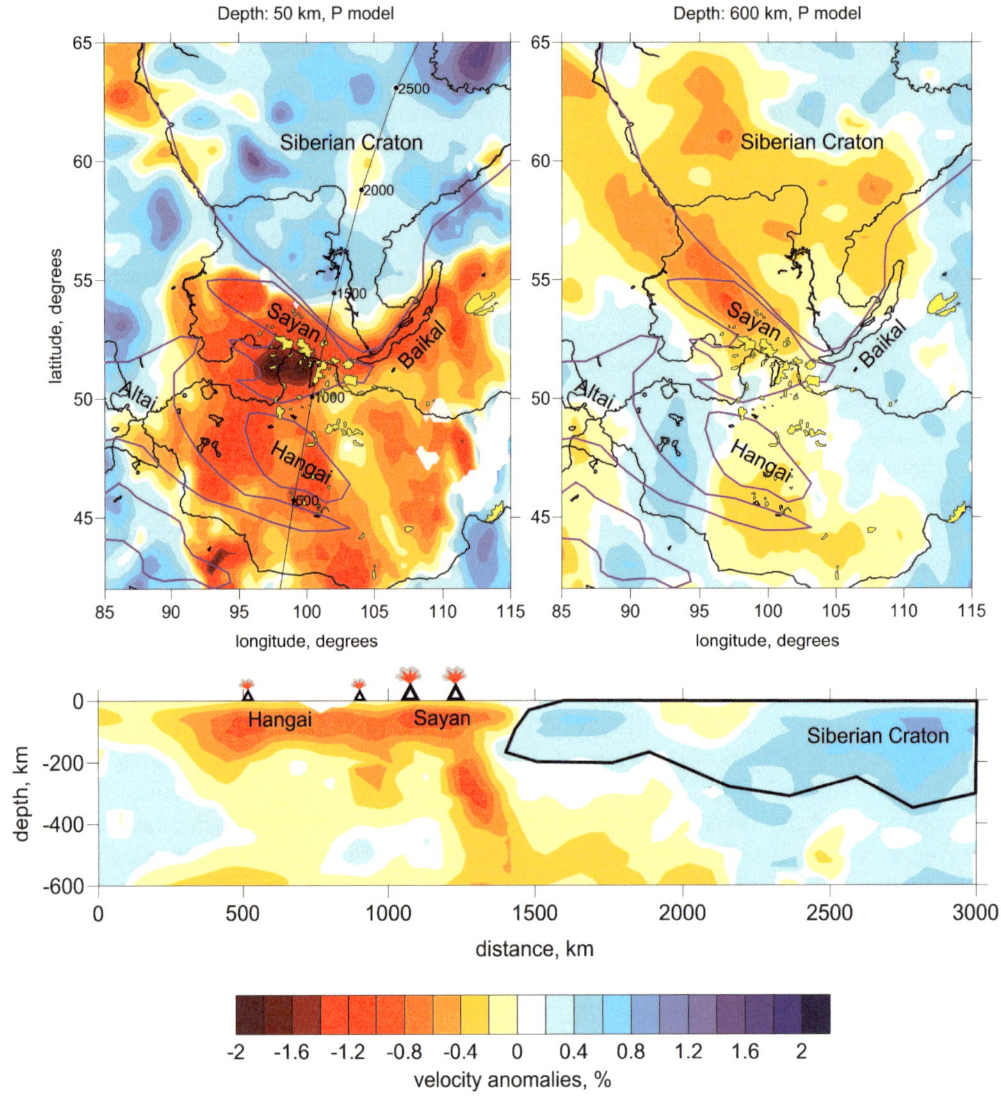

Fig. 7. P-velocity anomalies beneath Siberia and Mongolia from tomographic inversion of global P and PP data. Upper plots show two horizontal sections at 50 and 600 km depth. Yellow area depict the areas of Cenozoic basaltic volcanism. Violet lines highlight the main tectonic units. Lower plot show the resulting model in vertical section (location of the profile is shown in 50 km map). Hypothetical shape of the Siberian Craton is shown with black line. Locations of the main volcanic fields are indicated with volcano symbols.

4. Local earthquake tomography can reveal the link between subducting slab and volcanic arc

4.1 Toba Caldera: Can we see deep traces of super eruptions?

The Toba volcanic complex, located in northern Sumatra, Indonesia, is part of a 5,000 km long volcanic chain along the Sunda arc. Toba volcano produced the largest known volcanic eruption on the Earth during the past 2 million years (Smith & Bailey, 1968). About 74,000 years ago, around 2,800 cubic kilometers of magma were erupted and had a significant global impact on climate and the biosphere. The eruption led to the final formation of one of the largest calderas, the 35x100 km wide Toba caldera. Super scale eruptions at Toba have occurred several times (at least four eruptions of more than VEI 7 over the last 2 million years). In this sense, Toba seems to be a singularity in a chain of more than one hundred other volcanoes with explosive potential along the Sunda arc. The reasons for this unique behavior of the Toba volcanic activity are not yet clearly understood. As suggested by Bachmann and Bergantz, (2008), the explosive magma in super eruptions are produced by extracting interstitial liquid from long-lived "crystal mushes" (magmatic sponges containing >50 vol% of crystals) and collecting it in unstable liquid-dominated lenses. If there is a potential of a new super eruption, such signatures should be clearly seen in tomograms.

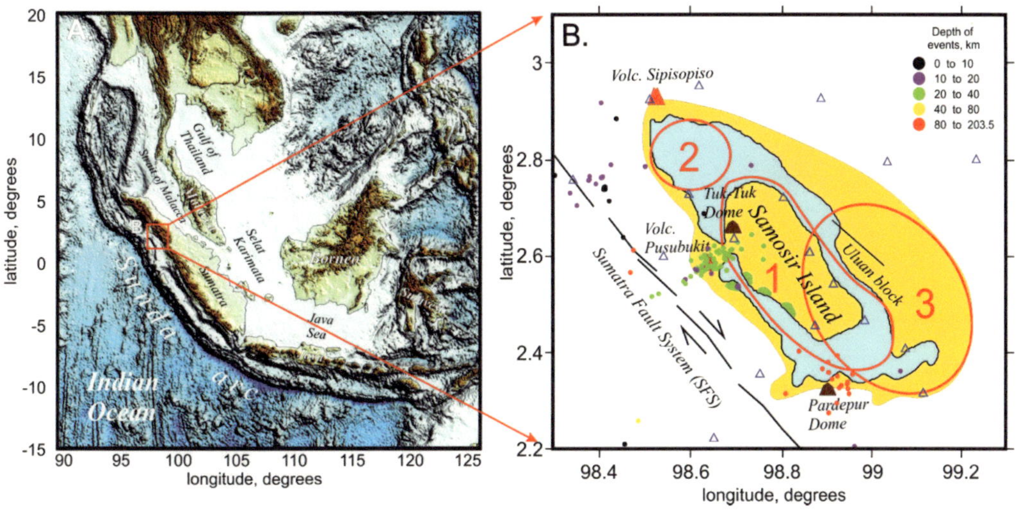

Fig. 8. Study region in the context of the Sunda Arc. *A*. Bathymetry and topography of the Sunda Arc and surrounding areas. B. The main structural elements of Toba caldera. Yellow areas are the present-day topographic depression; green indicates uplifted areas. Red ellipses mark different caldera units. 1 Sibandung caldera: created 74,000 years ago by the Toba YTT event (Young Toba Ash); 2 Haranggaol caldera: formed 500,000 years ago by the Toba MTT event (Middle Toba Ash); 3 Sibandung caldera: formed 800,000 years ago by the Toba OTT event (Old Toba Ash).Blue triangles depict stations; colored dots are the events used in this study.

Here I describe a tomographic model of the crust and uppermost mantle beneath the Toba Caldera constructed by Koulakov et al., (2009b). This work was based on a dataset corresponding to a rather old local seismic experiment and includes the travel time picks collected by seismic network around Toba which was operated for about 4 months (January-May, 1995) by Indonesian teams in cooperation with IRIS and PASSCAL. The network recorded ~1,500 local earthquakes; however, for this study only the 390 most reliable events were used (Figure 8 B). A special attention in this study was addressed to the verification of the model. In particular, a synthetic model with realistic configuration of anomalies was

constructed, so that the results of synthetic and real data inversion were almost identical both in shapes and amplitudes. This allowed assessing true magnitudes of anomalies.
Results of tomographic inversion are presented in Figure 9. Beneath the Toba Caldera and other volcanoes of the arc we observe relatively moderate (for volcanic areas) negative P and S velocity anomalies which reach 18% in the uppermost layer, 10-12% in the lower crust and about 7% in the uppermost mantle. Much stronger contrasts are observed for Vp/Vs ratio which is a possible indicator of dominant effect of melting in origin of seismic anomalies. At a depth of 5 km beneath active volcanoes we observe small patterns (7-15 km size) with high Vp/Vs ratio which might be an image of actual magmatic chambers filled with partially molten material feeding the volcanoes. In the mantle wedge we observe a vertical anomaly with low P and S velocities and high Vp/Vs ratio which link the cluster of events at 120-140 km depth with Toba caldera.

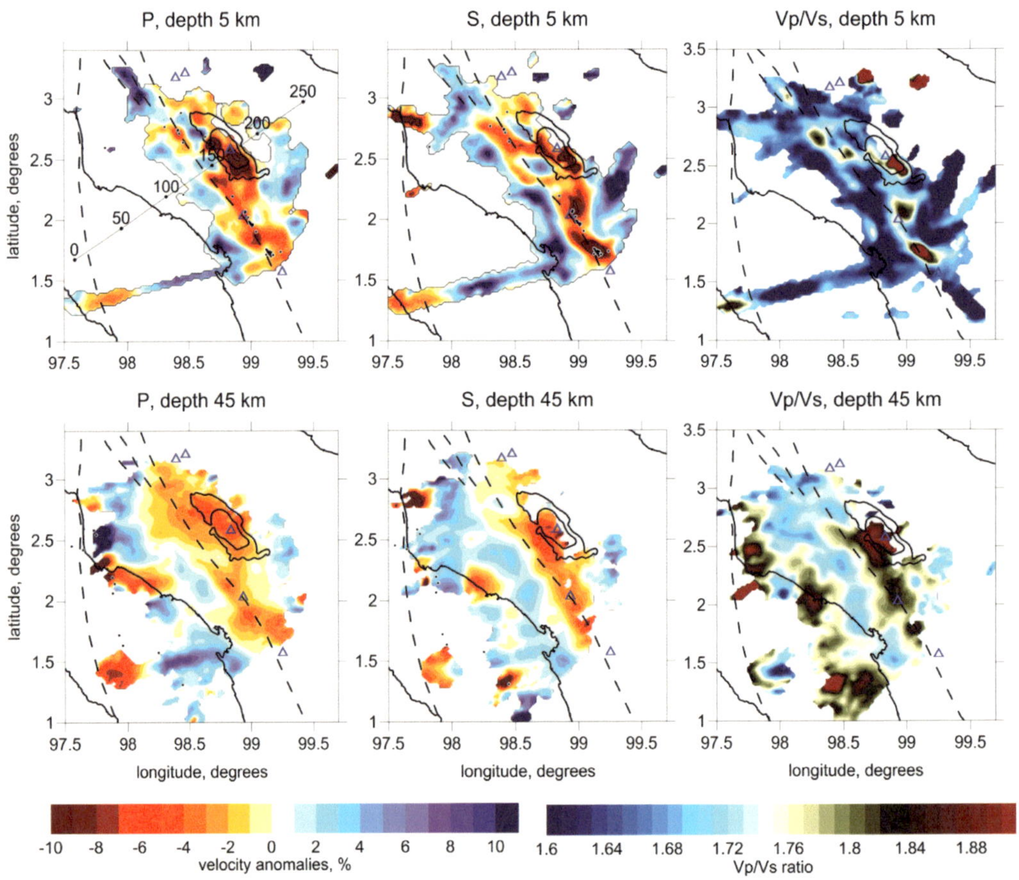

Fig. 9. P and S velocity anomalies and Vp/Vs ratio in the area of Toba Caldera in two horizontal sections. Black dots indicate the final locations of sources in a corresponding depth interval. Blue triangles show positions of active volcanoes. Dotted lines indicate the locations of the Mentawai and Sumatra fault zones. Positions of the profile for cross section in Figure 10 is given in the plots for P anomalies at 5 km depth.

The seismic structure beneath Toba allows quite clear interpretation which is illustrated in Figure 10. We observe a vertical low-velocity pattern which links the seismicity cluster at 130-150 km depth with the Toba Caldera. The most probable explanation for this link is that the phase transitions in the slab cause active release of fluids. Ascension of these fluids leads to decreasing melting temperature above the slab and the origin of diapirs and magmatic chambers (Poli and Schmidt, 1995). In this case, a prominent cluster of events at ~40 km depth beneath the western boundary of the caldera may delineate the border of an area with high content of molten material beneath Toba.

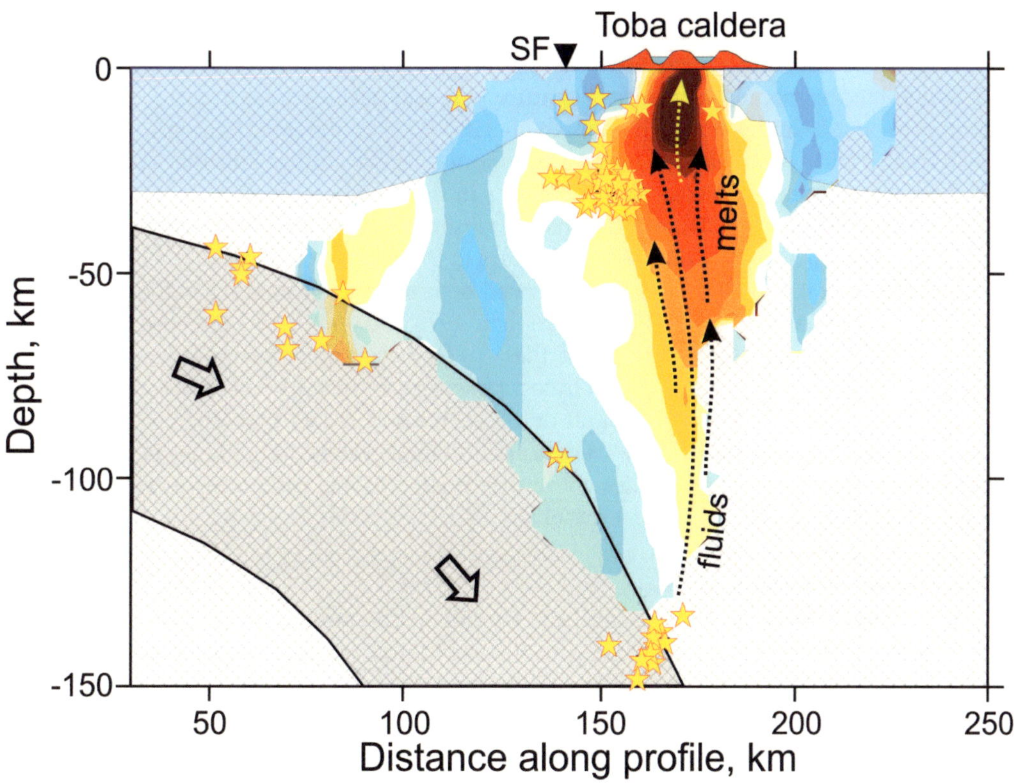

Fig. 10. Interpretation of the results along the vertical profile. Background is the distribution of P velocity anomalies. Yellow stars are the final location of events used in this study. Dotted arrows mark schematically the path of ascending fluids and partially molten material. Black triangle and SF indicates the Sumatra fault.

It is unexpected that beneath Toba we observe moderate values of P and S velocity anomalies. At a depth of 5 km the anomalies reach 16%-18%, however they may be partly related to young sediments which filled the Toba caldera. Some anomalies can also be related to fracturing of rocks along the Great Sumatra Fault. When considering only seismic velocities, it is not easy to distinguish these factors from volcano related anomalies (e.g. magma chambers). At 15 km depth beneath Toba and Helatoba volcanoes we observe anomalies which are slightly stronger than 10%. This is likely a signature of volcanic chambers and magma paths, however this value is not higher than observed in normal volcanic areas like Central Java (Koulakov et al., 2007, 2009c) and Central Chile (Koulakov et al., 2006) which will be discussed in next chapters. In the mantle the amplitude of anomalies is less than 7% that is relatively low for volcanic areas. In this sense, no significant signature of super volcanism is observed here.

Much more interesting features are observed at the distribution of Vp/Vs ratio which roughly reflects the content of melts and fluids. Variations of this parameter observed at different depths are much more significant than those of P and S velocity anomalies. At 5 km depth we obtain dominantly low values of Vp/Vs ratio with the value of about 1.62. At the same time, just beneath the presently active volcanoes we observe few local patterns with contrasted Vp/Vs ratio anomalies which reach 1.87. These patterns possibly indicate the magmatic reservoirs which feed the volcanoes. The concept of magma chamber is widely accepted in popular scientific literature (e.g. a typical cartoon in every school book where a volcano is connected with a large spherical body of magma chamber). At the same time, in many fairly robust and high resolution tomographic studies for other volcanoes, the chambers are not clearly detected (see examples in the next paragraphs). Beneath Toba and surrounding volcanoes at shallow depths we observe quite small bodies with extremely high Vp/Vs ratio which are quite clear indicators to the magma chambers.

The high Vp/Vs ratio patterns are also observed beneath the volcanic arc in the middle and lower crust (15 km and 25 km depth), where their sizes and amplitudes increase (up to 1.90). For the mantle wedge, the resolution of the Vp/Vs becomes rather poor, mostly due to trade-off with source depth and origin times. However, high values of Vp/Vs are robustly resolved beneath the Toba caldera in all depth sections. The fact that Vp/Vs contrasts are much more important beneath Toba caldera than P and S velocity anomalies is probably an indicator that the observed anomalies are mostly related to melting processes rather than to temperature, chemical or mechanical reasons.

4.2 Merapi volcano, Central Java

In this chapter I consider the tomographic studies in the area of Merapi in Central Java which is one of the most active volcanos in the world and represents a tremendous hazard to the local population. Mt. Merapi is a stratovolcano showing evidence of explosive eruptions over the last 7000 years (Newhall et al. 2000) and earlier (Camus et al. 2000). Besides volcano hazard, this part of Indonesia suffers of strong seismic activity. For example, on May 26, 2006 at 22:54:01 UTC a strong magnitude M_w=6.3 earthquake occurred in Central Java, Indonesia about 25 km SSW of Yogyakarta (May 27 at 5:54 AM local time in Java, Indonesia) and caused more than 6000 fatalities.

In 2004, combined amphibious seismological investigations have been performed in the framework of the MERAMEX (MERapi AMphibious EXperiment) project to study a volcanic arc system as part of an active continental margin. More than 100 seismic stations were operated continuously for more than 150 days (Figure 11). The local seismicity data recorded at these stations were used to perform a local tomographic inversion (Koulakov et al., 2007). Besides the passive data from local earthquakes, an active source experiment was performed at the same time. The airgun shots along three offshore lines were clearly registered and picked at onshore stations allowing the possibility to combine passive and active data to enhance the resolution in the overlapping area. The results of the combined active and passive data inversion are presented by Wagner et al., (2007).

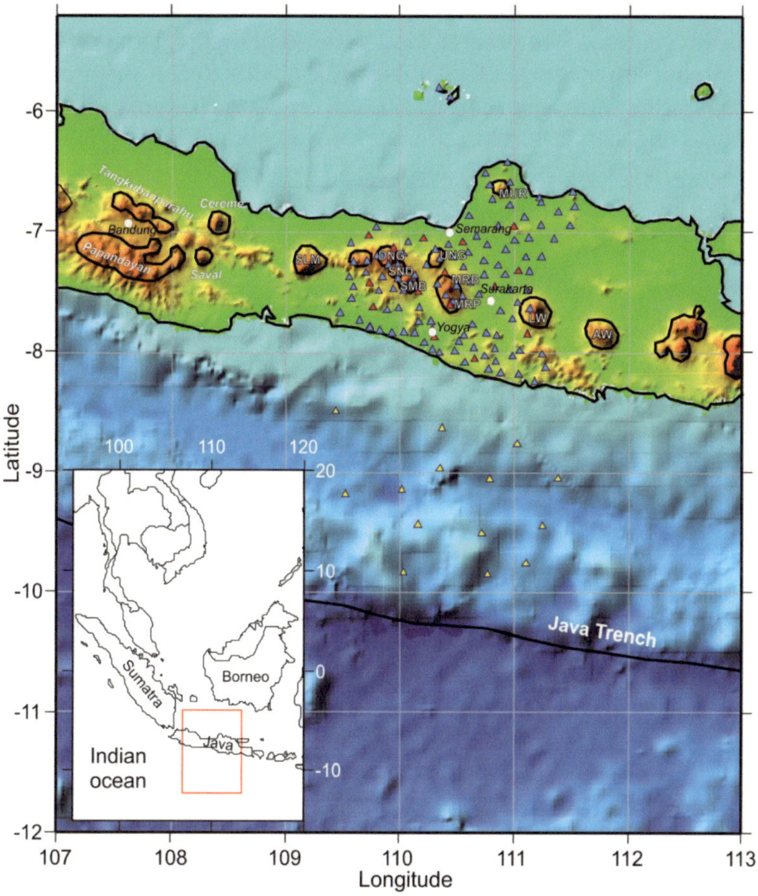

Fig. 11. Relief and bathymetry of the study area. The main volcanic complexes are highlighted with black contour lines. Volcanoes: SLM - Slamet, DNG - Dieng, SND - Sundoro, SMB - Sumbing, MRB - Merbabu, MRP - Merapi, MUR - Muria, LW - Lawu, AW - Arjuno-Welirang. The main cities are marked with white circles. Triangles show positions of the MERAMEX stations (red: broadband stations, blue: short-period stations, yellow: ocean bottom stations).

The dense ray coverage in this experiment made it possible to implement the anisotropic tomographic inversion for the P-velocity model. An orthorhombic anisotropy with one predefined direction oriented vertically is determined by four parameters in each point. Three of them describe slowness variations along three horizontal orientations with azimuths of 0°, 60° and 120°, and one is a perturbation along the vertical axis. Here I discuss the results of anisotropic inversion which are presented in Koulakov et al., (2009c).

The most important feature at shallow depths (Figure 12) is a strong low-velocity anomaly beneath the middle part of Central Java which is highlighted with violet dotted line. It consists of several segments which appear to be located between the main volcanic complexes. The strongest anomaly is observed between the Lawu and Merapi complexes and called MLA. The amplitude of this anomaly estimated from modeling with realistic patterns is about -30%. Another anomaly is located between the Sumbing and Merapi complexes, and the smallest anomaly is detected between the Sumbing and Slamet volcanoes. Beneath volcanoes we observe local patterns of higher anomalies which probably represent the location of channels or chambers filled with frozen magmatic rocks having higher seismic velocities. It is important that positive velocity patterns are observed beneath dormant volcanoes (e.g. Lawu, Sumbing, Sundoro, Merbabu, Ungaran). Beneath Merapi, the most active volcano in this region, such a feature is manifested less clearly.

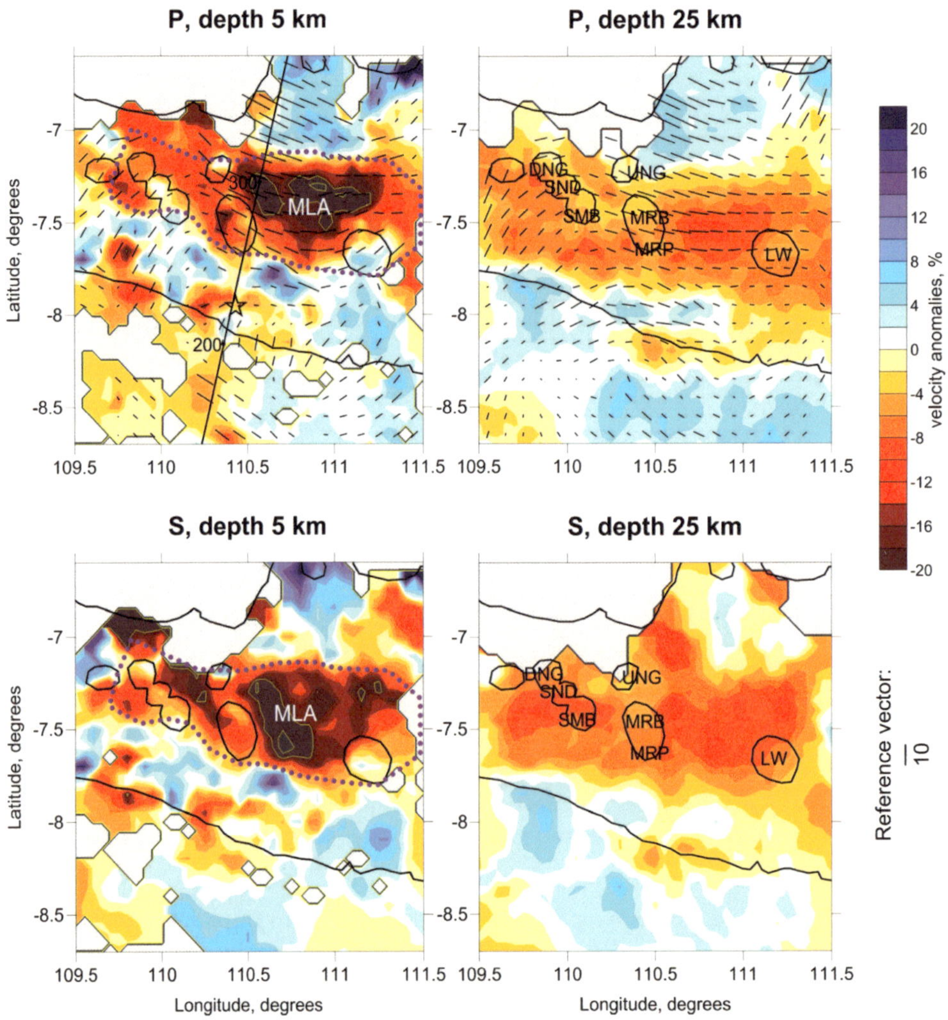

Fig. 12. Anisotropic P- and isotropic S-velocity distributions beneath Central Java in two horizontal sections. Colors indicate the isotropic velocity perturbations; yellow lines within "red" patterns are the contours of 20% and 25%; bars show directions of fast horizontal P velocities. Length of bars reflects the difference between fastest and slowest horizontal velocities. Position of the cross section presented in Figure 13 is marked in maps corresponding to 5 km depth. The star shows the hypocenter of Bantul M_w=6.5 earthquake (26.05.2006 UTC). Indications of volcanoes correspond to caption in Figure 11. Violet dotted line highlights the feeding area of volcanoes in Central Java.

In the forearc, between the coast and the volcanoes, the isotropic velocity structure is strongly heterogeneous. This complex structure can be explained by the presence of high velocity rigid crustal blocks separated with low velocity belts which represent folded areas.
Figure 13 shows the inversion result in a vertical section and presents our qualitative understanding of the link between the tomography results and the manifestation of surface tectonics and volcanism. In this section, an interesting combination of seismic velocity anomalies, anisotropy and seismicity distribution is observed. Beneath Merapi we can see an inclined low velocity anomaly which links the volcano with the seismicity cluster at 110-130 km depth marked by a pink ellipse in Figure 13A. This cluster can be explained by phase transitions and melting in the downgoing slab that cause significant fluids release (e.g. Poli and Schmidt, 1995). These fluids move upward (blue arrows in Figure 13), and decrease the melting temperature in the overlying mantle and crust. This might cause ascent of diapirs which reach the Earth surface and cause the origin of volcanism. We propose that the observed low-velocity anomaly represents the paths of migration of fluids ascending from the subducted slab. The natural question is why fluid migration does not occur vertically, as in the case of Toba Caldera (see previous chapter), but along an inclined zone with an angle of about 45°. Actually, several different processes (mechanical ascend due to Archimedes forces, corner flow, melting and chemical reactions) coexist in the mantle wedge. Summary effect of such multi component system is hardly predictable. One of the possible explanations can be an effect of drift of the ascending fluids and melts by corner flow in the mantle wedge.
The tomographic results show that crust and upper mantle beneath Central Java appear to be strongly anisotropic. The average magnitude of anisotropy is about 7%; however, due to uncertainty of damping determination and trade-off between isotropic and anisotropic parameters, these values should be interpreted with prudence.
In the crust beneath Central Java different anisotropic regimes are observed (Figure 12, left column). In the forearc between the southern coast of Java and the volcano chain the anisotropic properties seem to be chaotic and relatively weak: all directions of fast velocities coexist within this relatively small area. Variety of anisotropy orientations is a clear indicator to complex crustal structure in the forearc. This is also supported by mosaic structure of isotropic velocity anomalies. Note that the main patterns are similarly visible in P and S models computed independently. The strongest horizontal anisotropy in the crust is observed within a large low-velocity anomaly between Merapi and Lawu volcanoes (MLA). Inside this anomaly we observe an east-west trending of faster velocities, and the amplitude of anisotropy reaches 10%. In the vertical section (Figure 13B) we observe clear separation according to character of the vertical anisotropy. Beneath the volcanoes, in the southern part of MLA, faster vertical velocities are observed, while in the northern part, where maximal amplitudes of velocity anomalies were found (30-35%), the anisotropy tends to be horizontal. For the volcanic areas, we suppose that fast vertical velocities and arc parallel trending of horizontal anisotropy can be caused by existence of vertical dykes and channels beneath the volcanoes. Such structure with elongated vertical fine features causes higher effective velocity in vertical direction which is detected in the anisotropic results. Horizontal trending of anisotropy in the northern part of the MLA which is seen in the vertical section (Figure 13B) is probably due to layering of sediments in the uppermost part. Additionally, we suppose that it may be also due to penetration of horizontally oriented intrusions. Very low velocities in MLA and strong attenuation of S velocities are possible indicators that these intrusions are filled with molten magma. Another indirect argument for this

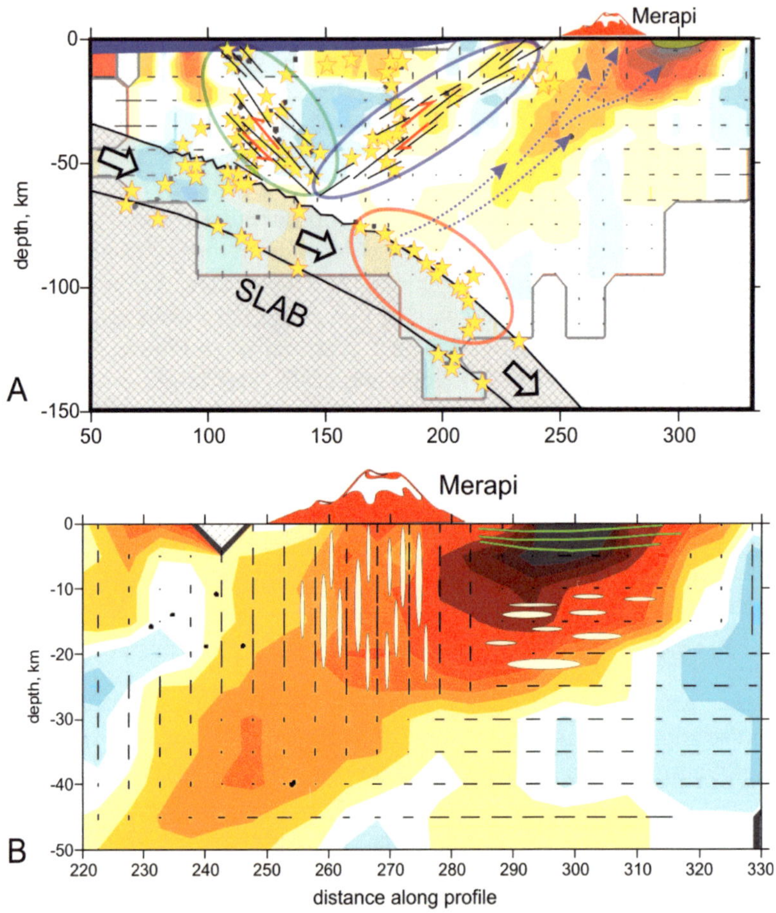

Fig. 13.a. Interpretation of the results. The background is the result for P velocities along the profile shown in Figure 12. Yellow stars show distribution of local seismicity recorded within this study. The pink ellipse highlights the events in the slab at 110 km - 130 km depth which could be related to phase transition and fluid release from the slab. Dotted blue arrows indicate schematic distribution of released fluids and partial melting zones according to our interpretation. Blue and green ellipses highlight the events which could be related to thrusting caused by increasing of the friction rate on the slab upper surface. Shallow green area to the right of Merapi indicates the sediment cover above the MLA. Black strokes above the slab indicate possible areas with maximum stress concentration. B.) Zoom for the area beneath Merapi volcano. Elongated ellipses show schematically distribution of channels, dykes and lenses filled with magmatic material. Green lines indicate schematically sediment layers.

hypothesis is large level of remnant residuals for the stations located within MLA. These unexplained residuals can be due to small high contrasted patterns (magma pockets) which are below the resolution capability of the inversion algorithm, but still affect the travel times. As for strong anisotropy in MLA observed in the map view, we believe that it can be controlled by the regional stress regime. Deviatoric stresses induce preferential opening and closing of cracks, potentially introducing seismic anisotropy in rocks. These effects are especially evident in cracked porous media (Gibson and Toksoz, 1990) as expected in MLA, where a lot of thermal fields are observed. In the case of tectonic deformations, fast seismic velocities are observed parallel to compressional stress direction (e.g., Lees and Wu, 1999).

For the deeper section, the horizontal orientation of fastest velocity axes seems to be fairly regular. At a depth of 25 km and deeper, which correspond to the uppermost mantle, the anisotropy beneath the coastal area is oriented parallel to the trench and roughly perpendicular to the direction of subduction. We explain the trench-parallel anisotropy in the mantle wedge by presence of B-type olivine (Jung and Karato 2001). This fabric is generated in special conditions which presume presence of water and melts. The tomographic results show an inclined low-velocity anomaly which connects the seismicity cluster at the depth of about 100 km with the active volcanoes on the surface (Figure 13). We interpret this anomaly as path of fluids which are released from the slab due to phase transitions. These fluids may initiate active melting in the mantle wedge and in the crust beneath the volcanoes. Both fluids and melts may take part at producing B-type olivine with orientation of the fast axis perpendicular to the flow direction.

In general the anisotropic tomographic study has provided a rich material for understanding the deep sources of feeding Merapi and other volcanoes of Central Java. The most important finding is a huge MLA anomaly with the amplitude of about 30%. At the same time, beneath the volcanoes we did not find any features which could be interpreted as magma chambers. So the classical concept of a large reservoir filled with liquid magma, which is often shown in popular and school books, is not valid for this case. However, the resolution of the tomographic model may appear to be not sufficient to resolve such a feature. To give a definitive answer about the existence of magma chamber, more dense networks around volcanoes are required.

4.3 Central Andes: Multiple paths of the arc feeding

The Andes are one of the largest active mountain ranges in the world, extending for over 8000 km along the western edge of South America (Figure 14). The subduction of the Nazca plate at a rate of about 8.4 cm/year (relative to South America (DeMets et al., 1990) under the South American continent has been going on for more than 200 Myr. The crustal structure of the Central Andes was influenced by the eastward migrating magmatic arc (Scheuber et al., 1994) due to strong erosion of the upper plate lithosphere in the subduction zone. This migration is expressed by a 200 km shift of the volcanic arc system to the east since the early Jurassic. Today the main volcanic arc is located along the Western Cordillera. The forearc region can be subdivided into four main morphological units from west to east (Figure 14): (1) Coastal Cordillera coinciding with the position of the volcanic arc in Jurassic times; (2) the Longitudinal Valley located at the place of the Jurassic back arc basin and the Mid-Cretaceous magmatic arc; (3) Precordillera located on the late Cretaceous magmatic arc; and (4) the Preandean depression containing the Salar Atacama block and bordered from the east by active volcanoes.

In the framework of the multidisciplinary international project SFB 267 several arrays of portable seismic stations were installed between 20°S and 25°S from the Pacific coast across the Andean mountain range (Figure 14C) in the time period from 1994 to 1997. In total, more than 100 seismic stations recorded about 1500 deep and crustal earthquakes and about 50 000 rays. This dataset was processed by several researchers and their results were published in many papers. Shurr et al., (1999) performed the preliminary location of sources and found the optimal 1D model. The results of travel-time tomographic inversion were presented as a PhD thesis by Schurr (2001). More than 15000 values of t* were inverted, and the corresponding model of P-wave attenuation was presented by Schurr et al., (2003) and Haberland et al., (2003). Koulakov et al., (2006) revised the same dataset and presented new models of P and S velocities, Vp/Vs ratio and Qp attenuation based on inhomogeneous 3D starting velocity model.

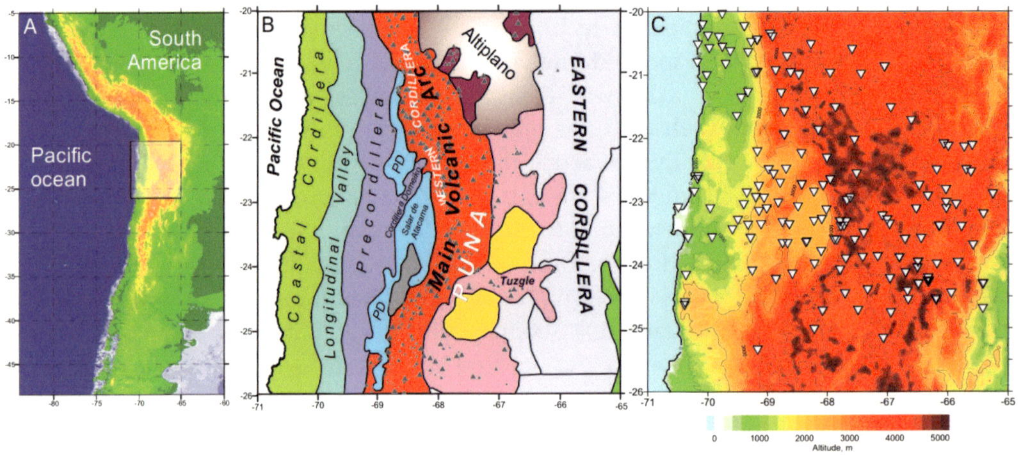

Fig. 14.a. Position of the study region in South America marked by rectangle. B: Simplified map of the main tectonic elements shown with different colors. PD: Preandean depression; APVC: Altiplano-Puna volcanic complex. The position of recent volcanoes (up to 5Ma) from the catalogue by de Silva & Francis, (1991) is shown by triangles. The size of the triangles reflects volume of the volcanoes. C: Smoothed topography and the distribution of seismic stations used for the inversion.

Here I present the most recent model constructed with the use of the LOTOS code (Koulakov, 2009a). The inversion results for P and S velocity anomalies are shown in horizontal and vertical sections in Figure 15. These new results show the link with the main geological structures much clearer than previously published models. At 25 km depth, we can see a very clear separation between high-velocity forearc and low-velocity volcanic arc. It was found that in Jurassic time the arc was located in the area of present location of Coastal Cordillera. After this time, the arc was replaced to more than 200 km due to erosion of the continental plate by the subducted lithosphere. Thus, the area between the coast and volcanic arc is composed of rigid igneous rocks erupted during eastward migration of the subduction complex. These strongly consolidated rocks are expressed as high-velocity patterns. Inside the high-velocity zone, we can distinguish several blocks which are visible in geological map (Figure 14B). For example, Salar de Atacama block is clearly separated from the Preandean Depression. Cordillera Domeiko, which marks the previous location of the volcanic arc, is expressed by locally lower velocities.

The crustal structure beneath the backarc appears to be more complex. Although, low-velocities dominate there, we do not observe direct correlation between seismic structures and the distribution of backarc volcanoes. On the contrary, in the mantle (below 70 km depth) a clear low-velocity anomaly is located beneath the Tuzgle volcano complex. Based on these results we can conclude that the origin of the backarc volcanism of the Tuzgle complex is significantly different of that of the arc volcanism. Beneath Tuzcle, the mantle seems to be overheated, and the volcanoes produce high-temperature and dry magmas. Beneath the arc, the temperature in the mantle wedge is normal, but high content of fluids lowers the melting temperature. As a result, the magma products in arc volcanoes are of lower temperature, but rich with fluids.

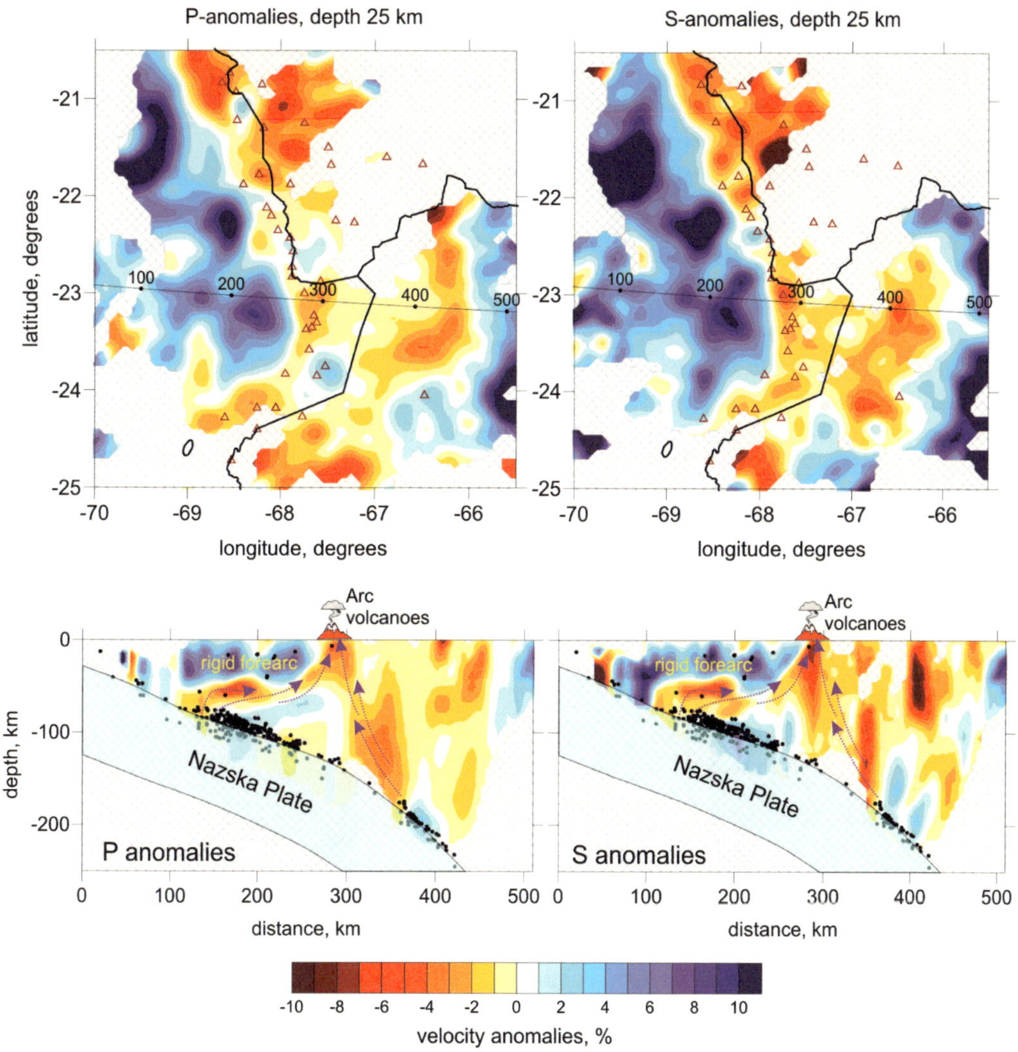

Fig. 15. P and S-velocity anomalies beneath Central Andes at 25 km depth and in a vertical section. Location of the profile is indicated in maps. Triangles in maps depict the locations of recent (1 Ma) volcanoes. Black dots in vertical sections are the locations of earthquakes at distances of less than 20 km from the profile. Violet dotted arrows indicate hypothetical migration of fluids from the slab.

The paths of feeding the arc volcanoes can be seen in vertical sections in lower row of Figure 15. It can be seen that the deep seismicity clearly delineates the upper boundary of the subducting slab. We observe two clusters: around the levels of 100 and 200 km depth. Between these clusters we see a clear gap. It is interesting that both in P and S-velocities, these clusters seem to be linked with the volcanic arc through low-velocity patterns. We propose that both of these clusters are due to phase transitions on the subducting slab which cause the release of fluids (Poli & Schmidt, 1995). In the case of the upper cluster at 100 km depth, the fluids ascend upward, but meet the bottom of the rigid forearc block. Then they migrate eastward and reach the surface at the eastern border of the forearc, where the present arc is located. Another cluster seem also to produce some fluids which also ascend, but with a westward bias. These fluids reduce the melting temperature of rocks in the mantle wedge and in the crust. The melting products are delivered to the volcanoes of the arc. Note that the trenchward orientation of the fluid path appears to be different of that observed beneath Merapi volcano in Central Java and Toba in Sumatra.

4.4 Kluchevskoy volcano group: Multi-level feeding determines the variety of eruption regimes

In subduction zones, the volcanic arcs are represented by complex distribution of volcanoes of considerably different eruption regimes and magma compositions. One of the brightest examples of volcano variability within one local complex is the Kluchevskoy volcano group located in Kamchatka Peninsula (Russia). In this chapter, I present results of tomographic inversion for seismic crustal structure beneath the Kluchevskoy group that were recently published by Koulakov et al., (2011a).

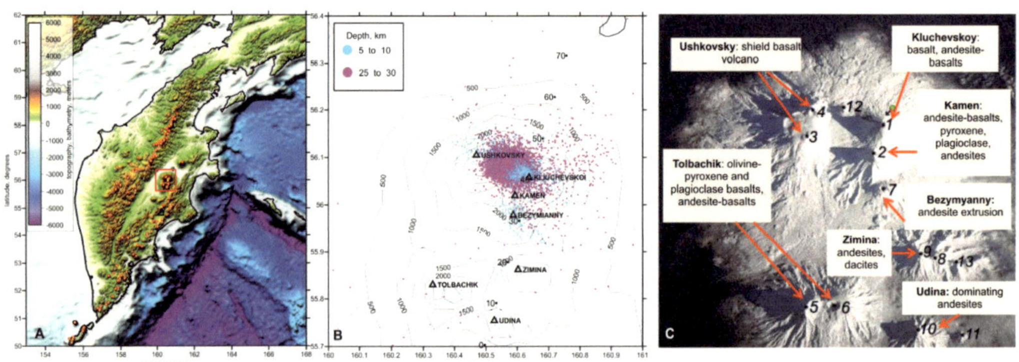

Fig. 16. Kluchevskoy Volcano Group: location, data distribution and information on volcanoes. A: Topography and bathymetry of the Kamchatka Peninsula. Red dots depict Cenozoic volcanoes; red rectangle marks the study area. B: Location of seismic stations (blue diamonds) and events in two depth intervals (colored dots). C: Satellite image and general information on volcanoes of the Kluchevskoy Group.

The Kluchevskoy volcano group covers an area of about 100x60 km size and includes 13 active and dormant volcanoes. Three of them, Kluchevskoy, Tolbachik and Bezymyanny, are some of the most productive and intensive volcanoes of the world. Kluchevskoy volcano, which is about 4800 meters high, is the largest active volcano in continental Eurasia. The composition and the activity regimes within the Kluchevskoy group (Figure 16C) vary very strongly from explosive andesitic to fissure basalt Hawaiian type eruptions (e.g., Fedotov et al., 2010). Furthermore, there are some evidences that, in some volcanoes of the group, the regimes and composition of eruptions might abruptly change during a relatively short periods of time (Laverov, 2005). For example, the major products of current eruptions of the Kluchevskoy volcano consist of the high-alumina basalts, (Khrenov et al., 1989), while older lavas consisted of high-magnesium basalts with up to 12% MgO and rocks of intermediate composition (Ozerov et al., 1997). The Kluchevskoy volcano acts with moderate eruptions alternated with silent periods lasting several years (Ozerov et al., 2007, Ivanov, 2008). The other volcanoes of the group are strongly variable in composition and activity regimes (Laverov, 2005). Bezymianny, which is located at about ten kilometers from Kluchevskoy, is an explosive andesite volcano with magmas containing 54.5 to 62.5% of SiO2 (Gorshkov and Bogoyavlenskaya, 1965, Ozerov et al., 1997). Before the years of fifties of the last century, Bezymyanny («Nameless », in Russian) was not considered as an active volcano. The catastrophic eruption in 1955-1956 which ejected more than five cubic kilometers of rocks was one of the largest volcanic events in twentieth century. The dormant Udina volcano, located in the southeastern part of the group, is composed of alternating lavas and pyroclastic layers consisting of basalts and andesites (Laverov, 2005). Zimina is a complex of three dormant stratovolcanoes consisting of andesite extrusions, dacite and basalts lavas (Braitseva et. al., 1994). The Ushkovsky volcano, located in the northwestern side of the Kluchevskoy group, has a very complex structure (Braitseva et. al., 1994). Initially, it was constructed as a shield volcano of the Hawaiian type and consisted primarily of basaltic lavas. In more recent stages, a series of stratovolcanoes appeared that consist of alternating pyroclastic andesite-basalts and lavas of similar composition (Laverov, 2005). Finally, the Tolbachik complex, located in the southwestern part of the Kluchevskoy group, contains several shield and stratovolcanoes of very different structures and compositions. The presently active Plosky Tolbachik volcano is a typical Hawaiian-type volcano with fissure eruptions and calderas. The fissure type eruption of Tolbachik in 1975-1976 appears to be the largest historical basalt eruption in Kamchatka (e.g., Fedotov et al., 2010). This diversity of composition and eruption regimes in volcanoes at distances of dozens of kilometers from each other shows that their feeding occurs both directly from mantle sources and through complex system of intermediate chambers in the crust (e.g., Khubunaya et al., 2007).

Another feature of this region which is very important for tomographic inversion is the major part of seismic energy is released at great depths (Figure 16B). Most seismic events occur at depths between 23 and 28 km. This enables favorable illumination of the crustal structures by seismic rays and creates better conditions for tomographic inversion compared to most other volcanoes, for example, Etna (Patanè et al., 2006). It is interesting that the depth distribution of seismicity over time seems to correlate with major eruptions of the Kluchevskoy volcano. During the activity periods, the seismicity clusters tend to be shallower.

The particular behavior of the Kluchevskoy group might be caused by its location near the junction of the Kuril-Kamchatka and Aleutian subduction zones (Figure 16 A) which is considered by many authors as a singular point with unusually active geodynamical processes. According to the distribution of deep seismicity and the results of regional and global tomographic studies (e.g., Gorbatov et al., 2001, Bijwaard et al., 1998, Koulakov et al, 2011b), between the Kuril-Kamchatka and Aleutian slabs there is a clear gap which is seen as a low-velocity anomaly. The horizontal distance from the Kluchevskoy volcanic group to the edge of the subducting Pacific plate is less than 100 km, whereas the vertical distance to the upper surface of the plate is approximately 150 km. This unique location above the end of the subduction zone may cause mantle flows that control a specific activity of the volcano group.

The tomographic inversion by Koulakov et al., (2011a) was based on data provided by the Kamchatka Branch of the Geophysical Survey of the Russian Academy of Sciences. The available dataset includes half millions travel times from about 80000 local earthquakes in the area of the Kluchevskoy volcano group recorded by seventeen permanent seismic stations in a time period from 1999 to 2009. The events are widely distributed in space, and most of them are located below 25 km deep, which provides good ray coverage for studying the crustal structure beneath the volcanoes. In addition, for most events, both P and S phases were handpicked by highly experienced specialists. The use of approximately equal numbers of P and S phases enables the determination of high quality source locations.

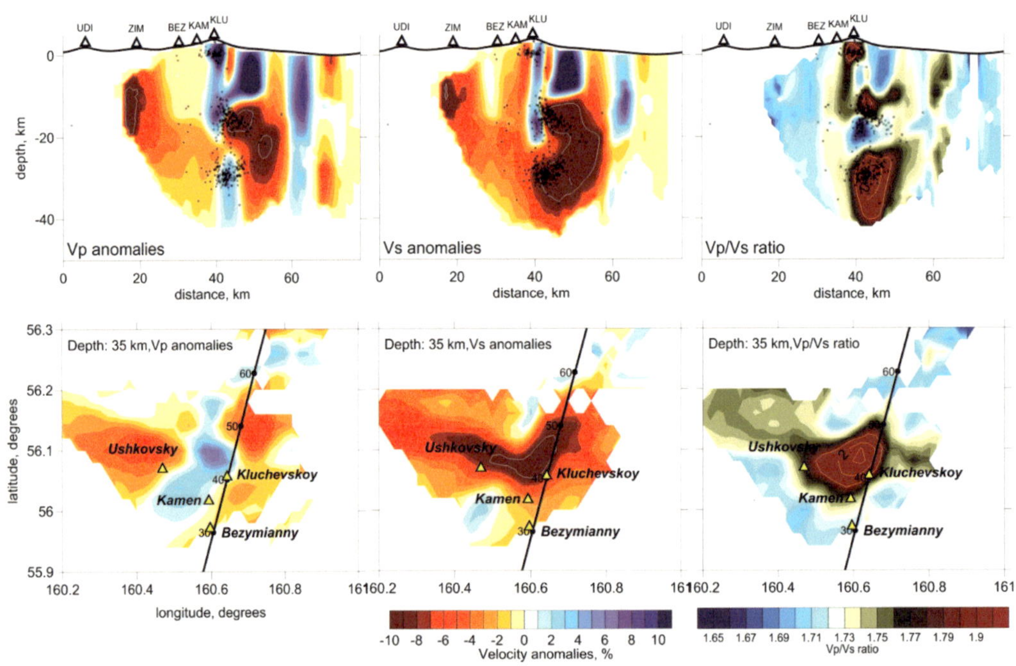

Fig. 17. P and S velocity anomalies and Vp/Vs ratio beneath the Kluchevskoy Volcano Group presented in vertical (upper) and horizontal (lower) sections. Location of the profile is indicated in maps. Black dots represent seismicity nearly the vertical section. Volcanoes are show by triangles above the vertical section plots.

Approximately same dataset has already been used by other authors in (Nizkous et al., 2006, Khubunaya et al., 2007) where they report the distribution of P velocity beneath the

Kluchevskoy group. We have also performed the inversion for the full dataset and obtained rather similar results as in the previous studies. However, a rather low value of variance reduction (about 15%) makes us to doubt in robustness of this solution. This means that it is hardly possible to build a single model which satisfies all the data for more than ten years. Koulakov et al., (2011a) performed the inversion for one year in 2004 and obtained much more robust solution than in the case of the entire data inversion. This fact indirectly shows strong variability of seismic structure beneath Kluchevskoy volcano group over time.

Here I present the result of inversion for the year 2003 in vertical and horizontal sections (Figure 17) which look rather similar to the results presented by Koulakov et al., (2011a) corresponding to 2004. The most prominent feature of the obtained seismic structure is a large anomaly located beneath the Kluchevskoy volcano at depths below 25 km. In this pattern, we observed positive P-velocity and negative S-velocity anomalies that result in very high Vp/Vs ratios, reaching 2.2. Very high Vp/Vs ratio below 25 km depth can be explained by both compositional and rheological properties of rocks. The P-velocity is more sensitive to the composition, and its higher values may be an indicator of rocks that came from lower depths. At the same time, very low values of the S-velocity indicate a high content of fluids and partial melting. We interpret this pattern as the top of a small plume (red area in Figure 20) that probably starts on the upper surface of the slab and reached the bottom of the crust. Active seismic clusters were observed within this high Vp/Vs anomaly only above ~30-35 km in depth. These depths correspond to the lowermost parts of the crust, whose thickness reaches the values of about 30-35 km, based on the results of receiver function analysis (Nikulin et al., 2010). We propose that these earthquakes in the lowermost crust, which fit to high values of Vp/Vs, mark the first level of magma storage between 25 and 30 km in depth. The seismicity at this level is probably due to strong thermal, chemical and mechanical effects of the ascending flow in the mantle channel which reaches the brittle crust.

The distribution of seismicity and velocity patterns in the crust forms a regular pattern which is marked in Figure 18 with S-shaped orange arrows. This cluster might indicate the paths of fluids and melts that ascend from the deep magma source at 25-30 km in depth to Kluchevskoy volcano at the surface. At intermediate depths between ~10 and 15 km beneath the Kluchevskoy volcano, we observed another anomaly of high Vp/Vs ratio, which probably marks the second level of magma storage. Clear records of shear waves do not support an idea of the existence of large chambers filled with liquid magma. We propose that these zones of high Vp/Vs ratio represent either sponge-structured areas with small blobs of partially molten material, or fracturing zones with systems of cracks filled with fluids and/or melts. In any case, these zones should play an important role in feeding the volcano system. Here, the magma migration can cause fracturing of crustal rocks, mixing and differentiation of molten material in magma storages. Thus, the properties of magma in these intermediate chambers might be considerably different from those of the initial magma sources in the lowermost crust. The coexistence of deep and intermediate sources can explain the compositional variability of the eruption products of the volcanoes of the Kluchevskoy group.

Just beneath the Kluchevskoy volcano, we observed another shallow anomaly with a high Vp/Vs ratio, which might reflect the existence of a third level of magma storage just beneath the volcano. This small pattern coexisting with shallow seismicity probably represent the third level of magma storage which is directly responsible for eruption of the Kluchevskoy volcano.

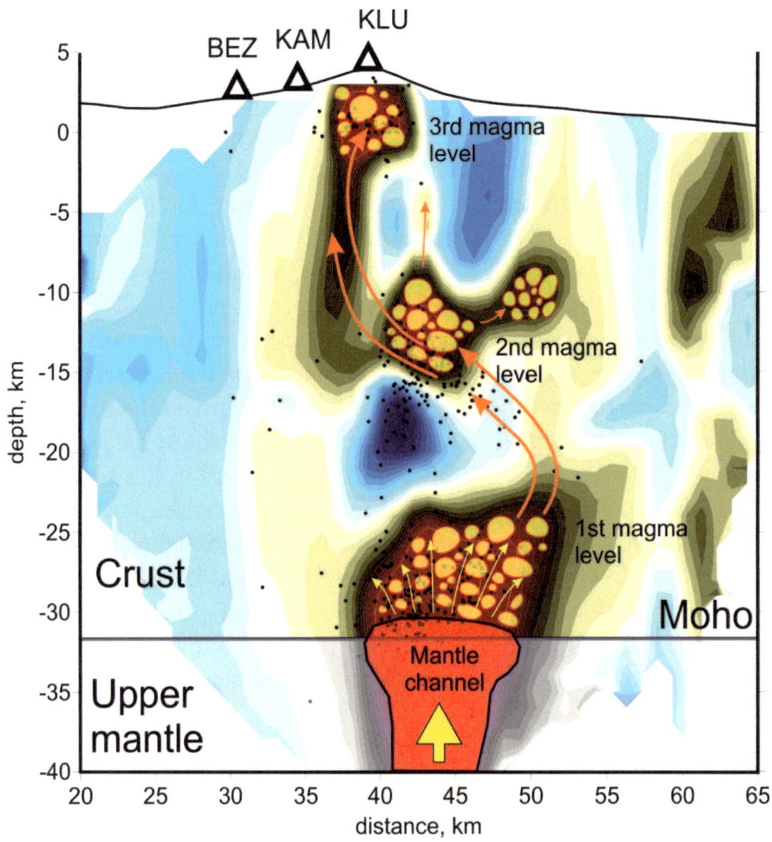

Fig. 18. Interpretation of the results of tomography inversion for the Kluchevskoy volcano group. The background is the distribution of the Vp/Vs ratio in the vertical profile (Figure 17). The dots indicate the distribution of earthquakes. The red area in the mantle depicts the main feeding channel that transports upward partially molten deep mantle material. Blobs in the crust that coincide with areas of high Vp/Vs ratio mark three levels of magma chambers in the crust. Orange arrows indicate possible paths of magma transport.

The existence of several layers of magma storage beneath the volcanoes of the Kluchevskoy group can explain their different composition and eruption behavior. When overheated liquid material of deep chambers reaches the intermediate depth reservoirs, mixing with crustal rocks, differentiation and decompression result in a very wide variety of magma composition and a high content of fluids and gases. These intermediate magma storages may feed, for example, Bezymianny, which is a caldera-forming explosive dacite-andesite volcano (Bogoyavlenskaya et al., 1991). On the other hand, Kluchevskoy and Kamen volcanoes are basalt stratovolcanoes that are feed directly from the deep magma storages. This is consistent with the results of petrochemical analysis by Ozerov et al., (1997) who observed different feeding regimes of Kluchevskoy and Bezymyanny volcanoes through direct channels and intermediate chambers, respectively.

All presented results of detailed tomographic studies in volcanic areas show that each volcanic complex appears to be particular and quite different of others. It shows that shallow structures which are directly responsible for volcano eruptions are in a close link with deep processes extending down to hundreds kilometers depth. It is worth noting that most of volcanic systems are very dynamic, and considerable changes of seismic parameters may occur during years, months and even days. We propose that some of the most important factors which control these changes are variations of stresses and defornations, as well as fast fluid migration.

5. Conclusions

In this chapter I made an overview of recent multiscale studies by seismic tomography related to volcanoes. I do not pretend presenting all variety of different approaches and results. Most of the presented models were produced and published in collaboration with colleagues from different scientific centers. A multiscale look to the volcanic systems allows us to reveal some common and particular features in different regions.

In case of intracontinental volcanism, all manifestations of Cenozoic magmatism are related to low-velocity seismic anomalies in the uppermost mantle. In Europe, the distribution of all known Cenozoic basaltic fields, such as French Massif Central, Rhine and Bohemian Massifs, Pannonian basin, well correlate with anomalous zones below the Moho interface. At the same time, neither in our works, nor in tomographic studies by other authors, no clear evidences of mantle plumes, thin contrasted channels, were revealed. Although, low-velocity anomalies are observed in most cases, their lateral size is compatible with height. Such pattern cannot be called plume. It can represent overheated zones either around a plume or in areas of ascending convective flows in the mantle. It should be noted that incapacity of tomography to reveal clear traces of the mantle plumes might be due to fundamental problems of the method. Even in cases of very high data density, as in the case of Europe, the seismic rays pass around contrasted low-velocity anomalies and accumulate very few information about it. So the plumes are fundamentally very hard objects to detect with tomography, and their robust studying is hardly possible. That is why, even having a lot of data in Europe, we still cannot answer a question if the volcanic fields are attributed to plumes or to ascending convective flows. Same conclusion can be made for two other considered areas, Southern Siberia and Eastern Africa, where the data coverage is poorer than in Europe.

In contrast to the plume-related volcanism, the arc volcanism processes in subduction complexes can be studied in details by seismic tomography thanks to favorable coverage of rays generated by deep seismicity. Examples presented in this chapter show different cases of the slab-arc interaction in various subduction zones. It was shown that the volcanoes of the arc appear to be connected through low-velocity patterns with seismicity clusters in the slab. At the same time, the configurations of these links, as well as depth of seismicity clusters in the slab, are considerably different.

In this chapter I have underlined the capacity of some tomographic schemes for studying the sources of volcano activity in different areas on different scales. The information derived from tomographic inversions should be considered in a multidisciplinary context together with various geological, geophysical and geochemical data. Note that tomography is a very young method which still is actively developed. This gives us a hope that in the future we will see much more detailed images of deep processes beneath volcanoes than those presented in this chapter.

6. Acknowledgments

This study is supported by the Helmholtz Society and RFBR Joint Research Project 09-05-91321-SIG a, Multidisciplinary Projects SB RAS #21, and Project ONZ RAS #7.4.

7. References

Aki, K., Christoffersson, A., Husebye, E.S., (1977). Determination of the three-dimensional seismic structure of the lithosphere. J. Geophys. Res. 82, 277–296.

Aoki Y., M. Takeo, H. Aoyama, J. Fujimatsu, S. Matsumoto, H. Miyamachi, H. Nakamichi, T. Ohkura, T. Ohminato, J. Oikawa, R. Tanada, T. Tsutsui, K. Yamamoto, M. Yamamoto, H. Yamasato, and T. Yamawaki. (2009). P-wave velocity structure beneath Asama volcano, Japan, inferred from active source seismic experiment. J. Vol. Geo. Res., 187:272–277, 2009.

Ashepkov I.V., Yu.D.Litasov and K.D.Litasov, (1996). Xenolites of granet peridotites from melanevelinit of the Khantey range (Southern Zabaikalie): evidences for the mantle diapir upwelling, Russ Geol. Geophys., v. 37, n. 1, P. 130-147.

Bachmann O. & G. Bergantz, (2008). The Magma Reservoirs That Feed Supereruptions, Elements, 4(1): 17 - 21.

Bell, K., F. Castorina, G. Lavecchia, G. Rosatelli, and F. Stoppa (2004), Is There a Mantle Plume Below Italy?, Eos Trans. AGU, 85(50), doi:10.1029/2004EO500002.

Bell K, Castorina F, Rosatelli G, Stoppa F. (2006). Plume activity, magmatism, and the geodynamic evolution of the eastern Mediterranean. Annals of Geophysics, 49/1, 357-372. ISSN: 1593-5213.

Benoit, M.H., Nyblade, A.A., VanDecar, J.C., (2006). Upper mantle P-wave speed variations beneath Ethiopia and the origin of the Afar hotspot. Geology 34, 329–332.

Bijwaard, H., W. Spakman, and E. R. Engdahl, (1998). Closing the gap between regional and global travel time tomography, J. Geophys. Res., v.103, p.30,055- 30,078

Bogoyavlenskaya G.E., Braitseva O.A., Melekestsev I.V., Maksimov A.P., Ivanov B.V., (1991). Bezymianny Volcano. Active Volcanoes of Kamchatka : In 2 vol. . Moscow: Nauka, 1991. V. 1. P.195-197.

Braitseva, O., I. Melektsev, V. Ponomareva, L. Sulezhitskiy, and S. Litasova, (1994). Ages of active volcanoes in the Kuril-Kamchatka region, Volcanology and seismology, 4-5, p. 5-32 (In Russian)

Camus, G. A., P.-C. Gourgaud, Mossand-Berthommier, and P.M. Vincent, (2000), Merapi (Central Java, Indonesia): An outline of the structural and magmatological evolution, with a special amphasis to the major pyroclastic events. J. *Volcanol. Geotherm. Res., 100,* 139-163.

Chang, S.-J., Van der Lee, S., (2011). Mantle plumes and associated flow beneath Arabia and East Africa, Earth Planet. Sci. Lett. doi:10.1016/j.epsl.2010.12.050

Daly, E., D. Keir, C. J. Ebinger, G. W. Stuart, I. D. Bastow, and A. Ayele, (2008). Crustal tomographic imaging of a transitional continental rift: the Ethiopian rift, *Geophys. J. Int. 172, 1033–1048, doi: 10.1111/j.1365-246X.*2007.03682.x.

Dercourt, J., L. Zonenshain, L.E. Ricou, G. Kazmin, X. Le Pichon, A.L. Knipper, C. Grandjacquet, I.M. Sbortshikov, J. Geyssant, C. Lepvrier, D.H. Pechersky, J. Boulin, J.C. Sibuet, L.A. Savostin, O. Sorokhtin, M. Westphal, M.L. Bazhenov, J.P. Lauer and B. Biju-Duval (1986): Geological evolution of the Tethys belt from the Atlantic to the Pamirs since the Lias,Tectonophysics, 123, 241-315.

DeMets, C., R. Gordon, D. Argus, and S. Stein, (1990). Current plate motions, *Geophys. J. Int., 101,* 425-478.

De Silva, S.L. and Francis, P.W., (1991). Volcanoes of the Central Andes, Springer Verlag, 199-212 (Appendix II).

Di Stefano R. and C. Chiarabba. Active source tomography at Mt. Vesuvius: Constraints for the magmatic system. J. Geophys. Res., 107:2278, doi:10.1029/2001JB000792, 2002.

Eskin A.C., A.A.Bukharov, Yu.A.Zorin, 1978. Cenozoic magmatizm of lake Baikal, Doklady AN SSSR, v. 239, n.4, p. 926-929.

Evans, J., Achauer, U., 1993. Teleseismic velocity tomography using the ACH method: theory and application to continental-scale studies. In: Iyer, H.M., Hirahara, K. (Eds.), Seismic Tomography: Theory and Practice. Chapman and Hall, London, pp. 319–360.

Faccenna, C., Jolivet, L., Piromallo, C. & Morelli, A., (2003). Subduction and the depth of convection in the Mediterranean mantle, J. Geophys. Res., 108(B2), 2099, doi:10.1029/2001JB001690.

Fedotov S.A., Zharinov N.A., Gontovaya L.I., (2010). The magmatic system of the Klyuchevskaya Group of volcanoes inferred from data on its eruptions, earthquakes, deformation and deep structure. Volcanology and Seismology, 1, p. 3-35 (in Russian).

Gibson, R.L., and M.N.Toksoz, (1990). Permeability estimation from velocity anisotropy in fractured rocks, *J.Geophys.Res.,* 95, 15643-15655.

Gorbatov A, Fukao Y, Widiyantoro S, Gordeev E, (2001). Seismic evidence for a mantle plume oceanwards of the Kamchatka-Aleutian trench junction, Geophysical Journal International, Volume 146, Issue 2, pp. 282-288.

Gorshkov G.S., Bogoyavlenskaya G.E., (1965). The Bezymianny Volcano and peculiarity of the last eruption (1955-1956). Moscow, Nauka, 165 p.

Granet, M., Wilson, M., and Achauer, U., (1995), Imaging a mantle plume beneath the French Massif Central: Earth and Planetary Science Letters, v. 136, p. 281-296.

Haberland, C., A. Riebrock, B. Schurr, and H. Brasse, (2003). Coincident anomalies of seismic attenuation and electrical resistivity beneath the southern Bolivian Altiplano plateau, Geophys. Res. Lett., 30(18), 1923, doi: 10.1029/2003GL017492.

Hansen, S., Schwartz, S., Al-Amri, A., Rodgers, A., (2006). Combined plate motion and density-driven flow in the asthenosphere beneath Saudi Arabia: evidence from shear-wave splitting and seismic anisotropy. Geology 34, 869–872. doi:10.1130/ G22713.1. International Seismological Centre, 2001, Bulletin Disks 1-9 [CD-ROM], Internatl. Seis. Cent., Thatcham, United Kingdom.

Ivanov, V.V. (2008). Current cycle of the Kluchevskoy volcano activity in 1995-2008 based on seismological, photo, video and visual data. in Proceedings of Conference, Petropavlovsk-Kamchatsky, 27-29 march, 100-109.

Jakovlev A., G. Rümpker, M.Lindenfeld, I.Koulakov, A.Schumann, N. Ochmann, (2011), Crustal seismic velocities from local travel-time tomography: a case study from the Rwenzori Mountains of the East African Rift, BSSA; v. 101; no. 2; p. 848-858; DOI: 10.1785/0120100023

Jung, H., and S.Karato, (2001). Water-induced fabric transitions in olivine. Science 293 (5534), 1460–1463.

Khrenov A.P., Antipin V.S., Chuvashova L.A., Smirnova E.V. (1989). Petrochemical and geochemical peculiarity of basalts of the Kluchevskoy volcano. Volcanology and seismology. 3, p.3-15.

Khubunaya S., L.Gontovaya, A. Sobolev, and I. Nizkous (2007). Magmatic sources beneath the Kluchevskoy volcano group (Kamchatka). Volcanology and seismology, 2, P 22-42

Koulakov, I. (2011). High-frequency P and S velocity anomalies in the upper mantle beneath Asia from inversion of worldwide traveltime data, J. Geophys. Res., 116, B04301, doi:10.1029/2010JB007938

Koulakov, I., E. I. Gordeev, N. L. Dobretsov, V. A. Vernikovsky, S. Senyukov, and A. Jakovlev (2011a), Feeding volcanoes of the Kluchevskoy group from the results of local earthquake tomography, Geophys. Res. Lett., 38, L09305, doi:10.1029/2011GL046957

Koulakov I.Yu., N.L. Dobretsov, N.A. Bushenkova , A.V. Yakovlev, (2011b). Slab shape in subduction zones beneath the Kurile–Kamchatka and Aleutian arcs based on regional tomography results, Russian Geology and Geophysics 52, 650–667

Koulakov I., and N. Bushenkova, (2010), Upper mantle structure beneath the Siberian craton and surrounding areas based on regional tomographic inversion of P and PP travel times, Tectonophysics, 486, 81-100.

Koulakov I., M.K. Kaban, M. Tesauro, and S. Cloetingh, (2009a). P and S velocity anomalies in the upper mantle beneath Europe from tomographic inversion of ISC data, Geophys. J. Int. 179, 1, p. 345-366. doi: 10.1111/j.1365-246X.2009.04279.x

Koulakov I., T. Yudistira, B.-G. Luehr, and Wandono, (2009b). P, S velocity and VP/VS ratio beneath the Toba caldera complex (Northern Sumatra) from local earthquake tomography, Geophys. J. Int., 177, p. 1121-1139, doi: 10.1111/j.1365-246X.2009.04114.x

Koulakov, I., A. Jakovlev, and B. G. Luehr (2009c). Anisotropic structure beneath central Java from local earthquake tomography, Geochem. Geophys. Geosyst., 10, Q02011, doi:10.1029/2008GC002109.

Koulakov I., (2009a). LOTOS code for local earthquake tomographic inversion. Benchmarks for testing tomographic algorithms, Bulletin of the Seismological Society of America, Vol. 99, No. 1, pp. 194-214, doi: 10.1785/0120080013

Koulakov I., (2009b). Out-of-network events can be of great importance for improving results of local earthquake tomography, Bulletin of the Seismological Society of America, Vol. 99, No. 4, pp. 2556–2563, doi: 10.1785/0120080365

Koulakov I.Y., (2008). Upper mantle structure beneath southern Siberia and Mongolia, from regional seismic tomography, Russian Geology and Geophysics Volume 49, Issue 3, Pages 187-196

Koulakov, I. (2007). Structure of the Afar and Tanzania plumes based on the regional tomography using ISC data, *Doklady Earth Sciences* 417 No. 8 1287–1292

Koulakov I., M. Bohm, G. Asch, B.-G. Lühr, A.Manzanares, K.S. Brotopuspito, Pak Fauzi, M. A. Purbawinata, N.T. Puspito, A. Ratdomopurbo, H.Kopp, W. Rabbel, E.Shevkunova (2007), P and S velocity structure of the crust and the upper mantle beneath central Java from local tomography inversion, J. Geophys. Res., 112, B08310, doi:10.1029/2006JB004712.

Koulakov, I., S.V.Sobolev, and G. Asch, (2006). P- and S-velocity images of the lithosphere-asthenosphere system in the Central Andes from local-source tomographic inversion, Geophys. Journ. Int., 167, 106-126.

Lavecchia G. & Stoppa F. (1996) - The tectonic significance of Italian magmatism: an alternative view. Terra Nova, 8, 435-343

Lavecchia G., Stoppa F. & Creati N. (2006) Carbonatites and kamafugites in Italy: mantle-derived rocks that challenge subduction. Annals of Geophysics, supplement to vol 49, 389-402.

Laverov N.P., (2005). Modern and Holocene volcanism in Russia, Moscow, Nauka, 604 p.

Lees J.M. and H. Wu, (1999). P wave anisotropy, stress, and crack distribution at Coso geothermal field, California, Journal of Goephys. Res., V. 104, N B8, p. 17955-17973.

Logachev, N.A., (2005). History and geodynamics of the Baikal rift, in: Goldin, S.V., Mazukabzov, A.M., Seleznev, B.C. (Eds.), Actual Problems of Modern Geodynamics of Asia [in Russian]. Izd. SO RAN, Novosibirsk, pp. 9–32.

Logachev, N.A., S.V.Rasskazov, A.V.Ivanov, K.G.Levi, A.A.Bukharov, S.A.Kashik, and S.I.Sherman, (1996). Cenozoic rifting in continental lithosphere, in: Logachev, N.A. (Ed.), Lithosphere of Central Asia [in Russian], Nauka, Novosibirsk, pp. 57–80.

Lustrino, M. (2000): Volcanic activity during the Neogene to Present evolution of the Western Mediterranean area: a review, Ofioliti, 25, 87-101.

Montelli, R., Nolet, G., Dahlen, F.A., Masters, G., Engdahl, E.R., and Hung, S.-H., 2004, Finite-Frequency Tomography Reveals a Variety of Plumes in the Mantle: Science, v. 303, p. 338-343.

Newhall, C.G., Bronto, S., Alloway, B., Banks, N.G., Bahar, I. & Marmol, M.A.D., (2000). 10000 Years of explosive eruptions of Merapi volcano, Central Java: Archaeological and Modern Implications, Journal of Volcanological and Geothermal Research, 100, 9-50

Nikulin, A., V. Levin, A. Shuler, and M. West (2010), Anomalous seismic structure beneath the Klyuchevskoy Group, Kamchatka, Geophys. Res. Lett., 37, L14311.

Nizkous, I., Sanina, I., Kissling, E., Gontovaya, L. (2006): Velocity properties of ocean-continent transition zone lithosphere in Kamtchatka region according to seismic tomography data. Physics of the solid Earth 42 (4): 286-296

Ozerov A.Y., Ariskin A.A., Kyle P., Bogoyavlenskaya G.E., Karpenko S.F. (1997). A petrological-geochemical model for genetic relationships between basaltic and andesitic magmatism of Klyuchevskoi and Bezymyannyi volcanoes, Kamchatka, Petrology, 5, p. 550-569.

Ozerov A.Yu., Firstov P.P., Gavrilov V.A. (2007) Periodicities in the Dynamics of Eruptions of Klyuchevskoi Volcano, Kamchatka. in Volcanism and Subduction: The Kamchatka Region. Geophysical Monograph Series, v.172, 2007, P. 283-291.

Park, Y. and A. A. Nyblade (2006), P-wave tomography reveals a westward dipping low velocity zone beneath the Kenya Rift, *Geophys. Res. Lett., vol. 33, L07311, doi:10.1029/2005GL025605.*

Park, Y., Nyblade, A.A., Rodgers, A.J., Al-Amri, A., (2007). Uppermantle structure beneath the Arabian Peninsula and northern Red Sea from teleseismic body wave tomography: implication for the origin of Cenozoic uplift and volcanism in the Arabian Shield. Geochem. Geophys. Geosyst. 8, Q06021. doi:10.1029/2006GC001566.

Park, Y., Nyblade, A.A., Rodgers, A.J., Al-Amri, A., (2008). S wave velocity structure of the Arabian Shield upper mantle from Rayleigh wave tomography. Geochem. Geophys. Geosyst. 9, Q07020. doi:10.1029/2007GC001895.

Patanè D, Barberi G, Cocina O, De Gori P, Chiarabba C, (2006). Time-resolved seismic tomography detects magma intrusions at Mount Etna. Science, 313(5788):821-3

Paulatto M., T.A. Minshull, B. Baptie, S. Dean, J.O.S. Hammond, T. Henstock, C.L. Kenedi, E.J. Kiddle, P. Malin, C. Peirce, G. Ryan, E. Shalev, R.S.J. Sparks, and B. Voight. (2010). Upper crustal structure of an active volcano from refraction/reflection tomography, Montserrat, Lesser Antilles. Geophys. J. Int., xxx:X–XX.

Piromallo, C., and Morelli, A., (2003). P wave tomography of the mantle under the Alpine Mediterranean area: Journal of Geophysical Research, v. 108, B2, 2065, doi:10.1029/2002JB001757.

Poli, S., & M.W. Schmidt, (1995). H2O transport and release in subduction zones: Experimental constrains on basaltic and andesitic systems, J.Geophys. Res., 100, 22299-22314.

Rasskazov S.V., (1993). Magmatism of the Baikal rift zone, Novosibirsk, Nauka, 288 p. (in Russian).

Ritter, J.R.R., Jordan, M., Christensen, U., Achauer, U., (2001). A mantle plume beneath the Eifel volcanic field, Germany, Earth Planet. Sci. Lett., 186, 7-14.

Scheuber, E., T. Bogdanic, A. Jensen, and K.-J Reutter, (1994). Tectonic development of the north Chilean Andes in relation to plate convergence and magmatism since the Jurassic, *in Tectonics of the Southern Central Andes, edited by K.-J. Reutter, E. Scheuber, and P. Wigger, pp. 121-139, Springer-Verlag, New York, 1994.*

Schurr, B., G.Asch, A.Rietbrock, R.Kind, M.Pardo, B.Heit, and T.Monfret, (1999). Seismicity and average velocity beneath the Argentine Puna, Geophys. Res. Lett., 26, 3025-3028.

Schurr, B., (2001). Seismic structure of the Central Andean Subduction Zone from Local Earthquake Data, STR01/01, PhD thesis, GFZ-Potsdam.

Schurr, B., G. Asch, A. Rietbrock, R. Trumbull, and C. Haberland, (2003). Complex patterns of fluid and melt transport in the central Andean subduction zone revealed by attenuation tomography, Earth and Planetary Science Letters 215 105-119.

Smith, R. L. & Bailey, R. A., (1968). Resurgent cauldrons. In: Coats, R. R., Hay, R. L. & Anderson, C. A. (eds) Studies in Volcanology. Geological Society of America, Memoir 116, 613-662.

Sobolev, S.V., Zeyen, H., Granet, M., Achauer, U., Bauer, C., Werling, F., Altherr, R., and Fuchs, K., (1997). Upper mantle temperatures and lithosphere-asthenosphere system beneath the French Massif central constrained by seismic, gravity, petrologic and thermal observations: Tectonophysics, v. 275, p. 143-164.

Stankiewicz, J., T. Ryberg, C. Haberland, Fauzi, and D. Natawidjaja, (2010). Lake Toba volcano magma chamber imaged by ambient seismic noise tomography, Geophys. Res. Lett., 37, L17306, doi:10.1029/2010GL044211.

Tongue, J. A., (1992). Tomographic study of local earthquake data from the Lake Bogoria region of the Kenya Rift Valley, *Geophys. Journ. Int., 109, 249 – 258.*

VanDecar, J. C., and R. S. Crosson, (1990). Determination of teleseismic relative phase arrival times using multi-channel cross-correlation and least-squares, Bull. Seismol. Soc. Am., 80, 150–169.

Wagner D., I. Koulakov, W. Rabbel, B.-G. Luehr, A. Wittwer, H. Kopp, M. Bohm, G. Asch, MERAMEX Scientists, (2007) Joint inversion of active and passive seismic data in Central Java // Geophysical Journal International.V 170, p. 923-932, doi:10.1111/j.1365-246X.2007.03435.x

Yarmoluk, V.V., V.I.Kovalenko, and O.A.Bogatikov, 1990. South-Baikal "Hot spot" in the mantle and its role in development of the Baikal rift zone, Doklady AN SSSR, v. 312, n. 1, p. 187-191.

Zandomeneghi D., A. H. Barclay, J. Almendros, J. M. Ibáñez, W. S. D. Wilcock, and T. Ben-Zvi. (2009). Crustal structure of Deception island volcano from p-wave seismic tomography: Tectonic and volcanic implications. J. Geophys. Res., 114:B06310.

Zollo A., L. D'Auria, R.De Matters, A. Herrero, J. Virieux, and P. Gasparini. (2002). Bayesia estimation of 2d p-velocity models from active seismic arrival time data: Imaging of the shallow structure of mt. vesuvius (southern italy). Geophys. J. Int., 151:566–582.

Zollo A., S. Judenherc, E. Auger, L. D'Auria, J. Virieux, P. Capuano, C. Chiarabba, R. de Franco, J. Makris, A. Michelini, and G. Musacchio. (2003). Evidence for the buried rim of Campi Flegrei caldera from 3D active seismic imaging. Geophys. Res. Letters, 30:10.1029,

Zorin, Yu.A., V.V. Mordvinova, E.Kh. Turutanov, B.G. Belichenko, A.A. Artemyev, G.L. Kosarev, and S.S. Gao, (2002). Low seismic velocity layers in the Earth's crust beneath Eastern Siberia (Russia) and Central Mongolia: receiver function data their possible geological implication, Tectonophysics, v. 359, p. 307 - 327.

Zorin Y. A., E. Kh. Turutanov, V. V. Mordvinova, V. M. Kozhevnikov, T. B. Yanovskaya, and A. V. Treussov, (2003). The Baikal rift zone: the effect of mantle plumes on older structure, Tectonophysics, v. 271, p. 153 - 173.55-75.